Methods in Molecular Biology

Series Editor
John M. Walker
School of Life and Medical Sciences
University of Hertfordshire
Hatfield, Hertfordshire, UK

For further volumes:
http://www.springer.com/series/7651

For over 35 years, biological scientists have come to rely on the research protocols and methodologies in the critically acclaimed *Methods in Molecular Biology* series. The series was the first to introduce the step-by-step protocols approach that has become the standard in all biomedical protocol publishing. Each protocol is provided in readily-reproducible step-by-step fashion, opening with an introductory overview, a list of the materials and reagents needed to complete the experiment, and followed by a detailed procedure that is supported with a helpful notes section offering tips and tricks of the trade as well as troubleshooting advice. These hallmark features were introduced by series editor Dr. John Walker and constitute the key ingredient in each and every volume of the *Methods in Molecular Biology* series. Tested and trusted, comprehensive and reliable, all protocols from the series are indexed in PubMed.

Aspergillus fumigatus

Methods and Protocols

Edited by

Natalia Mach

Health University of Applied Sciences Tyrol, FH Gesundheit Tirol, Innsbruck, Austria

Humana Press

Editor
Natalia Mach
Health University of Applied Sciences Tyrol
FH Gesundheit Tirol
Innsbruck, Austria

ISSN 1064-3745 ISSN 1940-6029 (electronic)
Methods in Molecular Biology
ISBN 978-1-0716-5221-3 ISBN 978-1-0716-5222-0 (eBook)
https://doi.org/10.1007/978-1-0716-5222-0

This Humana imprint is published by the registered company Springer Science+Business Media, LLC, part of Springer Nature.
The registered company address is: 1 New York Plaza, New York, NY 10004, U.S.A.

Preface

This book is a practical guide to current molecular biology methods for *Aspergillus fumigatus*. It brings together validated procedures designed to perform reliably at the bench and to reduce troubleshooting. It is intended for researchers in mycology, fungal genetics, and medical microbiology who want to deepen their understanding of *A. fumigatus* at the molecular level.

The book is organized into five sections that follow common workflows in ongoing *A. fumigatus* research. Part I presents current approaches to CRISPR-based genome editing and offers practical protocols for Cas9 ribonucleoproteins, plasmid-assisted strategies, and microhomology-mediated repair. Part II covers nucleic acid isolation, PCR-based typing, and sequencing workflows together with the bioinformatic steps needed for routine single-nucleotide polymorphism and mitochondrial analyses. Part III focuses on assays for gene regulation and contains detailed protocols for electrophoretic mobility shift assays and chromatin immunoprecipitation. Part IV describes models to study host-pathogen interactions, including micropatterned cell culture systems, bronchoalveolar lavage collection in murine models, and *Galleria mellonella* infection assays with imaging readouts. Part V addresses methods to investigate *A. fumigatus* cell-wall and secreted components, including protocols for luciferase-based sensing, metabolic activity assays, and extracellular vesicle isolation.

Each chapter is written by an active practitioner and provides step-by-step instructions, a materials list, expected results, and troubleshooting notes. Four chapters also include documented bioinformatic pipelines that accompany the wet lab protocol.

I thank all contributors for sharing their methods and valuable practical tips. My hope is that this volume will shorten the time spent on routine problems at the bench and help researchers move more quickly from idea to reproducible results.

Innsbruck, Austria *Natalia Mach*

Contents

Contributors

VISHUKUMAR AIMANIANDA • *Department of Biochemistry, Kasturba Medical College Manipal, Manipal Academy of Higher Education (MAHE), Manipal, India*
FAUSTO ALMEIDA • *Department of Biochemistry and Immunology, Ribeirão Preto Medical School, University of São Paulo, Ribeirão Preto, São Paulo, Brazil*
MARK ARENTSHORST • *Fungal Genetics and Biotechnology, Institute of Biology, Leiden University, Leiden, The Netherlands*
VIJENDRA ARYA • *Department of Molecular Physiology and Biophysics, Carver College of Medicine, University of Iowa, Iowa City, IA, USA*
ULRIKE BINDER • *Institute of Hygiene and Medical Microbiology, Medical University Innsbruck, Innsbruck, Austria*
MICHAEL J. BROMLEY • *Manchester Fungal Infection Group, University of Manchester, Manchester, UK*
MARGHERITA CACACI • *Dipartimento di Scienze Biotecnologiche di Base, Cliniche Intensivologiche e Perioperatorie, Università Cattolica del Sacro Cuore, Rome, Italy; Dipartimento di Scienze di Laboratorio ed Ematologiche, Fondazione Policlinico Universitario A. Gemelli IRCCS, Rome, Italy*
JONATAS ERICK MAIMONI CAMPANELLA • *Departamento de Genética e Evolução, Centro de Ciências Biológicas e da Saúde, Universidade Federal de São Carlos, São Carlos, SP, Brazil*
MARÍA DEL ROCÍO REYES-MONTES • *Departamento de Microbiología y Parasitología, Facultad de Medicina, Universidad Nacional Autónoma de México, Mexico City, Mexico*
THAILA FERNANDA DOS REIS • *Faculdade de Ciências Farmacêuticas de Ribeirão Preto, Universidade de São Paulo, São Paulo, Brazil; National Institute of Science and Technology in Human Pathogenic Fungi, São Paulo, Brazil*
ESPERANZA DUARTE-ESCALANTE • *Departamento de Microbiología y Parasitología, Facultad de Medicina, Universidad Nacional Autónoma de México, Mexico City, Mexico*
MELBA ELIZABETH-BIJU • *Department of Biochemistry, Kasturba Medical College Manipal, Manipal Academy of Higher Education (MAHE), Manipal, India*
SAM EL-KAMAND • *Children's Cancer Institute, Lowy Cancer Research Centre, UNSW Australia, NSW, Australia*
JARROD R. FORTWENDEL • *Department of Clinical Pharmacy and Translational Science, College of Pharmacy, University of Tennessee Health Science Center, Memphis, TN, USA; Department of Microbiology, Immunology, and Biochemistry, College of Medicine, University of Tennessee Health Science Center, Memphis, TN, USA*
MARÍA GUADALUPE FRÍAS-DE-LEÓN • *Unidad de Investigación Biomédica, Hospital Regional de Alta Especialidad de Ixtapaluca, Servicios de Salud del Instituto Mexicano de Seguro Social para el Bienestar (IMSS-BIENESTAR), Ixtapaluca, Mexico*
GUSTAVO HENRIQUE GOLDMAN • *Faculdade de Ciências Farmacêuticas de Ribeirão Preto, Universidade de São Paulo, São Paulo, Brazil; Departamento de Genética e Evolução, Centro de Ciências Biológicas e da Saúde, Universidade Federal de São Carlos, São Carlos, SP, Brazil*
YAOJIE GUO • *Zhejiang Key Laboratory of Precision Medicine Research on Head & Neck Cancer, Hangzhou, China; Department of Microbiology, School of Basic Medical Sciences, Zhejiang University School of Medicine, Hangzhou, People's Republic of China*

MARIANA HANDELMAN • *Department of Clinical Microbiology and Immunology, Sackler School of Medicine, Tel-Aviv University, Ramat-Aviv, Tel-Aviv, Israel*
LEA HEINRICH • *Research and Innovation Unit, Health University of Applied Sciences Tyrol, FH Gesundheit Tirol, Innsbruck, Austria*
NATALIA MACH • *Research and Innovation Unit, Health University of Applied Sciences Tyrol, FH Gesundheit Tirol, Innsbruck, Austria*
IRAN MALAVAZI • *National Institute of Science and Technology in Human Pathogenic Fungi, São Paulo, Brazil; Departamento de Genética e Evolução, Centro de Ciências Biológicas e da Saúde, Universidade Federal de São Carlos, São Carlos, SP, Brazil*
ADELA MARTIN-VICENTE • *Department of Clinical Pharmacy and Translational Science, College of Pharmacy, University of Tennessee Health Science Center, Memphis, TN, USA*
CHARLES OLIVER MORTON • *Western Sydney University, School of Science, Campbelltown Campus, NSW, Australia*
W. SCOTT MOYE-ROWLEY • *Department of Molecular Physiology and Biophysics, Carver College of Medicine, University of Iowa, Iowa City, IA, USA*
LUCAS F. B. NOGUEIRA • *Department of Biochemistry and Immunology, Ribeirão Preto Medical School, University of São Paulo, Ribeirão Preto, São Paulo, Brazil*
NIR OSHEROV • *Department of Clinical Microbiology and Immunology, Sackler School of Medicine, Tel-Aviv University, Ramat-Aviv, Tel-Aviv, Israel*
SANJOY PAUL • *Department of Molecular Physiology and Biophysics, Carver College of Medicine, University of Iowa, Iowa City, IA, USA*
SUSANNE PERKHOFER • *Research and Innovation Unit, Health University of Applied Sciences Tyrol, FH Gesundheit Tirol, Innsbruck, Austria*
JULIEN POLLEUX • *Research and Innovation Unit, Health University of Applied Sciences Tyrol, FH Gesundheit Tirol, Innsbruck, Austria; Department of Ophthalmic Optics, Health University of Applied Sciences Tyrol, FH Gesundheit Tirol, Innsbruck, Austria*
BRUNELLA POSTERARO • *Dipartimento di Scienze Biotecnologiche di Base, Cliniche Intensivologiche e Perioperatorie, Università Cattolica del Sacro Cuore, Rome, Italy; Unità Operativa "Medicina di Precisione in Microbiologia Clinica", Direzione Scientifica, Rome, Italy*
ARTHUR F. J. RAM • *Fungal Genetics and Biotechnology, Institute of Biology, Leiden University, Leiden, The Netherlands*
MAURIZIO SANGUINETTI • *Dipartimento di Scienze Biotecnologiche di Base, Cliniche Intensivologiche e Perioperatorie, Università Cattolica del Sacro Cuore, Rome, Italy; Dipartimento di Scienze di Laboratorio ed Ematologiche, Fondazione Policlinico Universitario A. Gemelli IRCCS, Rome, Italy*
DANIEL H. SCHARF • *Department of Microbiology, School of Basic Medical Sciences, Zhejiang University School of Medicine, Hangzhou, People's Republic of China; The Fourth Affiliated Hospital, International School of Medicine, Zhejiang University School of Medicine, Yiwu, China*
JAKOB SCHELER • *Institute of Hygiene and Medical Microbiology, Medical University Innsbruck, Innsbruck, Austria*
CHAITHANYA SHETTY • *Department of Biochemistry, Kasturba Medical College Manipal, Manipal Academy of Higher Education (MAHE), Manipal, India*
SAHO SHIBATA • *Medical Mycology Research Center, Chiba University, Chiba, Japan*
DAMIANO SQUITIERI • *Dipartimento di Scienze Biotecnologiche di Base, Cliniche Intensivologiche e Perioperatorie, Università Cattolica del Sacro Cuore, Rome, Italy*

MARK A. STAMNES • *Department of Molecular Physiology and Biophysics, Carver College of Medicine, University of Iowa, Iowa City, IA, USA*
HIROKI TAKAHASHI • *Medical Mycology Research Center, Chiba University, Chiba, Japan; Faculty of Science, Chiba University, Chiba, Japan; Molecular Chirality Research Center, Chiba University, Chiba, Japan; Plant Molecular Science Center, Chiba University, Chiba, Japan*
VERONICA THORN • *Deparment of Biology, McMaster University, Hamilton, ON, Canada*
JAMIE TINDALE • *Manchester Fungal Infection Group, University of Manchester, Manchester, UK*
MOMOTAKA UCHIDA • *Medical Mycology Research Center, Chiba University, Chiba, Japan*
CLARA VALERO • *Manchester Fungal Infection Group, University of Manchester, Manchester, UK*
NORMAN VAN RHIJN • *Manchester Fungal Infection Group, University of Manchester, Manchester, UK; Microbial Evolution Research at Manchester, University of Manchester, Manchester, UK*
CEES A. M. J. J. VAN DEN HONDEL • *Fungal Genetics and Biotechnology, Institute of Biology, Leiden University, Leiden, The Netherlands*
HILA WERNER • *Department of Clinical Microbiology and Immunology, Sackler School of Medicine, Tel-Aviv University, Ramat-Aviv, Tel-Aviv, Israel*
SARAH SZE WAH WONG • *Institut Pasteur, Université de Paris, Immunology of Fungal Infections, Paris, France*
JIANPING XU • *Deparment of Biology, McMaster University, Hamilton, ON, Canada*

Part I

Methods for CRISPR-Based Genome Editing

Chapter 1

Cas9 Ribonucleoproteins (RNPs) for Gene Deletion

Adela Martin-Vicente and Jarrod R. Fortwendel

Abstract

Genetic manipulations in pathogenic microorganisms, like *Aspergillus fumigatus*, allow us to study the function of genes in pathogenicity or other important traits. Classic genetic engineering tools in filamentous fungi rely on DNA fragment sub-cloning or fusion PCR to build a gene deletion cassette containing extended flanking regions of homology to promote efficient recombination. However, the introduction of CRISPR technology into fungal genetics has greatly simplified and accelerated the genome editing process. Here, we describe a simple and universal, one-step CRISPR-Cas9-mediated genetic tool employing repair templates containing microhomology regions. This gene targeting system displays high homologous recombination efficiency rates and can be easily implemented across different genetic backgrounds.

Key words *Aspergillus fumigatus*, CRISPR-Cas9, Gene deletion, Ribonucleoprotein complex, gRNA

1 Introduction

Genetic manipulations in *Aspergillus fumigatus*, as well as in other pathogenic organisms, allow us to study the function of different genes and their signaling pathways, so that we can better understand fungal biology, pathogenesis, and stress adaptation. Classical gene targeting techniques in *A. fumigatus* offer low rates of homologous recombination and often involve time-intensive steps like sub-cloning or fusion PCR for the generation of the gene deletion cassette. For efficient targeting in many filamentous fungi, long flanking regions (~750–1500 bp) of homology are usually required. The discovery of CRISPR (clustered regularly interspaced short palindromic repeat)-Cas9 about a decade ago revolutionized the genetic engineering field, simplifying the process of genome editing while achieving increased mutational efficiency in both prokaryotes and eukaryotes, and has become an indispensable tool in biological research [1]. Cas9 is a programmable endonuclease that produces double-strand breaks in specific DNA regions after forming a ribonucleoprotein (RNP) complex with a guide RNA (gRNA). The gRNA is a two-component complex formed

Natalia Mach (ed.), *Aspergillus fumigatus: Methods and Protocols*, Methods in Molecular Biology, vol. 3028,
https://doi.org/10.1007/978-1-0716-5222-0_1,

by a CRISPR RNA (crRNA), which contains a region complementary to the 20-base protospacer sequence at the target site that directs specific binding of the RNP complex, and a transactivating CRISPR RNA (tracrRNA), which is required for crRNA maturation and DNA cleavage by crRNA-bound Cas9 enzyme [2]. In addition, for Cas9 to cleave the target site, the protospacer must be immediately followed by a protospacer adjacent motif (PAM). In the commonly used Cas9 from *Streptococcus pyogenes*, the PAM usually corresponds to any nucleotide followed by two guanines (i.e., NGG) [2]. CRISPR-Cas9 gene editing systems have been demonstrated to be faster, more efficient, and more reliable than conventional methods, and they have already been implemented in *A. fumigatus* [3, 4], as well as in other fungal organisms [5].

Here, we describe a protocol for gene deletion in *A. fumigatus*, using in vitro-assembled Cas9 RNPs composed of two different crRNAs recognizing PAM sites located 5′ and 3′ of the gene of interest. We have found the use of two crRNAs to be necessary because efficient homologous integration of a repair template in wild-type *A. fumigatus* genetic backgrounds requires targeting recombination to occur within ~100 base pairs of the Cas9 RNP-induced double-strand break. However, we note that this same technique can be used employing only a single crRNA to achieve other genetic manipulations, including gene disruption, promoter replacements, or in situ gene fusions. For example, using a single crRNA, we were able to successfully generate a targeted disruption library of protein kinases [6]. It is also worth mentioning that the gene editing system described here has proved to be efficient, not only in *A. fumigatus* laboratory strains but also in isolates of different origin, including clinical isolates [7], and, due to its high efficiency, it can also be used to generate marker-free mutations [8]. In the protocol below, we utilize an example employing the antibiotic hygromycin for selection coupled with a repair template carrying the *hph* gene as a resistance marker, since it is commonly used in *A. fumigatus* wild-type strains [9]. However, the same process can be performed in wild-type strains with other dominant drug resistance markers such as phleomycin or pyrithiamine, as well as using the orotidine 5′-phosphate decarboxylase (*pyrG*) gene as a selective marker in uridine/uracil auxotrophic strains [10–12].

2 Materials

2.1 Growth Media

1. Salt solution (20×): Dissolve 10.4 g of KCl, 10.4 g of $MgSO_4{\cdot}7H_2O$, and 30.4 g of KH_2PO_4 in 1 L of sterile deionized water (SDW). Store at room temperature.

2. $NaNO_3$ (100×): Dissolve 21.3 g of $NaNO_3$ in 250 mL of SDW. Store at room temperature.
3. Trace elements solution (1000×): Dissolve the following reagents in 80 mL of SDW, ensuring that every compound is completely dissolved before adding the next one: 2.2 g of $ZnSO_4{\cdot}7H_2O$, 1.1 g of H_3BO_3, 0.5 g of $MnCl_2{\cdot}4H_2O$, 0.5 g of $FeSO_4{\cdot}7H_2O$, 0.16 g of $CoCl_2{\cdot}5H_2O$, 0.16 g of $CuSO_4{\cdot}5H_2O$, 0.11 g of $(NH_4)_6Mo_7O_{24}{\cdot}4H_2O$, and 5 g of Na_4EDTA. Heat the solution until boiling and adjust the pH to 6.5 with KOH pellets. Cool to room temperature, and make up to 100 mL with SDW. Store at room temperature (*see* **Note 1**).
4. Glucose Minimal Media (GMM) agar: Dissolve 50 mL of 20 × salt solution, 10 mL of 100 × $NaNO_3$, 1 mL of 1000 × trace elements solution, and 10 g of glucose in 800 mL of SDW. Adjust the pH to 6.5 and make up to 1 L with SDW. Add 15 g of agar, autoclave.
5. Glucose Minimal Media broth with yeast extract (GMM + YE): Dissolve 50 mL of 20 × Salt solution, 10 mL of 100 × $NaNO_3$, 1 mL of 1000 × trace elements solution, 10 g of glucose, and 5 g of yeast extract in 800 mL of SDW. Adjust the pH to 6.5, make up to 1 L with SDW, and autoclave.
6. Stabilizing Minimal Media (SMM) agar (also called Sorbitol Minimal Media): Dissolve 50 mL of 20 × salt solution, 10 mL of 100 × $NaNO_3$, 1 mL of 1000 × trace elements solution, 10 g of glucose, and 218.6 g of sorbitol in 800 mL of SDW. Adjust the pH to 6.5, make up to 1 L with SDW. Add 15 g of agar, and autoclave.
7. SMM Top agar: Dissolve 50 mL of 20 × salt solution, 10 mL of 100 × $NaNO_3$, 1 mL of 1000 × trace elements solution, 10 g of glucose, and 218.6 g of sorbitol in 800 mL of SDW. Adjust the pH to 6.5, make up to 1 L with SDW. Add 7.5 g of agar and autoclave.
8. 100 mm sterile Petri dishes.
9. Glass Erlenmeyer flasks (250–1000 mL, for fungal culture incubation).
10. Incubator set at 32–37 °C, with and without shaking capability.

2.2 Protoplasts Generation and Transformation

1. Sterile deionized water (SDW).
2. Sterile Miracloth filter material (Merk Millipore).
3. Sterile spatula.
4. Sterile 50 mL plastic tubes.
5. Sterile 10 mL pipettes.
6. Sterile microcentrifuge tubes.

7. 2 M sodium phosphate buffer: Dissolve 9.9 g of Na_2HPO_4 in 80 mL of SDW. Add 16.34 g of NaH_2PO_4 and adjust the pH to 6.5 with 1 M Na_2HPO_4. Make up to 100 mL with SDW and filter-sterilize (0.22 μm PES filter, e.g., Millipore Steri-flip).
8. Osmotic Media (OM): Add 147.88 g $MgSO_4 \cdot 7H_2O$ to 250 mL of SDW. Add 2.5 mL of 2 M sodium phosphate buffer, pH 6.5. Make up to 500 mL with SDW, filter-sterilize (0.22 μm PES filter), and store at 4 °C.
9. Protoplasting solution: Dissolve 2 g of ExtraLyse (e.g., Laffort) in 40 mL of warm OM (*see* **Note 2**). Shake well; filter-sterilize (0.22 μm PES filter).
10. Trapping buffer: Dissolve 109.3 g of sorbitol in 700 mL of SDW. Add 100 mL of 1 M Tris-HCl, pH 7.0. Make up to 1 L with SDW, filter-sterilize (0.22 μm PES filter), and store at 4 °C.
11. STC buffer: Dissolve 218.6 g of sorbitol in 700 mL of SDW. Add 1.11 g of $CaCl_2$ and 10 mL of 1 M Tris-HCl, pH 7.5, make up to 1 L with SDW, filter-sterilize (0.22 μm PES filter), and store at 4 °C.
12. PEG-$CaCl_2$ solution: Dissolve 6 g of PEG 3350 in 4.5 mL of Tris-HCl, pH 7.5, and add 0.5 mL of 1 M $CaCl_2 \cdot 2H_2O$. Incubate at 50 °C until dissolved, filter-sterilize (0.22 μm PES filter), and store at room temperature for no more than 2 weeks.
13. DNA repair template (*see* Subheading 3.2 for details).

2.3 CRISPR-Cas9 Components

The CRISPR-Cas9 components (tracrRNA, Cas9 enzyme, and custom-designed crRNA) for the system described here are typically purchased from a commercial vendor (e.g., Integrated DNA Technologies [IDT]). The presentation and concentrations of each component provided may vary from vendor to vendor.

1. Cas9 enzyme working buffer: Mix 0.2 mL of 1 M HEPES, pH 7.5, with 0.75 mL of 2 M KCl and 9.05 mL of nuclease-free water (*see* **Note 3**). Verify pH 7.5, store at room temperature.
2. tracrRNA: The tracrRNA is a universal 67mer that can hybridize to any custom-designed crRNA (*see* **item 4** in Subheading 2.3) to form the gRNA duplex. The tracrRNA can usually be ordered to be delivered as a lyophilized material. Resuspend the lyophilized material with nuclease-free duplexing buffer to a concentration of 100 μM.
3. Nuclease-free duplexing buffer: 30 mM HEPES, pH 7.5, 100 mM potassium acetate. When purchasing the tracrRNA commercially, the vendor often provides a duplexing buffer. Store at 4 °C for short periods of time or at −20 °C for long storage.

4. crRNA: The crRNA is custom-designed for each target (*see* Subheading 3.1). Resuspend the lyophilized material with nuclease-free duplexing buffer to a concentration of 100 μM.
5. Cas9 nuclease: Prepare a working solution by diluting 1:10 with Cas9 enzyme working buffer to achieve a final concentration of 1 μg/mL, store at −20 °C.
6. Heat block set to 95 °C.
7. Sterile microcentrifuge tubes, sterile, RNAase-free filter tips.

3 Methods

3.1 Design of crRNAs

To delete the entire coding sequence of a specific gene, design two crRNAs (cr5' and cr3') that direct Cas9 cleavage upstream of the START and downstream of the STOP codons (Fig. 1). crRNAs may be designed manually or with online tools (e.g., Eukaryotic Pathogen CRISPR guide RNA/DNA Design Tool or "EuPaGDT" [13]). In our example, the complete gene coding sequence will be replaced by a repair template carrying a hygromycin resistance cassette, allowing for facile identification of positive transformant colonies. To design the crRNAs:

1. Acquire the genomic sequence of the target gene to be deleted, to include ~200 bp upstream and downstream as flanking regions.

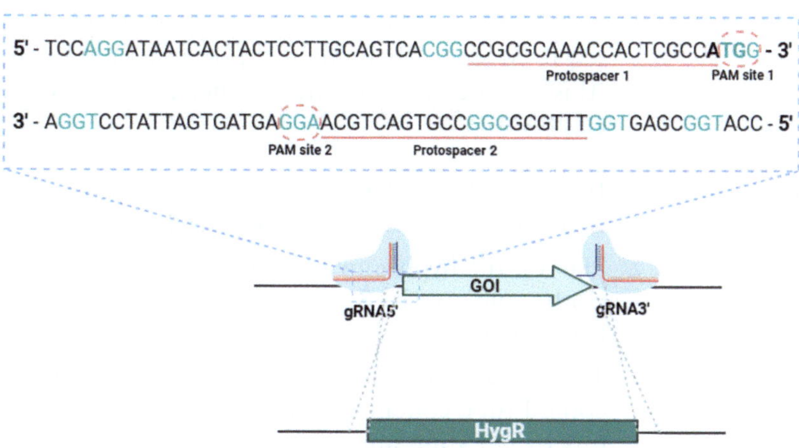

Fig. 1 Schematic representation of a gene deletion by CRISPR-Cas9 gene targeting employing two gRNAs and hygromycin as a selectable marker. The two RNPs, formed by the duplex crRNA-tracrRNA (gRNA5′ and gRNA3′) and Cas9 nuclease, are shown near the START and STOP codons of the gene of interest (GOI). Zoomed in, the DNA sequence (and the complemented strand) corresponding to −50 to +4 nucleotides from the START codon (marked in bold), is shown. All the available PAM sites in that area are denoted in green, and two examples of PAM sites and corresponding protospacer sequences (one in each DNA strand) are denoted with orange lines. (Created with BioRender.com by Fortwendel, J)

2. Identify all putative protospacer adjacent motif (PAM) sites, each consisting of any nucleotide [N], followed by two guanines [GG] (i.e., NGG), in the areas of −50 to +4 nucleotides upstream of the START codon and −4 to +50 nucleotides downstream of the STOP codon.
3. Identify the putative protospacers adjacent to each PAM. Protospacer sequences are composed of the 20 nucleotides immediately upstream (5′) of the selected PAM sites (Fig. 1) (*see* **Note 4**).
4. To ensure specific Cas9 cleavage and to avoid off-target mutations, perform a BLAST search of each protospacer against the reference genome sequence. Avoid protospacers that display greater than 15 bp of identity to off-target genomic loci (*see* **Note 5**).

3.2 Design of Oligonucleotides for the Amplification of Repair Template with Microhomology Regions

The repair template used for gene deletion in *A. fumigatus* consists of a resistance cassette containing a gene that confers resistance to an antibiotic (here, we use hygromycin), together with a promoter that drives its expression and a terminator. This resistance cassette (HygR, carrying the *hph* gene) is stored in the plasmid pUCGH [14] for easy amplification.

1. Using standard parameters, design oligonucleotides for amplification of the selection cassette.
2. Complete the repair template design by including a microhomology flanking region of 40 bp in length to the 5′ end of each oligonucleotide. These microhomology regions should be complementary to the genomic sequence upstream of the chosen PAM site(s), beginning at position −2 relative to the PAM (this is the site of the predicted DSB induced by the Cas9 RNP).

3.3 Generation of Protoplasts

1. Inoculate 1.2×10^8 conidia into 100 mL of GMM + YE. Incubate in a shaking incubator, 250 rpm at 32 °C for 16 h.
2. Harvest the hyphae by pouring the culture over a sterile Miracloth, and wash 3–4 times by pouring SDW over the hyphal mat.
3. Transfer approximately 5 mL of wet hyphal mat to a 50 mL conical tube containing 20 mL of protoplasting solution, and mix by pipetting up and down.
4. Incubate tubes horizontally for 2 h at 32 °C and shake incubator at 75 rpm. After 1 h, remove tubes and pipette up and down with a 10 mL pipette to break up clumps.
5. Monitor protoplast formation microscopically by sampling a 10 μL aliquot on a slide. Protoplasts begin to form at 1 h but are much more abundant after 1.5–2 h.

6. When cultures are full of protoplasts, pipette up and down to break remaining clumps and overlay the protoplast mixture with 10 mL of trapping buffer by allowing the buffer to slowly trickle down the side of the tube (*see* **Note 6**).
7. Centrifuge for 15 min at 3500 rpm, 4 °C with no centrifuge brake. A thin, white layer of protoplasts should be present at the interface of the two layers.
8. Carefully remove the protoplasts with a pipette and transfer to a 15 or 50 mL conical (depending on the volume). Typically, 2–4 mL of protoplasts can be obtained from each tube.
9. Add 3× the volume of STC buffer to the isolated protoplasts, and centrifuge for 10 min at 3500 rpm and 4 °C, with full brake.
10. Discard supernatant and resuspend the protoplasts in STC buffer. Count protoplasts with a hemocytometer, and adjust the concentration to 5×10^7 protoplasts/mL. Keep protoplasts on ice until use.

3.4 Cas9 RNP-Based Transformation

To prepare all the CRISPR-Cas9 components, use sterile, RNAse-free filter tips.

3.4.1 Guide RNA (gRNA) Duplex Generation

1. Mix 5 μL of duplexing buffer, 5 μL of 100 μM tracrRNA, and 5 μL of 100 μM crRNA in a sterile microcentrifuge tube (Fig. 2a).
2. Mix by pipetting up and down and gently spin down.
3. Incubate at 95 °C for 5 min.
4. Let the gRNA cool to room temperature for 10 min, place on ice until use, or store at −20 °C for long-term storage (*see* **Note 7**).

3.4.2 Ribonucleoprotein (RNP) Complexes Assembly

1. Mix 4.5 μL of Cas9 working buffer, 0.5 μL of gRNA (*see* **Note 8**), and 0.5 μL of 1 μg/mL Cas9 enzyme in a microcentrifuge tube to get a final volume of 5 μL (Fig. 2b).
2. Incubate at room temperature for 5 min to allow the assembly of the RNP complexes. Place on ice until use.

3.4.3 A. fumigatus Transformation

1. In a microcentrifuge tube, combine 10 μL of protoplasts (5×10^7 protoplasts/mL), 5 μL of each RNP complex, 200–500 ng of repair template, and 3 μL of PEG-$CaCl_2$ solution. Make up to 30 μL with STC buffer. Mix by gently tapping the tube (*see* **Note 9**).
2. Incubate on ice for 50 min.
3. Add 57 μL of PEG-$CaCl_2$ solution and mix lightly by gently tapping the tube.

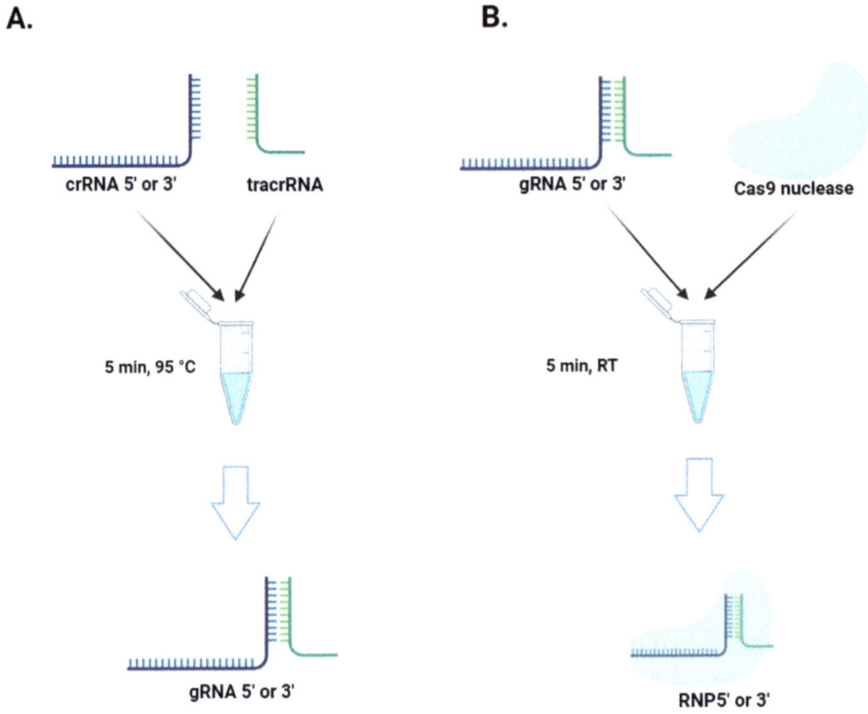

Fig. 2 Schematic representation of gRNA duplexes and Cas9 ribonucleoprotein (RNP) complex in vitro assembly. (**a**) Equimolar amounts of crRNA (targeting the 5′ or 3′ end of the target locus) and tracrRNA are mixed with duplexing buffer to a final concentration of 33 μM. The mixes are boiled at 95 °C for 5 min and then cooled down to room temperature for about 10 min. (**b**) For the generation of the RNP complexes, the gRNA duplexes are mixed with Cas9 nuclease in Cas9 working buffer and incubated at room temperature for 5 min. (Created in BioRender.com by Fortwendel, J)

4. Incubate at room temperature for 20 min.
5. Add 113 μL of STC buffer to obtain a total volume of 200 μL.
6. Spread the whole transformation mixture on a single SMM plate.
7. Incubate the plates at room temperature overnight.
8. The next day, add 10 mL of Top Agar containing 150 μg/mL of hygromycin.
9. Incubate the plates at 37 °C until colonies are observed.
10. With the help of a toothpick or a pipette tip, carefully transfer individual colonies to fresh GMM supplemented with 150 μg/mL of hygromycin.
11. Incubate the plates at 37 °C for 2–3 days or until enough growth or conidiation.
12. Extract genomic DNA and screen for correct integration by PCR or Southern blot.

4 Notes

1. If solid precipitation is observed, mix the solution thoroughly before use.
2. Additional protocols are available that utilize varying combinations of α- and β-glucanases, β-glucosidases, chitinases, and proteases to generate *A. fumigatus* protoplasts from germlings or mature hyphae (e.g., VinoTaste (Novozymes) or lytic enzymes from *Trichoderma harzianum*).
3. Use nuclease-free HEPES and KCl solutions.
4. When purchasing the crRNA from a commercial vendor, the 20 bp protospacer sequence is typically the only information requested. The vendor will convert the DNA sequence to RNA and will add the additional bases required to hybridize with the tracrRNA.
5. If it is not possible to identify a protospacer with less than 15 bp of homology to an off-target target locus, the protospacer can still be selected as long as it is not immediately followed by a PAM in the genome.
6. It is very important that the trapping buffer forms a layer on top of the protoplasting solution. Do not allow the layers to mix.
7. The crRNA:tracrRNA gRNA complex can be used for up to 3 months with no loss in activity when stored at −20 °C. Longer storage times have not been tested.
8. Typically, 2 gRNAs are used for complete gene deletions, so two RNP complexes will be generated separately.
9. Do not pipette up and down to mix. The protoplasts are very delicate.

Acknowledgments

This work was supported by NIH grants R01AI158442, R01AI143197, and R21AI178148 to JRF. Figures were created with Biorender.com.

References

1. Wang JY, Doudna JA (2023) CRISPR technology: a decade of genome editing is only the beginning. Science 379:eadd8643. https://doi.org/10.1126/science.add8643
2. Sander JD, Joung JK (2014) CRISPR-Cas systems for editing, regulating and targeting genomes. Nat Biotechnol 32(4):347–355. https://doi.org/10.1038/nbt.2842
3. Al Abdallah Q, Ge W, Fortwendel JR (2017) A simple and universal system for gene manipulation in *Aspergillus fumigatus*: *in vitro*-assembled Cas9-guide RNA ribonucleoproteins

coupled with microhomology repair templates. mSphere 2(6). https://doi.org/10.1128/mSphere.00446-17

4. Al Abdallah Q, Souza ACO, Martin-Vicente A, Ge W, Fortwendel JR (2018) Whole-genome sequencing reveals highly specific gene targeting by *in vitro* assembled Cas9-ribonucleoprotein complexes in *Aspergillus fumigatus*. Fungal Biol Biotechnol 5:11. https://doi.org/10.1186/s40694-018-0057-2
5. Wang D, Jin S, Lu Q, Chen Y (2023) Advances and challenges in CRISPR/Cas-based fungal genome engineering for secondary metabolite production: a review. J Fungi 9(3):362. https://doi.org/10.3390/jof9030362
6. Souza ACO, Martin-Vicente A, Nywening AV, Ge W, Lowes DJ, Peters BM, Fortwendel JR (2021) Loss of Septation Initiation Network (SIN) kinases blocks tissue invasion and unlocks echinocandin cidal activity against *Aspergillus fumigatus*. PLoS Pathog 17(8): e1009806. https://doi.org/10.1371/journal.ppat.1009806
7. Rybak JM, Ge W, Wiederhold NP, Parker JE, Kelly SL, Rogers PD, Fortwendel JR (2019) Mutations in *hmg1*, challenging the paradigm of clinical triazole resistance in *Aspergillus fumigatus*. MBio 10(2). https://doi.org/10.1128/mbio.00437-19
8. Van Rhijn N, Furukawa T, Zhao C, McCann BL, Bignell E, Bromley MJ (2020) Development of a marker-free mutagenesis system using CRISPR-Cas9 in the pathogenic mould *Aspergillus fumigatus*. Fungal Genet Biol 145: 103479. https://doi.org/10.1016/j.fgb.2020.103479
9. Punt PJ, Oliver RP, Dingemanse MA, Pouwels PH, van den Hondel CA (1987) Transformation of *Aspergillus* based on the hygromycin B resistance marker from *Escherichia coli*. Gene 56(1):117–124. https://doi.org/10.1016/0378-1119(87)90164-8
10. Weidner G, D'Enfert C, Koch A, Mol PC, Brakhage AA (1998) Development of a homologous transformation system for the human pathogenic fungus *Aspergillus fumigatus* based on the *pyrG* gene encoding orotidine 5′-monophosphate decarboxylase. Curr Genet 33(5):378–385. https://doi.org/10.1007/s002940050350
11. Kubodera T, Yamashita N, Nishimura A (2002) Transformation of *Aspergillus* sp. and *Trichoderma reesei* using the pyrithiamine resistance gene (*ptrA*) of *Aspergillus oryzae*. Biosci Biotechnol Biochem 66(2):404–406. https://doi.org/10.1271/bbb.66.404
12. Austin B, Hall RM, Tyler BM (1990) Optimized vectors and selection for transformation of *Neurospora crassa* and *Aspergillus nidulans* to bleomycin and phleomycin resistance. Gene 93(1):157–162. https://doi.org/10.1016/0378-1119(90)90152-h
13. Peng D, Tarleton R (2015) EuPaGDT: a web tool tailored to design CRISPR guide RNAs for eukaryotic pathogens. Microb Genom 1(4):e000033. https://doi.org/10.1099/mgen.0.000033
14. Al Abdallah Q, Norton TS, Hill AM, LeClaire LL, Fortwendel JR (2016) A fungus-specific protein domain is essential for RasA-mediated morphogenetic signaling in *Aspergillus fumigatus*. mSphere 1(6). https://doi.org/10.1128/mSphere.00234-16

Chapter 2

Generation of Seamless Point Mutations with Cas9 RNP and pTel-hygR Plasmid in *Aspergillus fumigatus*

Mariana Handelman, Hila Werner, and Nir Osherov

Abstract

The environmental mold *Aspergillus fumigatus* is a common human fungal pathogen that causes a wide range of diseases. The antifungal triazoles that inhibit the Cyp51 enzyme involved in ergosterol biosynthesis are used to treat *A. fumigatus* infections. However, triazole resistance is an increasing concern due to mutations in the genes *cyp51A*, *hmg1*, and others, and efflux pumps overexpression. The process of verifying mutations is time-consuming, even with CRISPR-Cas9 methods, as it still requires constructing repair templates with selectable markers. This study presents a faster and more efficient method to introduce mutations conferring triazole resistance in *A. fumigatus* by using in vitro assembled CRISPR-Cas9, along with a recyclable selectable marker. With this approach, we successfully introduced triazole resistance-conferring mutations in *A. fumigatus* genes (*cyp51A*, *cyp51B*, and *hmg1*), both individually and in combination. The technique has the potential to introduce mutations for resistance to other antifungals, toxic metals, and environmental stressors, thus enhancing the ability to generate dominant mutations in *A. fumigatus*.

Key words *Aspergillus fumigatus*, CRISPR-Cas9, ARMS-PCR, pTel, Genetic manipulation

1 Introduction

Introducing point mutations in *A. fumigatus* is a time-consuming process, requiring the creation of a repair template that includes the target gene with the mutation, a resistance marker cassette, and homologous flanking regions of approximately 1500 bp on each side [1]. While CRISPR-Cas9 gene-editing has reduced the size of the homologous flanking regions needed to about 50 bp on each side [2, 3], a selectable marker, such as hygromycin B (hyg), is still necessary to isolate transformants from the protoplast population seeded on the transformation plate.

The pTel-hyg^R plasmid [4], which contains human telomeric sequences at both ends, linearizes upon introduction into the fungal cell and is maintained as an episome. Similar pTel plasmids, in

Natalia Mach (ed.), *Aspergillus fumigatus: Methods and Protocols*, Methods in Molecular Biology, vol. 3028, https://doi.org/10.1007/978-1-0716-5222-0_2,

combination with CRISPR-Cas9 methods, have proven effective for transforming filamentous fungi [2, 4, 5].

In this method's protocol, we present a novel application of the pTel-hyg^R plasmid as an efficient recyclable marker, combined with CRISPR-Cas9 for seamless gene-editing and isolation of single and multiple resistance-conferring point-mutations in the pathogenic mold *A. fumigatus*. Here, we present the introduction of an F262del point mutation into the *hmg1* gene (AFUB_020770) as an example to demonstrate this method, coupled with ARMS PCR [6] for the isolation of mutated colonies, and validation of the removal of the pTel-hyg^R plasmid.

2 Materials

2.1 Preparing A. fumigatus for Transformation

1. Trace elements solution: 0.08% $FeCl_2{\cdot}4H_2O$, 0.71% HCl, 0.04% H_3BO_3, 0.04% $CuSo_4{\cdot}5H_2O$, 0.08% $McCl_2$, 0.08% $NaMoO_4{\cdot}2H_2O$, 0.8% $ZnSo_4{\cdot}7H_2O$. Dissolve $FeCl_2$ in HCl until fully dissolved. Dissolve all other components in double-distilled water (DDW), then add the $FeCl_2$/HCl solution. Filter-sterilize the final mixture.
2. Vitamin mix solution: 0.1% p-aminobenzoic acid, 0.1% niacin, 0.1% pyridoxine HCl, 0.1% riboflavin, 0.1% thiamine HCl, 0.1% choline HCl, 0.02% D-biotin.
3. YAG plates: 0.5% yeast extract, 1% dextrose, 0.01 M $MgSO_4$, 0.1% trace elements solution, 0.2% vitamin mix solution, 1.5% agar. Prepare both hygromycin B (hyg) non-supplemented and supplemented variants with 350 μg/mL hyg.
4. Tween 20 solution: 0.02% Tween in DDW.
5. Liquid Sabouraud medium (SAB).

2.2 Transformation

In the following section, "transformation" is equivalent to performing a single gene-editing event.

1. Sterilized Miracloth filter material (Merck Millipore, Cat # 475855), cut into 8 × 8 cm squares, 2 per transformation.
2. Liquid SAB: 30 mL per transformation.
3. TRAFO 1 solution: 0.6 M KCl, 50 mM $CaCl_2$, 5 mM Tris-HCl, pH 7.5, autoclave. Prepare 200 mL in DDW, sufficient for 4–8 transformations.
4. TRAFO 2 solution: 40% PEG 3350 in TRAFO 1, autoclave. Prepare 100 mL in DDW, sufficient for 9–10 transformations.
5. Fungal cell wall protoplasting solution: 10 mL TRAFO 1, 5% VinoTaste FCE solution (Novozymes), filter-sterilized, prepare less than 30 min before use.

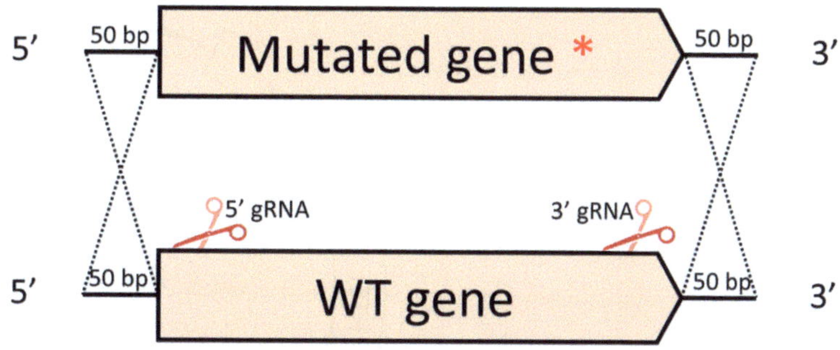

Fig. 1 Outline of the repair template. To introduce the point mutation, the mutated gene was PCR-amplified and used as a repair template. Transformation was performed with pTel-hyg^R and two gRNAs designed for both 5′ and 3′ ends of the gene

6. Cas9 working buffer: 20 mM HEPES, pH 7.5, 150 mM KCl, pH 7.5, filter-sterilize through 0.22 μm filter, prepare 9 mL in DDW.
7. Alt-R™ S.p. Cas9 Nuclease 3NLS.
8. Cas9 guide RNAs (gRNA): Mix equal volumes of 100 μM Alt-R CRISPR-Cas9 tracrRNA (trans-activation CRISPR RNA) and 100 μM Alt-R CRISPR-Cas9 custom crRNA (CRISPR RNA) in Nuclease-Free Duplex Buffer. Heat at 95 °C for 5 min, then gradually cool down to room temperature (20–25 °C). Use two crRNAs for each gene, targeting the 5′ and 3′ areas of the gene. Design using EuPaGDT—a web tool tailored to design CRISPR guide RNAs for eukaryotic pathogens.
9. For *A. fumigatus hmg1*, use the following crRNAs: GCACCC TATACACACCATTG and GCATGGCGAAACATGAAGTA. Use 0.175 μL of each crRNA and Cas9 (1 μg/μL in Cas9 working buffer) per transformation plate.
10. Repair template. For each gene, the repair template is a copy of the target gene containing the point mutation within its sequence. Design flanks to be at least 50 bp before and after the 5′ and 3′ crRNAs, respectively. For *hmg1* F262del, use primers: CAGCATCGAGTCGAGAGAATTT and CTGCGT TACTCGGTCTTGGTAC, to amplify genomic DNA isolated from a mutant strain containing this mutation. Use 0.5–1 μg per transformation plate (*see* Fig. 1).
11. pTel-hyg^R plasmid: Use 0.5–1 μg per transformation plate (*see* Fig. 2).

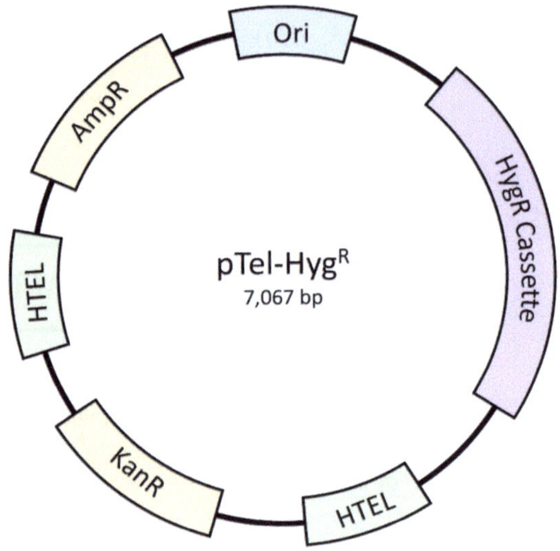

Fig. 2 Scheme of the pTel-hyg^R plasmid

12. YPGS agar plates: 2% yeast extract, 0.5% peptone, 2% D-glucose, 1 M sucrose, 1.5% agar, 250 μg/mL hyg, pH 6.0. Pour 15 mL per plate.
13. Top YPGS agar plates: 2% yeast extract, 0.5% peptone, 2% D-glucose, 1 M sucrose, 0.7% agar, 250 μg/mL hyg, pH 6.0.

2.3 Mutation Selection, Isolation, and Verification

1. YAG plates (*see* **item 3** in Subheading 2.1).
2. Selection plates: YAG plates supplemented with a selection compound of choice. (e.g., voriconazole).
3. Tween 20 solution (*see* **item 4** in Subheading 2.1).
4. Gene-specific primers for the amplification and verification of the mutation. For *hmg1* F262del, use primers TATGCTGCCA TATTTGCTGATG and ACGAGACAGTAGAGGTAGGCC.

2.4 Equipment and Consumables

1. Benchtop centrifuge.
2. Incubator set at 30 °C and 37 °C.
3. Orbital shaker (settable to 70 × rpm).
4. Water bath (capable of holding 48 °C).
5. Sterile microcentrifuge tubes (1.5 mL and 2.0 mL); sterile 50 mL conical tubes.
6. Sterile Petri dishes (90 mm diameter).
7. Hemocytometer.
8. PCR thermocycler, gel electrophoresis system.
9. Sterile cell scrapers.
10. Parafilm or plate-sealing tape.

3 Methods

3.1 Preparing A. fumigatus for Transformation

1. Plate *A. fumigatus* conidia (10–50 μL of frozen −80 °C stock, counting is not required) onto YAG plates, and incubate at 37 °C for 48–72 h (*see* **Note 1**). For hyg sensitivity control, plate a sample of conidia on YAG with 350 μg/mL hyg.
2. Collect conidia using 1–1.5 mL of 0.02% Tween 20 and a sterile scraper, until the solution is murky green.
3. Inoculate 30 mL of liquid SAB medium with 1 × 10^6/mL conidia (total 3 × 10^7) of the recipient strain.
4. Transfer to two sterile, empty Petri dishes. Seal with parafilm or tape. Incubate overnight at 37 °C without shaking.

3.2 Transformation

1. Filter the mycelium from both Petri dishes through sterilized Miracloth, and discard the flow-through.
2. Wash the retained mycelium with ~20 mL of fresh SAB medium.
3. Transfer the mycelium to 10 mL of freshly prepared, filter-sterilized fungal cell wall protoplasting solution in a sterile 50 mL tube.
4. Incubate tubes horizontally at 30 °C for 2 h while shaking at 70 rpm.
5. Chill the TRAFO 1 solution at 4 °C and place on ice.
6. Boil the TRAFO 2 solution briefly and then place one ice.
7. Boil Top YPGS agar and hold in water bath at 48 °C.
8. Filter the protoplasts through sterile Miracloth and transfer to ice. Keep the flow-through containing protoplasts.
9. Centrifuge protoplasts at 3000 rpm for 5 min at room temperature.
10. Resuspend the pellet in 10 mL of cold TRAFO 1 by gently pipetting to lightly disperse the pellet.
11. Centrifuge again at 3000 rpm for 5 min at room temperature.
12. Resuspend the pellet in 1.0 mL of cold TRAFO 1 and keep on ice.
13. Count the protoplasts using a hemocytometer (typically ~1–3 × 10^7/mL), and adjust the concentration to 5 × 10^6 protoplasts/mL in TRAFO 1 on ice.
14. Prepare the Cas9 RNP complex. For each transformation plate, mix 0.175 μL of gRNA1, 0.175 μL of gRNA2, and 0.175 μL of Cas9 nuclease (Cas9 nuclease concentration is 1 μg/μL in Cas9 working buffer). Incubate at room temperature for 5 min to allow the assembly of the RNP complexes. Use immediately.

15. For each transformation plate, combine 105 μL protoplast suspension, 0.5–1 μg of repair template, 0.5–1 μg of pTel-hyg^R plasmid, RNP complex (from **step 14**), and 25 μL of TRAFO 2. Mix gently by pipetting 3–4 times.
16. Incubate on ice for 50 min.
17. Add 250 μL of TRAFO 2, and mix gently by pipetting 3–4 times.
18. Incubate at room temperature for 20 min.
19. Add 6 mL of Top YPGS agar (48 °C, from **step** 7) containing 250 μg/mL hyg, mix gently, and immediately pour onto YPGS plates supplemented with 250 μg/ml hyg (*see* **Note 2**).
20. Repeat **steps 14–19** of transformation for the following controls: for the positive growth control (to verify the viability of the protoplasts), use transformation with protoplasts only, no repair template, no pTel-hyg^R, no RNP complex, and no hyg. For the negative control (to verify full susceptibility of the recipient strain to hyg), use transformation with protoplasts and with 250 μg/ml hyg, no repair template, no pTel-hyg^R, and no RNP complex.
21. Let plates rest at room temperature for 1 h, then seal with parafilm.
22. Incubate at 37 °C for 3–5 days, inverted (agar side up).
23. During this time, determine the minimal inhibitory concentration of the compound used for selection. The ideal concentration is the lowest concentration at which the WT strain fails to grow. For example, for voriconazole, a concentration of 0.5 μg/mL was used, as it blocks WT growth but not the growth of WT transformed with *hmg1* F262Del, which is eightfold more resistant.

3.3 Mutation Selection, Isolation, and Verification

1. After 3–5 days, verify growth patterns on plates: full growth on the positive control plate without hyg, no growth on the negative control plate with hyg, and several hyg-resistant colonies on the experimental plate with hyg, pTel-hyg^R, and repair template (*see* Figs. 3 and 4).
2. Use 1.5–3 mL of 0.02% Tween 20 to harvest as many conidia as possible from all transformation plates using a scraper. Pool conidia from identical transformation plates where appropriate (*see* **Note 2**).
3. Count the harvested conidia using a hemocytometer.
4. Plate 10^3–10^5 conidia from transformation and control plates onto prepared selection plates containing the determined concentration of compound of choice (*see* **step 23** in Subheading 3.2) (e.g., 0.5 μg/mL voriconazole). Include replicates.

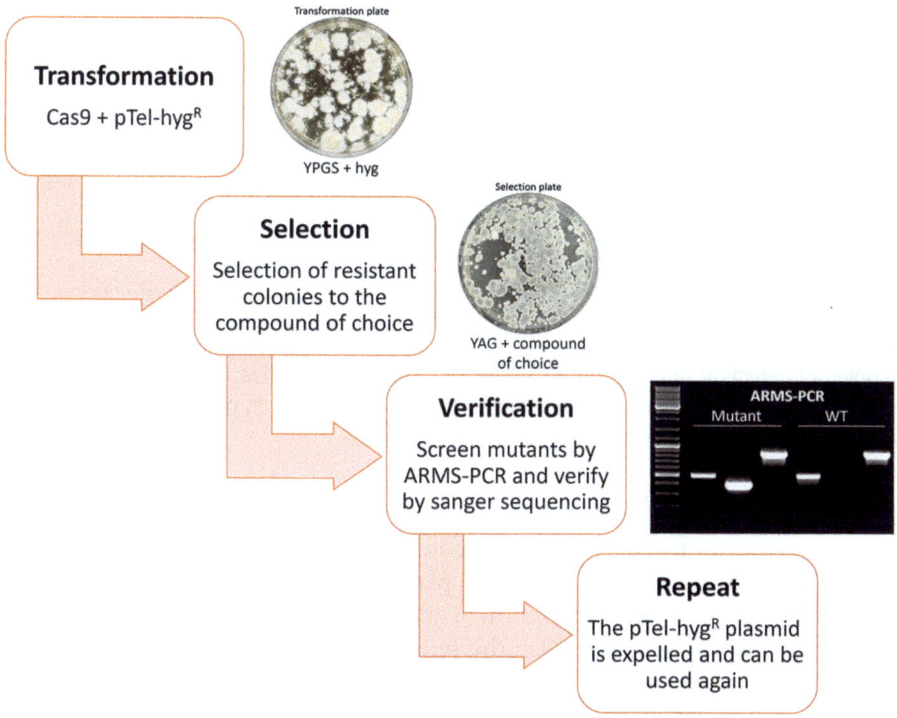

Fig. 3 Layout of the experimental steps used to generate seamless point mutations. *A. fumigatus* protoplasts are transformed using CRISPR-Cas9 transformation methodology in the presence of the pTel-hyg^R recyclable plasmid and selected on hyg. Conidia are then further selected on VRC-containing plates (or compound of choice), isolated, and screened by ARMS-PCR. Candidate isolates were Sanger-sequenced to verify the insertion of the mutation

Fig. 4 Examples of transformation plates after 3–5 days of incubation. Positive control plate shows full growth, negative control plate shows no growth, and experimental plate, on which seeded protoplasts were transformed with Cas9, gRNA, pTel-hyg^R, and repair template, shows growth of several hyg-resistant colonies that potentially contain the point mutation in the gene of interest

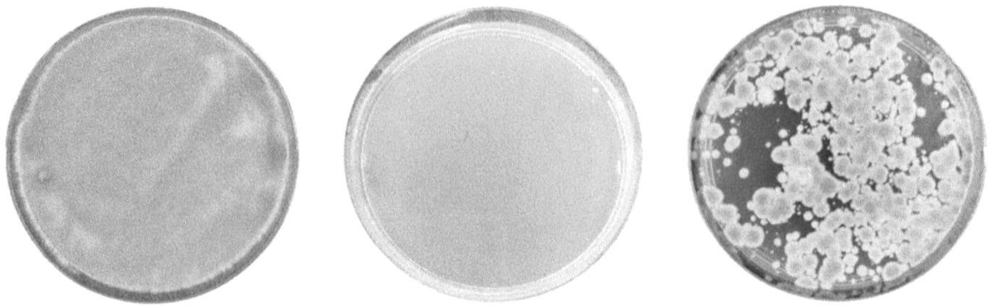

Fig. 5 Examples of selection plates after 3–5 days of incubation. Positive control plate shows full growth, negative control plate shows no growth, and selection plate shows growth of several VRC-resistant colonies that potentially contain the point mutation in the gene of interest

5. Incubate the plates at 37 °C for 3–5 days, inverted.
6. After 3–5 days, confirm expected outcomes: full growth on the positive control plate without compound, no growth on the negative control plate with compound, and several resistant colonies on the selection plate (*see* Fig. 5).
7. Pick single resistant colonies from the selection plates, and streak on YAG without any selection compounds for two rounds. After two rounds of isolation on non-selectable media, the pTel-hyg^R plasmid is lost from the cells and recycled.
8. Verify mutants by your method of choice. For example, perform ARMS-PCR (*see* **Note 3**), PCR-amplify the mutation region or full gene, check product size by gel electrophoresis, purify (e.g., use Wizard SV Gel and PCR Clean-Up System), and submit for Sanger sequencing (*see* **Note 4**).
9. This CRISPR-Cas9 method with pTel-hyg^R and ARMS-PCR validation allows rapid editing of *A. fumigatus* with high success rates of up to 100%, depending on how strongly the mutation affects susceptibility to the compound of choice (*see* **Note 5**).

4 Notes

1. It is recommended to use the *akuB*^{KU80} strain for transformation, to ensure a correct single integration of the repair template.
2. It is advisable to prepare 3–6 duplicates of the experimental transformation plates (those with Cas9, pTel-hyg^R, and repair template), to ensure sufficient spore collection for the selection step.

3. Once colonies resistant to your compound of choice are growing, it is recommended to use ARMS-PCR [6] to quickly screen for the mutated colonies, before sending them to sequencing.
4. Once verified via Sanger sequencing, ~4 isolates of the same mutation are tested and expected to present the same phenotype, to confirm correct integration of the repair template.
5. This method of selectable marker recycling allows the introduction of several point mutations one by one, by using the same selectable marker (pTel-hyg^{R}) again in each transformation. For example, we have used this plasmid to introduce the G457S point mutation in *cyp51B*, and after recycling of the plasmid, we used it again to introduce the F262del point mutation in *hmg1*. After that, we used the pTel-hyg^{R} for the third time to introduce the G448S point mutation in *cyp51A*, to create a triple mutant [7].

References

1. Krappmann S, Sasse C, Braus GH (2006) Gene targeting in aspergillus fumigatus by homologous recombination is facilitated in a nonhomologous end- joining-deficient genetic background. Eukaryot Cell 5:212–215. https://doi.org/10.1128/EC.5.1.212-215.2006
2. Al Abdallah Q, Ge W, Fortwendel JR (2017) A simple and universal system for gene manipulation in aspergillus fumigatus: in vitro -assembled Cas9-guide RNA ribonucleoproteins coupled with microhomology repair templates. mSphere 2. https://doi.org/10.1128/mSphere.00446-17
3. Wang Q, Coleman JJ (2019) Progress and challenges: development and implementation of CRISPR/Cas9 technology in filamentous fungi. Comput Struct Biotechnol J 17:761–769. https://doi.org/10.1016/j.csbj.2019.06.007
4. Barreau C, Iskandar M, Turcq B et al (1998) Use of a linear plasmid containing telomeres as an efficient vector for direct cloning in the filamentous fungus Podospora anserina. Fungal Genet Biol 25:22–30. https://doi.org/10.1006/fgbi.1998.1064
5. Leisen T, Bietz F, Werner J et al (2020) CRISPR/Cas with ribonucleoprotein complexes and transiently selected telomere vectors allows highly efficient marker-free and multiple genome editing in Botrytis cinerea. PLoS Pathog 16:e1008326. https://doi.org/10.1371/journal.ppat.1008326
6. Newton CR, Graham A, Heptinstall LE et al (1989) Analysis of any point mutation in DNA. The amplification refractory mutation system (ARMS). Nucleic Acids Res 17:2503–2516. https://doi.org/10.1093/nar/17.7.2503
7. Handelman M, Osherov N (2023) Efficient generation of multiple seamless point mutations conferring Triazole resistance in aspergillus fumigatus. JoF 9:644. https://doi.org/10.3390/jof9060644

Chapter 3

Marker-Free CRISPR/Cas9 RNP Transformation for Gene Disruption in *Aspergillus fumigatus*

Clara Valero, Jamie Tindale, Michael J. Bromley, and Norman van Rhijn

Abstract

CRISPR-based genetic engineering has resulted in the possibility of far more efficient and challenging genetic engineering in filamentous fungal pathogens. In this respect, selection marker-free CRISPR-Cas9 transformation has been possible in fungi such as *Aspergillus fumigatus*, one of the most prevalent fungal pathogens of humans. Here, we describe a protocol to perform this technique, which can be widely adapted to use multiple strains and other species.

Key words CRISPR, Genetic engineering, Aspergillus, Protoplasts, Transformation

1 Introduction

Clustered Regularly Interspaced Short Palindromic Repeats (CRISPR)-based genome editing relies on utilizing the CRISPR-Cas system, originally found to be an adaptive immune system in bacteria [1, 2]. In this system, an endonuclease, most commonly Cas9 or Cas12a, is used to make a double-stranded break (DSB) [3, 4]. This endonuclease targets specific DNA sequences, being guided by a guide RNA. This can either be via a single guide RNA or a tracRNA and crRNA. Upon binding to the specific DNA sequence, the endonuclease generates a DSB. The cell then undergoes DNA repair, resulting in either mutations from the nonhomologous end joining pathway or integration of a provided DNA repair template, via heterologous recombination [5].

Traditionally, genetic engineering in *Aspergillus* spp. relied heavily on selectable markers and homologous recombination, processes that were often time-consuming and limited by low efficiency [6]. The introduction of CRISPR-Cas9, and more recently Cas12a-based systems, has enabled targeted genome editing by introducing site-specific DSBs [7, 8]. Early work adapted CRISPR for *Aspergillus nidulans* and *Aspergillus fumigatus*,

Natalia Mach (ed.), *Aspergillus fumigatus: Methods and Protocols*, Methods in Molecular Biology, vol. 3028, https://doi.org/10.1007/978-1-0716-5222-0_3,

demonstrating that direct transformation of Cas9 ribonucleoprotein (RNP) complexes, together with short donor templates, could yield high-efficiency editing without the need for long homology arms [9, 10]. This method notably reduces the risk of random integration events, minimizing off-target effects [11].

Another advancement in CRISPR technology for *Aspergillus* species was the development of marker-free editing approaches [12]. Traditional selection markers, such as pyrG or hygromycin resistance cassettes, often complicate downstream applications, especially when multiple rounds of genome editing are required. In this context, we previously developed a highly efficient, markerless CRISPR-Cas9 system for *Aspergillus fumigatus* using an RNP-based method combined with microhomology-mediated end joining (MMEJ) repair, which allowed us to generate single-nucleotide polymorphism, promoter replacements, and introduction of paralogues into *A. fumigatus* [13–16]. By delivering preassembled RNP complexes directly into fungal protoplasts along with minimal donor DNA, we have achieved up to 10% transformation efficiency without the need for auxotrophic selection markers [12].

The increasing techniques using CRISPR tools in fungal species have also expanded into base editing, transcriptional modulation, and CRISPR interference (CRISPRi) systems [17–19]. Recent adaptations of catalytically dead Cas9 (dCas9) fused to transcriptional repressors or activators allow for precise control of gene expression without permanent genomic alterations [20]. Nonetheless, challenges remain, including the delivery of CRISPR components into certain recalcitrant strains and the mitigation of off-target activity, especially when multiplexed guides are employed.

2 Materials

Prepare all solutions using deionized and analytical grade reagents. Prepare and store all reagents at room temperature (unless indicated otherwise).

2.1 Fungal Culture Medium

1. YPS media: 2% yeast extract (20 g/L), 5 mM Tris base (0.6057 g/L), 0.5% peptone (5 g/L), 1 M sucrose (342.3 g/L), adjust the pH to pH 6.0 (with citric acid). Add 1.5% bacteriological agar.
2. Sabouraud dextrose agar: 40 g/L dextrose, 10 g/L mycological peptone, 15 g/L agar, pH 5.6.
3. Erlenmeyer flasks that at least fit 50 mL of medium. Glass flasks are preferable; make sure to wash properly and autoclave before usage.

2.2 Protoplasting Solutions

1. KCl, citric acid solution: Add 8.2 g of KCl and 2.1 g of citric acid monohydrate (molecular weight 210 g/mol) to approximately 50 mL of deionized or MilliQ water. Allow salts to dissolve. Adjust pH to 5.8 with 1.1 M KOH solution (*see* **Note 1**). Make the volume up to 100 mL with deionized water. This solution can be stored for short periods (less than 2 weeks) in a refrigerator or sterilized carefully by autoclaving (volume should not be reduced during autoclaving) and stored indefinitely at room temperature.
2. 0.6 M KCl solution: Dissolve 4.47 g of KCl in 100 mL (final volume) of deionized water. Autoclave the solution. Make sure that the volume is not reduced during autoclaving. Can be stored at room temperature.
3. 0.6 M KCl, 50 mM $CaCl_2$ solution: Dissolve 4.47 g KCl and 0.74 g of $CaCl_2$ in 100 mL (final volume) of deionized water. Can be stored indefinitely at room temperature.
4. 2 × Protoplasting solution: Make fresh on the day. To 10 mL of KCl, citric acid solution, add 1.28 g of VinoTaste FCE (Novo Nordisk). Sterilize the solution by filtering it through a sterile 0.22 μm pore size filter (*see* **Note 2**).

2.3 CRISPR-Cas9 Transformation

1. Cas9 enzyme working buffer: 20 mM HEPES, 150 mM KCl, pH 7.5. Add 200 μL HEPES (1 M, pH 7.5 nuclease-free grade) and 750 μL of KCl (2 M) to 9 mL of deionized water. Mix components, adjust pH to 7.5, and store at room temperature.
2. Nuclease-free duplex buffer: Commercially available buffer, included with Alt-R CRISPR-Cas9 reagents, store at −20 °C (short term at 4 °C).
3. PEG-$CaCl_2$ solution: 60% (W/V) PEG 3350, 50 mM $CaCl_2$, 450 mM Tris-HCl, pH 7.5 (*see* **Note 3**).
4. Alt-R CRISPR-Cas9 crRNA custom: 2 nmol.
5. Alt-R CRISPR-Cas9 tracrRNA: 5 nmol.
6. Alt-R S.p. Cas9 Nuclease 3NLS: 100 μg.

2.4 Equipment

1. Orbital shaking incubator capable of 130 rpm at 27–30 °C.
2. Benchtop centrifuge capable of 1800 × g and 2400 × g.
3. Microcentrifuge for 1.5 mL Eppendorf tubes.
4. Erlenmeyer flasks (minimum volume 50 mL), glass, autoclavable.
5. Miracloth or equivalent filtration material.
6. Thermal cycler for RNA duplex annealing.

7. Sterile filtration unit with 0.22 μm pore size filter.
8. Microscope for counting protoplasts.

3 Methods

3.1 Preparation of Guide RNA (gRNA)

1. Resuspend crRNA and tracrRNA to 100 μM stock concentrations in Nuclease-Free Duplex Buffer (or Nuclease-Free IDTE Buffer). Store resuspended RNA oligos at −20 °C. For each gRNA, prepare a 33 μM RNA duplex solution, by mixing crRNA and tracrRNA oligonucleotides in equimolar concentrations in a sterile microcentrifuge tube. For example, combine 3 μL of 100 μM Alt-R CRISPR-Cas9 crRNA, 3 μL of 100 μM Alt-R CRISPR-Cas9 tracrRNA, and 3 μL of Nuclease-Free Duplex Buffer to a total volume of 9 μL.
2. Heat the mixture at 95 °C for 5 min, then gradually cool down to room temperature (20–25 °C) using a thermal cycler. Before use, thoroughly mix the Alt-R S.p. Cas9 Nuclease 3NLS stock by inverting the tube several times and briefly centrifuge. Dilute Cas9 to a final concentration of 1 μg/μL using Cas9 working buffer. Store diluted Cas9 at −20 °C.

3.2 Preparation of Protoplasts

1. Inoculate two Erlenmeyer flasks, each containing 50 mL of Sabouraud Dextrose Agar, with 2×10^8 spores of *A. fumigatus*. Incubate for 16 h at 27–30 °C, 130 × rpm in shaking incubator (*see* **Note 4**).
2. Filter the cultures through miracloth, then transfer to flasks containing 16 mL of Sabouraud Dextrose Agar and 16 mL of 2 × protoplasting solution.
3. Incubate for up to 4 h at 30 °C, 130 rpm in shaking incubator (*see* **Note 5**).
4. Filter the residual undigested hyphae through Miracloth.
5. Top up the volume with 0.6 M of KCl to 50 mL and centrifuge at 1800 × g for 10 min.
6. Discard the supernatant and resuspend the pellet in 2 mL of 0.6 M KCl.
7. Transfer the resuspended pellet into two 1.5 mL Eppendorf tubes.
8. Centrifuge the pellets at 2400 × g for 3 min and remove the supernatant.
9. Wash each pellet by resuspending in 1 mL of 0.6 M KCl, pipetting thoroughly. Repeat this washing step twice more.

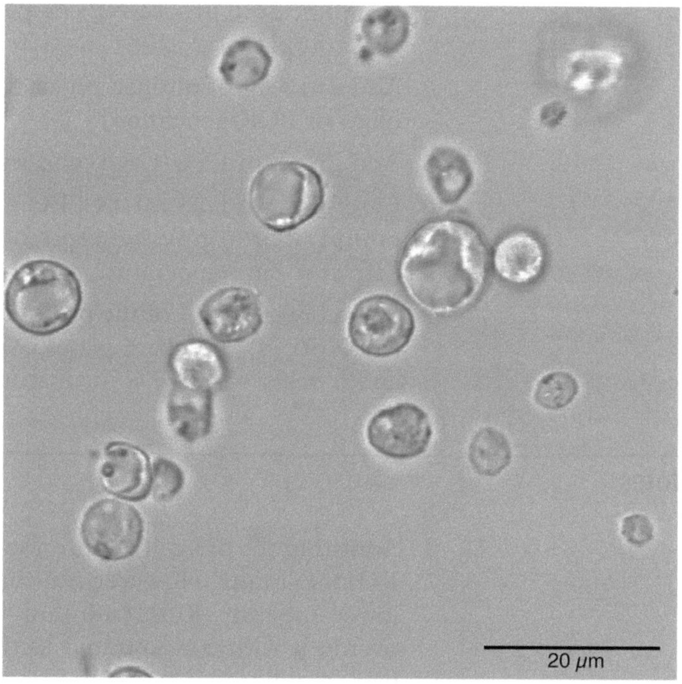

Fig. 1 *A. fumigatus* protoplasts after washing steps have been completed

10. Resuspend the final pellet in 0.5 mL of 0.6 M KCl, 50 mM $CaCl_2$ solution, and combine the two suspensions in a single Eppendorf tube.
11. Centrifuge at 2400 × g for 3 min and remove the supernatant.
12. Resuspend the final protoplast pellet in 1 mL of 0.6 M KCl, 50 mM $CaCl_2$ solution, and count the protoplasts using a microscope (*see* **Note 6**). Healthy protoplasts can vary in shape and size (Fig. 1 shows protoplasts).
13. Dilute the protoplast suspension with 0.6 M KCl, 50 mM $CaCl_2$ solution to a final concentration of 5×10^6 protoplast/mL (*see* **Note 7**).
14. Use 200 μL of the protoplast suspension for each transformation.

3.3 Transformation

1. In a 1.5 mL Eppendorf tube, combine 1.5 μL of 33 μM RNA duplex solution 1 (gRNA 1), 1.5 μL of 33 μM RNA duplex solution 2 (gRNA 2) (if only using one gRNA, add an extra 1.5 μL of Cas9 working buffer), 1.5 μL of Cas9 nuclease (1 μg/μL), and 22 μL of Cas9 working buffer to obtain a total volume of 26.5 μL.
2. Incubate the mixture at room temperature for 5 min to allow the assembly of the RNP complexes.

3. Combine 26.5 μL of RNP complex with 200 μL of *A. fumigatus* protoplasts.
4. Add 2–10 μg of purified repair template (can be bought as an oligo or PCR-generated).
5. Add 25 μL of PEG-$CaCl_2$ and leave on ice for 50 min.
6. Carefully add 1.25 mL of PEG-$CaCl_2$ solution (scale down to 600 μL if protoplast concentration is limited), mix, and incubate at room temperature for 20 min.
7. Spread 50–200 μL of the protoplast suspension on a YPS plate (*see* **Note 8**). Incubate plates overnight at room temperature and transfer to a 37 °C incubator the following day.

4 Notes

1. Adjusting the pH will require a substantial volume of the 1.1 M KOH solution. However, we found that it is better to use a lower molarity KOH solution, as it is easy to significantly overdo it with concentrated KOH, turning the solution more basic rapidly.
2. The protoplasting solution is turbid before filtering and may require multiple filters, as these can clog up quite easily. Another way to address this is to prepare the solution the day before in a bottle-top filtration unit, followed by storage in the fridge or a cold room. This allows it to filter overnight and keeps the protoplasting solution intact, which is especially useful if many transformations are planned in 1 day.
3. PEG-$CaCl_2$ solution is viscous. The PEG needs to be added slowly and dissolved before adding more. This can take a significant amount of time. Sometimes, it is useful to slightly heat the mixture, as this decreases the time needed to dissolve the PEG. The PEG solution is a balance of managing viscosity and ease of handling. We have had success with PEG6000 as well (at the same concentration as PEG3350), but handling this solution and dispensing it accurately is more difficult and requires wide-bore pipette tips.
4. If multiple strains are to be transformed in a single day, these can be grown overnight in a 50 mL centrifuge tube containing 25 mL of rich medium (Aspergillus Complete Medium or Sabouraud Dextrose Agar). This gives sufficient biomass for protoplasting. However, this culture should not be filtered prior to protoplasting. Protoplasting solution (25 mL per tube) should be added on top of the culture medium.
5. Regularly check for protoplast formation under the microscope. Incubating too long can lead to protoplasts bursting.

The time required to effectively protoplast differs per strain and species, so it is advisable to optimize this step first using a time course experiment.

6. As can be seen from the image provided (*see* Fig. 1), protoplasts can differ in shape and size. All protoplasts should be counted, as there does not seem to be a correlation between size and the number of nuclei present. However, a protoplast can contain from 0 to more than 10 nuclei.
7. Protoplast transformation can be performed with lower or higher concentrations of protoplasts, but the efficiency is much lower, and 5×10^6 seems to be the optimum concentration.
8. YPS is the normal medium used if no selection or hygromycin selection is used. However, for pyrithiamine selection, Aspergillus Minimal Medium plus 1% sorbitol is used as a pyrithiamine selection on YPS in noneffective. In a similar fashion, Aspergillus Minimal Medium plus 1% sorbitol is used when bleomycin/phleomycin selection is used.

Acknowledgments

This work has been funded through Wellcome Trust (grant: 226408/Z/22/Z).

References

1. Doudna JA, Charpentier E (2014) Genome editing. The new frontier of genome engineering with CRISPR-Cas9. Science 346:1258096. https://doi.org/10.1126/science.1258096
2. van Barrangou R, Fremaux C, Deveau H et al (2007) CRISPR provides acquired resistance against viruses in prokaryotes. Science 315: 1709–1712. https://doi.org/10.1126/science.1138140
3. Heler R, Samai P, Modell JW et al (2015) Cas9 specifies functional viral targets during CRISPR-Cas adaptation. Nature 519:199–202. https://doi.org/10.1038/nature14245
4. Chen JS, Ma E, Harrington LB et al (2018) CRISPR-Cas12a target binding unleashes indiscriminate single-stranded DNase activity. Science 360:436–439. https://doi.org/10.1126/science.aar6245
5. Xue C, Greene EC (2021) DNA repair pathway choices in CRISPR-Cas9-mediated genome editing. Trends Genet 37:639–656. https://doi.org/10.1016/j.tig.2021.02.008
6. Son Y-E, Park H-S (2021) Genetic manipulation and transformation methods for Aspergillus spp. Mycobiology 49:95–104. https://doi.org/10.1080/12298093.2020.1838115
7. Nødvig CS, Nielsen JB, Kogle ME et al (2015) A CRISPR-Cas9 system for genetic engineering of filamentous fungi. PLoS One 10: e0133085. https://doi.org/10.1371/journal.pone.0133085
8. Krappmann S (2017) CRISPR-Cas9, the new kid on the block of fungal molecular biology. Med Mycol 55:16–23. https://doi.org/10.1093/mmy/myw097
9. Zhang C, Meng X, Wei X et al (2016) Highly efficient CRISPR mutagenesis by microhomology-mediated end joining in Aspergillus fumigatus. Fungal Genet Biol 86: 47–57. https://doi.org/10.1016/j.fgb.2015.12.007
10. Al Abdallah Q, Ge W, Fortwendel JR (2017) A simple and universal system for gene manipulation in aspergillus fumigatus: in vitro-assembled Cas9-guide RNA ribonucleoproteins coupled with microhomology repair

templates. mSphere 2(6):e00446-17. https://doi.org/10.1128/mSphere.00446-17

11. Al Abdallah Q, Souza ACO, Martin-Vicente A et al (2018) Whole-genome sequencing reveals highly specific gene targeting by in vitro assembled Cas9-ribonucleoprotein complexes in Aspergillus fumigatus. Fungal Biol Biotechnol 5:11. https://doi.org/10.1186/s40694-018-0057-2
12. van Rhijn N, Furukawa T, Zhao C et al (2020) Development of a marker-free mutagenesis system using CRISPR-Cas9 in the pathogenic mould Aspergillus fumigatus. Fungal Genet Biol 145:103479. https://doi.org/10.1016/j.fgb.2020.103479
13. van Rhijn N, Storer ISR, Birch M et al (2024) Aspergillus fumigatus strains that evolve resistance to the agrochemical fungicide ipflufenoquin in vitro are also resistant to olorofim. Nat Microbiol 9:29–34. https://doi.org/10.1038/s41564-023-01542-4
14. van Rhijn N, Zhao C, Al-Furaji N et al (2024) Functional analysis of the Aspergillus fumigatus kinome identifies a druggable DYRK kinase that regulates septal plugging. Nat Commun 15:1–16. https://doi.org/10.1038/s41467-024-48592-8
15. van Rhijn N, Hemmings S, Storer ISR et al (2022) Antagonism of the Azoles to Olorofim and cross-resistance are governed by linked transcriptional networks in aspergillus fumigatus. MBio 13:e0221522. https://doi.org/10.1128/mbio.02215-22
16. van Rhijn N, Bromley M, Richardson M et al (2021) CYP51 paralogue structure is associated with intrinsic azole resistance in fungi. mBio 12:e0194521. https://doi.org/10.1128/mBio.01945-21
17. Zhao F, Sun C, Liu Z et al (2023) Multiplex base-editing enables combinatorial epigenetic regulation for genome mining of fungal natural products. J Am Chem Soc 145:413–421. https://doi.org/10.1021/jacs.2c10211
18. Wensing L, Sharma J, Uthayakumar D et al (2019) A CRISPR interference platform for efficient genetic repression in Candida albicans. mSphere 4(1):e00002-19. https://doi.org/10.1128/mSphere.00002-19
19. Mózsik L, Hoekzema M, de Kok NAW et al (2021) CRISPR-based transcriptional activation tool for silent genes in filamentous fungi. Sci Rep 11:1–13. https://doi.org/10.1038/s41598-020-80864-3
20. Song R, Zhai Q, Sun L et al (2019) CRISPR/Cas9 genome editing technology in filamentous fungi: progress and perspective. Appl Microbiol Biotechnol 103:6919–6932. https://doi.org/10.1007/s00253-019-10007-w

Chapter 4

CRISPR-Cas9 and Microhomology-Mediated End-Joining for Biosynthetic Pathway Reconstruction in *Aspergillus fumigatus*

Yaojie Guo and Daniel H. Scharf

Abstract

We have recently engineered *Aspergillus fumigatus* using an AMA1-based episomal expression CRISPR vector and a microhomology-mediated end-joining repair system. We adopted *A. fumigatus* as a heterologous expression host to investigate the biosynthetic pathways of epipolythiodioxopiperazine-type compounds. In the present chapter, we describe the assembly of the CRISPR-Cas9 vector and microhomology template plasmid, their transformation into *A. fumigatus*, and metabolite detection.

Key words *Aspergillus fumigatus*, CRISPR-Cas9, Microhomology-mediated end-joining, Pathway reconstruction, Epipolythiodioxopiperazine, Gliotoxin, *Chaetomium cochliodes*, Fungal biosynthesis

1 Introduction

Aspergillus fumigatus is an opportunistic human pathogen that the World Health Organization (WHO) included among the highest "critical" group in its first fungal priority pathogens list in 2022 [1] (*see* **Note 1**). This species can produce an array of bioactive compounds with a broad functional spectrum. So far, 19 biosynthetic gene clusters have been linked to their products in this species [2]. Among its various types of secondary metabolites, the epipolythiodioxopiperazine (ETP)-type compound, gliotoxin, is one of the most studied compounds [3]. The versatility of *A. fumigatus*' secondary metabolism makes it an interesting host for the reconstruction of pathways from organisms that might be inaccessible to genetic manipulation.

In addition to the classical nonhomologous end-joining (NHEJ) and homologous recombination (HR) pathways, an alternative DNA repair mechanism known as microhomology-mediated end-joining (MMEJ) in eukaryotic cells has attracted great attention recently. Briefly, an MMEJ process involves the sequence

Natalia Mach (ed.), *Aspergillus fumigatus: Methods and Protocols*, Methods in Molecular Biology, vol. 3028, https://doi.org/10.1007/978-1-0716-5222-0_4,

resection at both broken ends of the DNA double-strand breaks (DSBs), generating single-strand stretches. This is followed by the microhomologous annealing within the overhangs and the nucleolytic cleavage of the unpaired bases. After the DNA gap-filling and ligation, the DNA breaks are sealed [4]. MMEJ has been successfully applied in several fungal engineering attempts, including *A. fumigatus*, *A. melleus*, *A. nidulans*, and *A. aculeatus*, among others [5–7].

Here, we describe a strategy for reconstructing a biosynthetic pathway by replacing a gene in *A. fumigatus* with another gene from a different fungal species, utilizing CRISPR-Cas9 and MMEJ. To make this approach easier to understand we illustrate it with the replacement of the nonribosomal peptide synthetase (NRPS) encoding gene *gliP*, responsible for gliotoxin biosynthesis, with the gene *cheP* from *Chaetomium cochliodes* CGMCC3.17123 as an example in this protocol [8, 9]. We use the isothermal homology-based assembly (known as Gibson assembly) [10] method to build the CRISPR-Cas9 plasmid instead of the previously used USER cloning [11] strategy. In the repair template preparation step, we incorporate the microhomologous segments (50 bp) separately into two pairs of PCR primers, ensuring that each PCR product features a 20 bp overlap for the next vector assembly. For transformant screening, we utilize fungal colony PCR in a 10 μL volume directly using the fungal colonies (mycelia) as the DNA template for diagnostic PCR, eliminating the need for any DNA isolation.

We have demonstrated that this protocol can also be used for transferring other exogenous ETP-pathway genes (such as methyl transferase genes and transporter genes), and simultaneously replacing two genes of the gliotoxin BGC in *A. fumigatus*. We anticipate that our methodology can be used to explore the gene functions within predicted biosynthetic gene clusters (BGCs) in a well-characterized BGC whose host is suitable for genetic engineering. This approach provides a broader perspective on heterologous expression.

2 Materials

2.1 Strains

1. *A. fumigatus* CEA17 *ΔakuB pyrG*$^{+}$ (referred to as AfuB001) [12].
2. *Escherichia coli* DH5α competent cells.
3. Original fungal strain with target biosynthetic gene(s), e.g. *Chaetomium cochliodes* CGMCC3.17123 [9].

2.2 Plasmids

1. pFC332 (Addgene ID #87842) [13] (*see* **Note 2**).
2. pFC902 (Addgene ID #106904) [13].
3. pUC57 (Addgene ID #54338) (*see* **Note 3**).

2.3 Reagents and Solutions

1. Salt solution (50×): 26 g KCl and 76 g KH_2PO_4. Dissolve in ultrapure water, make up to 1 L, and autoclave.
2. Glucose (20%): dissolve 200 g of glucose in ultrapure water, make up to 1 L, and autoclave.
3. $MgSO_4$ (20%): dissolve 20 g of $MgSO_4$ in 60 mL ultrapure water, make up to 100 mL, and autoclave.
4. Clive Roberts Trace Elements solution (1 L): 1 g $FeSO_4{\cdot}7H_2O$, 8.8 g $ZnSO_4{\cdot}7H_2O$, 400 mg $CuSO_4{\cdot}5H_2O$, 150 mg $MnSO_4{\cdot}H_2O$, 100 mg $Na_2B_4O_7{\cdot}10H_2O$, and 50 mg $(NH_4)_6Mo_7O_{24}{\cdot}4H_2O$. Dissolve in ultrapure water, make up to 1 L, and autoclave.
5. Solution 1: dissolve 44.73 g KCl in ultrapure water, make up to 1 L, and autoclave.
6. Solution 2: dissolve 44.73 g KCl and 5.55 g $CaCl_2$ in ultrapure water, make up to 1 L, and autoclave.
7. *Aspergillus* transformation buffer (ATB, 1 L solution): 218.61 g sorbitol, 44.73 g KCl, 5.55 g $CaCl_2$, 20 mL 1 M Tris-HCl pH 8.0 in ultrapure water. Adjust to pH 7.2, make up to 1 L, and autoclave. Can be stored at 4 °C for up to 1 year.
8. PCT solution (200 mL): 100 g PEG 8000, 8.95 g KCl, 1.11 g $CaCl_2$, 4 mL 1 M Tris-HCl pH 8.0. Dissolve in ultrapure water, adjust to pH 7.5, make up to 200 mL, and autoclave. Can be stored at 4 °C for up to 1 year.
9. NaCl solution (0.9%): dissolve 9 g NaCl in ultrapure water, make up to 1 L, and autoclave.
10. Antibiotics for selection: final concentration in the medium 100 μg/mL for ampicillin and 200 μg/mL for hygromycin.

2.4 Kits and Enzymes

1. Restriction enzyme: PacI (e.g., FuniCut PacI from Yeasen Biotechnology).
2. Fungi genomic DNA extraction kit.
3. Homology-based cloning kit (e.g., the Hieff Clone Universal II One-Step Cloning Kit from Yeasen Biotechnology).
4. Phusion Flash High-Fidelity PCR Master Mix (Thermo Scientific), use for the amplification of CRISPR fragments (*see* **Note 4**).
5. AG12204 DNA polymerase kit (*see* Table 1) containing *ApexHF* HS DNA Polymerase CL (1 U/μL) and 2X *ApexHF* CL Buffer (with 2 mM Mg^{2+} and 400 μM dNTP) from

Table 1
PCR reaction mix (50 μL) using AG12204 polymerase

Components	Volume (μL)
ApexHF HS DNA Polymerase CL	1
2X *ApexHF* CL Buffer	25
Template (< 500 ng)	1
Primer F (10 μM)	2.5
Primer R (10 μM)	2.5
Nuclease-free water	18

Accurate Biotechnology (Hunan). Use for the PCR reactions including the amplification of repair template fragments, *E. coli* colony PCR, and diagnostic tissue PCR (fungi colony PCR).

6. Gel extraction kit.
7. Plasmid miniprep kit.
8. VinoTaste® Pro enzyme powder (Novozymes) for preparing fungal protoplasts.

2.5 Equipment and Consumables

1. Syringe and syringe-driven filter with a 0.22 μm polyethersulfone (PES) membrane.
2. Filtration materials: 40 μm and 70 μm cell strainers.
3. Sterile filter material (e.g., Miracloth from Merk Millipore).
4. T75 cell culture bottles.
5. 90 mm sterile petri dishes.
6. Gas-permeable film (e.g., EasyFilm nanopore PE film from Wuhan Biorun Biosciences) (*see* **Note 5**).
7. 2 mL and 50 mL sterile centrifuge tubes.
8. Sterile inoculation loops and Drigalski spatulas.
9. Microscope and cell counting chamber.
10. 250 mL shaking flasks with vented sealing film, autoclaved.
11. PCR tubes, and a thermocycler with the option to decrease temperatures at a specific cycle (e.g., Applied Biosystems ProFlex 3 × 32-well thermocycler).
12. Agarose gel, nucleic acid stain, loading dye, DNA marker, electrophoresis tank, combs, TAE buffer, power station, and gel imaging system.
13. Ethyl acetate, LC-MS-grade methanol, vials and glass inserts, LC-DAD-MS instrument.
14. Microvolume spectrophotometer (e.g., Nanodrop).

15. Orbital shaker.
16. Vortex mixer.

2.6 Media

1. Aspergillus Transformation Media (ATM): 182 g sorbitol, 6 g $NaNO_3$, 0.815 g KH_2PO_4, 1.045 g K_2HPO_4, 20 g agar, 10 g glucose. Dissolve in ultrapure water. Add 1 mL of Clive Roberts Trace Elements solution, make up to 1 L, and autoclave. Add 2.5 mL of 20% $MgSO_4$ after autoclaving.
2. Aspergillus Minimal Media (AMM): 20 mL of 50× salt solution, 1 mL of Clive Roberts Trace Elements solution, 6 g $NaNO_3$ in 950 mL ultrapure water. Adjust pH to 6.5 with KOH solution. Autoclave, and add 50 mL of 20% glucose and 2.5 mL of 20% $MgSO_4$ after cooling down.
3. Aspergillus Minimal Media Agar (1 L): prepare AMM (see Step 2). Add 15 g agar before autoclaving.
4. Luria-Bertani (LB) liquid and solid medium.
5. Malt Extract Medium (MAG): 20 g of malt extract, 2 g peptone, 10 g dextrose, 1 mL Clive Roberts Trace Elements solution, 20 g agar in ultrapure water. Make up to 1 L, and autoclave.
6. Czapek-Dox broth (e.g., Coolaber #MM1010).

2.7 Primers for the Construction of CRISPR Vector

1. 902F: TAGCTGTTTCCGCTGAGGGTTTAATGATCACATAGATGCTCGGTTGACA
2. 902R: TCTGCTGTCTCGGCTGAGGTCTTAATACCCTGAGAAGATAGATGTGAATGTG

Based on the genomic sequence of *Aspergillus fumigatus* AfuB001, we designed two double-strand breaks (DSBs) flanking the *gliP* gene open reading frame, in the proximity of its start and stop codons [14]. For each DSB, we selected a protospacer adjacent motif (PAM; 5’-NGG-3′; specifically, 5’-CGG-3′ or 5′-AGG-3′) and defined the 20 nucleotides directly upstream as the protospacer sequence (*see* Fig. 1).

3. gliPPS1R: **AGCGACGGACCGGTCAAAAA** TGCATCATCCGTGAATCGAAC,

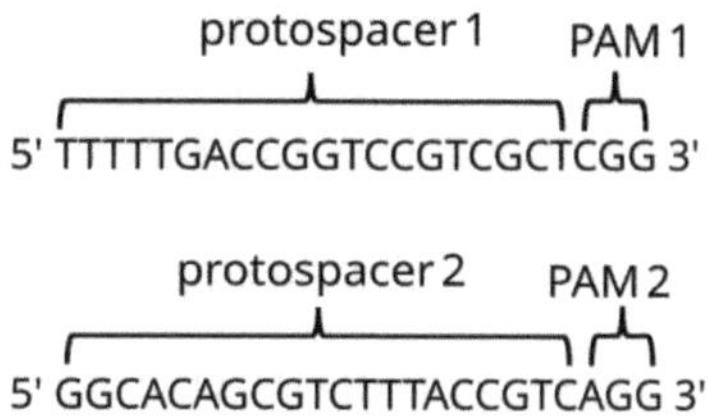

Fig. 1 The protospacer sequences in the genomic sequence of *gliP*

4. gliPPS1F: **TTTTTGACCGGTCCGTCGCT** GTTTTA GAGCTAGAAATAGCAAG,
5. gliPPS2R: GACGGTAAAGACGCTGTGCC TGCAT CATCCGTGAATCGAAC,
6. gliPPS2F: GGCACAGCGTCTTTACCGTC GTTTTAGAGC TAGAAATAGCAAG,

In the four primer sequences above, the **bold**-highlighted and underlined portions correspond to the first and second protospacers, respectively. The sequences TGCATCATCCGTGAATC GAAC and GTTTTAGAGCTAGAAATAGCAAG within the primers anneal to the tRNA and gRNA coding sequences of pFC902, respectively during PCR amplifications.

7. YJG078: CTCGTTTTCATGGCTTAGGGGC.
8. YJG079: TGCTCGGTTGACAGGACCAC.

Primers YJG078 and YJG079 are employed to validate the construction of CRISPR vectors.

2.8 Primers for the Construction of pUC57-Based Repair Template

We utilize the amplification of the hypothetical NRPS gene *cheP* from the genomic DNA (gDNA) of *C. cochliodes* CGMCC3.17123 as a representative example in this section.

1. 57ptailF: **AGCCAGACAATAATCGAGTAGATATT-CAAAA***TTAATTAA*aaccctggcgttacccaact
2. 57ptailR: TGTAGCCTTGCTTGAGGTTGTTCATCTGTC *TTAATTAA*aaagcctggggtgcctaatga.

This primer pair, 57ptailF and 57ptailR, consists of three components: a portion of the microhomology sequence (with the upstream segment highlighted in **bold** and the downstream segment underlined), a recognition site for a restriction enzyme (e.g., the *Pac*I recognition site *TTAATTAA* in *italics*), and a binding site for the plasmid pUC57 to facilitate PCR amplification (denoted in lowercase) (see *View-1* and *View-2* in Fig. 2).

3. 57chepF: **TACTCGATTATTGTCTGGCTCAACCC-TACGCTCTGCATC**atgactgctgaagctagcac
4. 57chepR: CAACCTCAAGCAAGGCTACAATGTA-TACCGTGTGGAAGAAcggaagccgaagtgagaag

The second pair of primers entails two components: a segment of the microhomology sequence (with the upstream portion highlighted in **bold** and the downstream portion underlined) and a binding site corresponding to the original fungal gDNA sequence (e.g., *C. cochliodes*), indicated in lowercase.

5. YJG075: AGTTGGGTAACGCCAGGGTT.
6. YJG076: TCATTAGGCACCCCAGGCTTT.

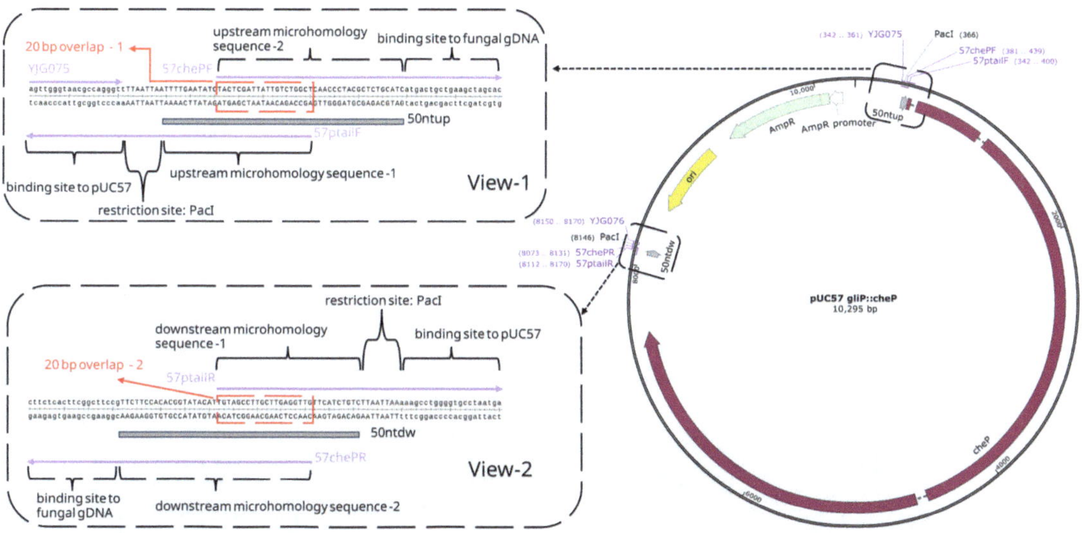

Fig. 2 Schematic representation of the amplification of the *cheP* sequence using a pUC57-based plasmid

Primers YJG075 and YJG076 are used for the verification of repair plasmid construction. A map for the complete repair template (*gliP*::*cheP*) is shown (*see* Fig. 2).

2.9 Primers for PCR Product Verification

1. 330seqckR: TGGGCGGTGATTCTGCTGT, use for sequencing.
2. glipckF: ACCTCGAAGTCAACCCCCAC, use for the verification of *gliP*::*cheP*.
3. glipckR: GTCGAATGGAACGATCTCCCAGTC, use for the verification of *gliP*::*cheP*.

3 Methods

3.1 Generation of DNA Fragments for CRISPR Vectors

1. Linearize the pFC332 plasmid backbone using FuniCut PacI. In a microcentrifuge tube, combine 1 μg of pFC332, 1 μL of FuniCut *Pac*I, 2 μL of 10× FuniCut Buffer, and add nuclease-free water to a final volume of 20 μL. Incubate at 37 °C for 15 min, and inactivate the reaction at 80 °C for 20 min.
2. Set up three PCR reactions to target two sites within a specific gene (here *gliP* gene) of AfuB001. For each 50 μL reaction, mix 25 μL of 2× Phusion Flash PCR Master Mix, 1 μL of DNA template (pFC902), 2.5 μL of forward and reverse primers, and add 19 μL of nuclease-free water to a final volume of 50 μL. To generate three CRISPR fragments, we use three pairs of primers: 902F + gliPPS1RT, gliPPS1FT + gliPPS2RT, and gliPPS2FT + 902R.

Table 2
PCR thermocycling steps for generating the CRISPR fragments

Cycle step	Temperature	Time	Cycles
Initial denaturation	98 °C	10 s	1
Denaturation	98 °C	1 s	25 to 34
Annealing	70 °C, with −0.5 °C/cycle	5 s	
Extension	72 °C	15 s/kb	
Final extension	72 °C	1 min	1

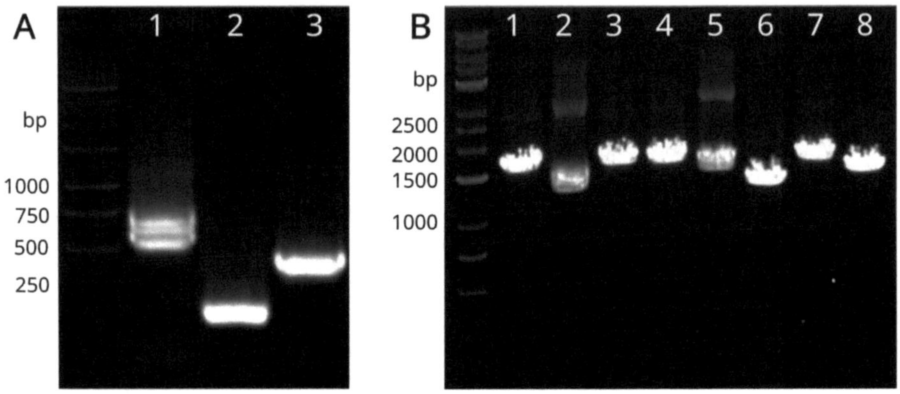

Fig. 3 Illustrations of (**a**) three CRISPR fragments, with the first PCR reaction yielding two fragments (calculated sizes 584, 755 bp), and (**b**) colony PCR results for CRISPR vector construction, with a 50% success rate (bands 1, 3, 4, and 7, positive calculated band size 1574 bp)

3. Perform the PCR reaction with a touchdown thermocycling procedure (*see* Table 2).
4. Verify the PCR products using agarose gel electrophoresis. If the fragments are correct (expected sizes 584, 191, and 423 bp, *see* Fig. 3a and **Note 6**), purify the DNA fragments with a DNA recovery kit.
5. Measure DNA concentrations.

3.2 Generation of DNA Fragments for Repair Templates

The construction of the plasmid repair requires the generation of two PCR products. The first product mainly contains the pUC57 plasmid backbone (e.g., using primers 57ptailF and 57ptailR) with the expected band size around 2584 bp. The second product consists primarily of the gene sequence of interest (e.g., using primers 57chepF and 57chepR), with the product size varying according to the specific gene of interest. We use the AG12204 DNA polymerase kit for PCR amplification (*see* Table 3). Notably, both products share two 20 bp overlapping regions at their respective ends (see *View-1* and *View-2* in Fig. 2).

Table 3
PCR thermocycling steps using AG12204 polymerase

Cycle step	Temperature	Time	Cycles
Initial denaturation	94 °C	6 min	1
Denaturation	98 °C	10 s	25 to 34
Annealing	68 °C, with −0.5 °C/cycle	30 s	
Extension	68 °C	1 min/kb	
Final extension	68 °C	Variable	1

1. For each 50 μL reaction (*see* Table 1), mix 1 μL of ApexHF HS DNA Polymerase CL, 25 μL of 2× ApexHF CL Buffer, 1 μL of DNA template, 2.5 μL of forward and reverse primers, and add 18 μL of nuclease-free water to 50 μL. We use primers 57ptailF and 57ptailR, and plasmid pUC57 as a DNA template to generate the DNA product bearing the pUC57 plasmid backbone. Further, we utilize primers 57chepF and 57chepR, and DNA template (gDNA of *C. cochliodes*) during PCR preparation to generate the gene to be introduced.
2. Perform the PCR reactions using a touchdown thermocycling procedure (*see* Table 3).
3. Verify the PCR products using agarose gel electrophoresis. If the fragments are correct (the first PCR product size is fixed at 2584 bp, while the size of the second PCR product depends on the gene to be inserted), purify the DNA fragments with a DNA recovery kit.
4. Measure DNA concentrations.

3.3 Construction of the CRISPR Vector and the Repair Template Plasmid

We build the vectors based on the Gibson DNA assembly strategy (*see* **Note 7**) using Hieff Clone Universal II One-Step Cloning Kit.

1. Combine in 200 μL PCR tube: 2× Hieff Clone Universal Enzyme Premix (5 μL), 1 μL linearized pFC332 (concentration > 10 ng/μL), 1 μL each of the three CRISP PCR fragments (concentration > 10 ng/μL) (*see* Figs. 3a and 4) or the two PCR fragments for repair template plasmid (*see* Fig. 2) (concentration > 10 ng/μL), and nuclease-free water to 10 μL. Mix gently by pipetting.
2. Incubate the mixture at 50 °C for 10 min.
3. Thaw 100 μL *E. coli* DH5α competent cells on ice for 10 min.
4. Add 50 μL of DH5α cells to the assembly reaction and mix by gently flicking. The remaining cells are used for a negative control.
5. Incubate on ice for 25 min.

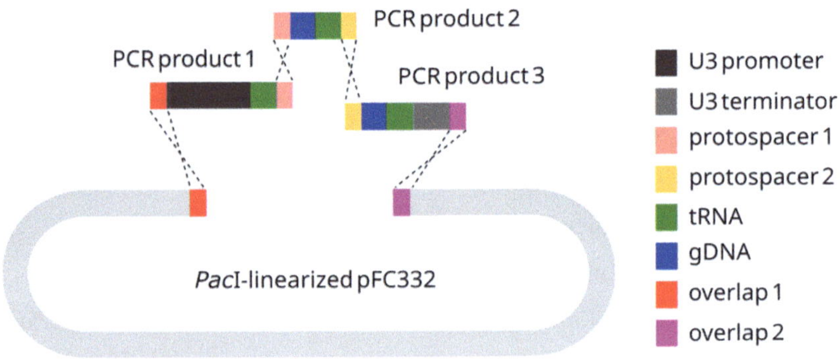

Fig. 4 Assembly of the CRISPR vector

6. Heat shock at 42 °C for 1 min.
7. Incubate on ice for 5 min.
8. Transfer the mixture (approximately 60 μL) to a sterile 2 mL tube, and add 750 μL of LB liquid.
9. Incubate at 37 °C with shaking at 200 rpm for 30 min.
10. Centrifuge at 5000 rpm in a microcentrifuge for 1 min, and discard 750 μL of supernatant.
11. Resuspend the pellet, and spread on an LB plate containing 100 μg/mL ampicillin.
12. Incubate the plate at 37 °C overnight.
13. Pick and transfer eight colonies into an eight-tube PCR strip using pipette tips, each tube containing 10 μL of sterile water.
14. Perform PCR using the AG12204 DNA polymerase kit to verify the successful plasmid construction. In the colony PCR reactions, which can be performed in a 10 μL reaction volume, we utilize 1 μL of the above cell suspension as the DNA template and employ primers YJG078 and YJG079 for CRISPR vector confirmation and YJG075 and YJG076 for repair plasmid verification.
15. The anticipated PCR fragment size for the CRISPR vector is 1574 bp. The repair template PCR product size varies based on the targeted gene sequence, and in this case, the calculated PCR product size for *cheP* amplification is 7829 bp.
16. Select PCR-verified colonies, inoculate in LB broth supplemented with 100 μg/mL ampicillin at 37 °C, and incubate at 200 rpm overnight.
17. Extract the assembled plasmids from *E. coli* DH5α using a commercial plasmid miniprep kit.
18. Send the plasmids for DNA sequencing (*see* **Note 8**). The sequencing primer for CRISPR plasmids is 330seqckR primer. Design appropriate sequencing primers to verify the plasmid

repair template. The double-confirmed plasmids will be used for the subsequent *A. fumigatus* transformation.

19. Linearize the pUC57-based repair plasmid with *Pac*I before use.

3.4 Transformation

1. Grow *A. fumigatus* AfuB001 on MAG agar in a T75 cell culture bottle a at 37 °C for 72 h (*see* **Note 9**). Harvest conidia using a 0.9% NaCl solution and an inoculation loop. Filter the suspension through a 40 μm cell filter into a new 50 mL tube. Count conidia under a microscope.
2. Inoculate 1×10^8 *A. fumigatus* conidia into 100 mL AMM medium, and incubate at 37 °C with 200 rpm shaking overnight.
3. Dissolve 2.2 g VinoTaste® Pro enzyme powder in 20 mL solution 1, stir using a magnetic stirrer for 10 min. Filter the mixture through a 0.22 μm syringe-driven PES filter into a sterile 250 mL flask.
4. Harvest the mycelium by filtering the AMM liquid through a 70 μm cell strainer.
5. Transfer a pea-sized portion of the mycelium into the enzyme solution and incubate at 28 °C with shaking at 60 rpm for 3 h. Check the protoplast production microscopically after 2 h.
6. Filter the enzyme mixture through a 40 μm cell strainer into a sterile 50 mL centrifuge tube.
7. Washing step: Centrifuge the filtrate at 1300 × g for 10 min at 4 °C, discard the supernatant, gently add 20 mL of Solution 2, and resuspend the pellet. Repeat the washing step two more times.
8. Discard the supernatant, add 1 mL of cold ATB to the pellet, and carefully mix by pipetting.
9. Measure the protoplast concentration under a microscope. The optimal protoplast concentration for transformation is between 1×10^8 and 1×10^9 protoplasts/mL (*see* **Note 10**).
10. Combine 100 μL of protoplasts, 10 μL of CRISPR plasmid (> 100 ng/μL) and 10 μL of linearized repair template plasmid (> 100 ng/μL) in a 2 mL centrifuge tube.
11. Add 150 μL of cold PCT solution, mix gently, and incubate at room temperature for 10 min.
12. Add 250 μL of ATB, mix gently, and plate on selective ATM plates (e.g., hygromycin). Prepare control plates. For the positive control, spread on ATM plates without antibiotics. For the negative control, spread on selective ATM plates. Air-dry plates in the biosafety cabinet for 20 min before putting the lid on. Seal plates with gas-permeable film.

13. Incubate all plates at 37 °C for 2–3 days.
14. Pick up 8–10 colonies from the transformation plates and transfer them to a new AMM agar plate supplemented with selective antibiotics using sterile pipette tips.
15. Incubate the selective AMM plate at 37 °C until new colonies appear.
16. Select the transformants for PCR verification: prepare the PCR reaction mix according to Table 1. For a single transformant check, a PCR reaction of 10 μL was shown to be sufficient for screening. Gently touch the colony surface of a transformant with a sterile 10 μL pipette tip, and dive the tip into the PCR mix, pipetting twice to release the tissue sample (*see* **Note 11**). We used glipckF+glipckR primers for the verification of *gliP*::*cheP*. Prepare a negative control with a fresh colony of the starting strain AfuB001.
17. Perform the PCR reaction according to the parameters in Table 3.
18. Verify the fragment size using agarose gel electrophoresis.
19. Transfer colonies showing correct PCR bands to nonselective AMM plates using a sterile pipette tip, seal the plate with a film, and incubate at 37 °C for 2–3 days. Removing the selection agent results in losing the *AMA1*-based CRISPR vector.
20. When colonies reach an approximate diameter of 5 mm (about 2 days), repeat colony PCR verification. Transfer the newly verified colony to another nonselective AMM plate. Culture the colonies on nonselective AMM for 2–4 rounds to confirm the CRISPR vector loss.
21. Inoculate the CRISPR vector-free mutant on MAG media in a T75 cell culture bottle at 37 °C for 3 days to harvest the mutant conidia.
22. Plate 10^6 conidia on selective and nonselective AMM, compare colony growth after 1–2 days of incubation at 37 °C. Lack of growth on the selective plate confirms vector loss. Alternatively, extract the genomic DNA of the mutant and perform PCR reactions using CRISPR vector-specific primers (e.g., YJG078 and YJG079) (*see* step 13 in Sect. 3.3). Absence of a 1574 bp band confirms vector loss.
23. Use vector-free mutants for subsequent transformations using the same selectable marker.

3.5 Metabolic Analysis

Compare the metabolic profiles of the AfuB001 strain and the *gliP*::*cheP* mutant under the gliotoxin production condition to assess the production of new ETP-type compounds.

1. Inoculate 1×10^8 spores of the *A. fumigatus* mutant in 100 mL Czapek-Dox broth, incubate at 28 °C with 200 rpm shaking for 7 days.
2. Filter the culture through Miracloth filter material.
3. Extract 40–100 mL of filtrate with an equal volume of ethyl acetate in a conical glass flask (250 mL) by liquid-liquid partitioning for 3 h at room temperature with occasional swirling (*see* **Note 12**).
4. Gently pour the upper organic layer into a new conical flask, and transfer 1.5 mL to a 2 mL microcentrifuge tube. Evaporate organic extracts to dryness under a stream of nitrogen gas, and reconstitute in 300 μL LC-MS-grade methanol by vortex mixing for 5 min.
5. Centrifuge the methanolic liquid at maximum speed for 5 min, and transfer 200 μL of the supernatant to a new HPLC vial with a glass insert.
6. Inject 1–10 μL of the sample into an LC-DAD-MS system to analyze the metabolic profile.

4 Notes

1. The handling of all *Aspergillus fumigatus* strains throughout this work was performed in a Biosafety Level 2 Laboratory. We recommend referring to the WHO and local regulations for working with this species.
2. The use of an auxotrophic marker, such as *pyrG*, *riboB*, or *argB*, requires the deletion of the corresponding gene within the host strain as the first step to obtain an auxotrophic strain.
3. Other high-copy-number-type plasmids can also be used for propagation.
4. Other DNA polymerases, such as Q5 High-Fidelity DNA Polymerase from New England Biolabs, also work.
5. We usually seal the plates in the incubator to avoid contamination. The widely used PARAFILM sealing film decomposes easily when incubated at 37 °C. EasyFilm stays intact for longer than 7 days at 37 °C.
6. It is worth noting that the first PCR reaction often yields two bands at 755 and 584 bp due to the presence of dual binding sites within the pFC902 sequence. We recommend performing gel electrophoresis at a low voltage (such as 120 V for a Tanon HE-120 electrophoresis tank) with 1.2% agarose gel for a duration of 30–40 min. Monitor the gel constantly to avoid overrun.

7. Compared to the previously reported CRISPR plasmid assembly using the USER cloning method [11, 13], the CRISPR vector construction based on the Gibson DNA assembly strategy has several advantages, including the elimination of the need for expensive deoxyuracil (dU)-containing primers, specialized DNA polymerase for PCR amplification (required for dU), and the nicking endonuclease Nt.BbvCI to generate single-stranded overhangs. Therefore, this protocol is more efficient and cost-effective. However, its success rate is approximately 50% due to the possible unwanted mismatch during recombination (*see* Figs. 3b and 4), which is lower than that of USER cloning. We recommend using Gibson assembly when less than three protospacers are designed, and USER cloning, when three or more DSBs are planned.
8. Analyzing the plasmids by restriction digestion is an option before DNA sequencing.
9. Harvesting spores grown in a cell culture bottle reduces the number of spores released into the air.
10. We use a Thoma chamber to count protoplasts and spores. Make sure that the chamber is properly cleaned, and eventually add 0.1% of Tween 80 to the spores to avoid aggregation.
11. The mycelia sample at the tip should be barely visible to the naked eye. Once you see the white mycelia, that might be too much for tissue PCR.

Acknowledgments

This work was supported by NSFC grant No. W2432017 to Daniel H. Scharf.

References

1. World Health Organization (ed) (2022) WHO fungal priority pathogens list to guide research, development and public health action, 1st edn. World Health Organization, Geneva
2. Seo H-W, Wassano NS, Amir Rawa MS et al (2024) A timeline of biosynthetic gene cluster discovery in aspergillus fumigatus: from characterization to future perspectives. J Fungi (Basel) 10(4):266. https://doi.org/10.3390/jof10040266
3. Huber EM (2022) Epipolythiodioxopiperazine-based natural products: building blocks, biosynthesis and biological activities. Chembiochem 23: e202200341. https://doi.org/10.1002/cbic.202200341
4. Seol J-H, Shim EY, Lee SE (2018) Microhomology-mediated end joining: good, bad and ugly. Mutat Res 809:81–87. https://doi.org/10.1016/j.mrfmmm.2017.07.002
5. Zhang C, Meng X, Wei X et al (2016) Highly efficient CRISPR mutagenesis by microhomology-mediated end joining in aspergillus fumigatus. Fungal Genet Biol 86: 47–57. https://doi.org/10.1016/j.fgb.2015.12.007
6. Jarczynska ZD, Garcia Vanegas K, Deichmann M et al (2022) A versatile in vivo DNA assembly toolbox for fungal strain engineering. ACS Synth Biol 11:3251–3263. https://doi.org/10.1021/acssynbio.2c00159

7. Yuan B, Grau MF, Murata RM et al (2023) Identification of the Neoaspergillic acid biosynthesis gene cluster by establishing an in vitro CRISPR-ribonucleoprotein genetic system in aspergillus melleus. ACS Omega 8: 16713–16721. https://doi.org/10.1021/acsomega.2c08104
8. Guo Y, Liu J-Z, Limwachiranon J et al (2024) Reconstitution of the early stage of Chetomin biosynthesis in aspergillus fumigatus leads to the production of Epipolythiodioxopiperazines. Org Lett 26:4469–4474. https://doi.org/10.1021/acs.orglett.4c01274
9. Wang M-H, Hu Y-C, Sun B-D et al (2018) Highly photosensitive poly-sulfur-bridged Chetomin analogues from *Chaetomium cochliodes*. Org Lett 20:1806–1809. https://doi.org/10.1021/acs.orglett.8b00304
10. Gibson DG, Young L, Chuang R-Y et al (2009) Enzymatic assembly of DNA molecules up to several hundred kilobases. Nat Methods 6: 343–345. https://doi.org/10.1038/nmeth.1318
11. Nour-Eldin HH, Geu-Flores F, Halkier BA (2010) USER cloning and USER fusion: the ideal cloning techniques for small and big laboratories. Methods Mol Biol 643:185–200. https://doi.org/10.1007/978-1-60761-723-5_13
12. Da Silva Ferreira ME, Kress MRVZ, Savoldi M et al (2006) The akuB(KU80) mutant deficient for nonhomologous end joining is a powerful tool for analyzing pathogenicity in Aspergillus fumigatus. Eukaryot Cell 5:207–211. https://doi.org/10.1128/EC.5.1.207-211.2006
13. Nødvig CS, Nielsen JB, Kogle ME et al (2015) A CRISPR-Cas9 system for genetic engineering of filamentous fungi. PLoS One 10: e0133085. https://doi.org/10.1371/journal.pone.0133085
14. Fedorova ND, Khaldi N, Joardar VS et al (2008) Genomic islands in the pathogenic filamentous fungus Aspergillus fumigatus. PLoS Genet 4:e1000046. https://doi.org/10.1371/journal.pgen.1000046

Part II

Methods for Identification, Genotyping, and Detection

Chapter 5

PCR Typing and Genome Sequencing of *Aspergillus fumigatus*

Hiroki Takahashi, Saho Shibata, and Momotaka Uchida

Abstract

Aspergillus fumigatus is a major opportunistic fungal pathogen, presenting a serious health risk, particularly to immunocompromised individuals. Comprehensive genomic analysis of both clinical and environmental isolates provides valuable insights into its population structure, antifungal resistance mechanisms, and evolutionary dynamics. This chapter outlines protocols for DNA extraction and typing PCR, including a step-by-step workflow for high-resolution genomic analysis of *A. fumigatus* using short-read sequencing data. This protocol enables researchers to perform reliable variant analysis and phylogenomic reconstruction, offering a reproducible protocol for genomic studies in *A. fumigatus* and related fungal species.

Key words *Aspergillus fumigatus*, Whole-genome sequencing, PCR typing, Bioinformatics pipeline, Phylogenomics, Variant analysis

1 Introduction

Aspergillus fumigatus is a filamentous fungus that is ubiquitously present in the environment and is one of the most common causative agents of invasive aspergillosis, particularly in immunocompromised individuals [1]. Its clinical significance, combined with the growing concern over antifungal resistance, highlights the critical need for genomic studies to elucidate its genetic diversity, population structure, and evolutionary adaptations [2–4].

Advancements in high-throughput sequencing technologies have enabled in-depth exploration of fungal genomes, allowing for the identification of genetic variants, structural rearrangements, and phylogenetic relationships. In recent years, an expanding array of publicly available genomic resources, such as reference genomes, variant datasets, and transcriptomic profiles, has been made accessible through dedicated databases [5]. These resources, alongside robust bioinformatics pipelines, have significantly enhanced our

Natalia Mach (ed.), *Aspergillus fumigatus: Methods and Protocols*, Methods in Molecular Biology, vol. 3028,
https://doi.org/10.1007/978-1-0716-5222-0_5,

understanding of fungal biology, pathogenicity, and resistance mechanisms.

This chapter describes an integrated experimental and computational pipeline for the genomic analysis of *A. fumigatus* strains. The workflow includes experimental procedures such as fungal culture, genomic DNA extraction, and typing PCR, as well as computational steps for SNP analysis and draft genome assembly. This protocol enables reproducible and scalable genomic studies of *A. fumigatus* or related fungal species, facilitating downstream applications in comparative genomics, epidemiology, antifungal resistance monitoring, and evolutionary research.

2 Materials

2.1 Media

1. Potato dextrose agar (PDA): 39 g PDA powder in 1 L of distilled water. Autoclave, dispense it into 90 mm Petri dishes, dry for 10 min, and store at 4 °C.
2. Potato dextrose broth (PDB): 24 g PDB powder in 1 L of distilled water. Autoclave, store at 4 °C.

2.2 Buffers and Solutions

1. Tween 80 buffer: 8 g NaCl, 1 mL Tween 80 in 1 L of distilled water. Autoclave, store at room temperature.
2. Lysis buffer: 4.85 g Tris base, 0.29 g NaCl, 1.86 g EDTA, 1.0 g SDS in 200 mL distilled water. Autoclave, and store at room temperature.
3. 10 × PBS: 8.0 g NaCl, 0.2 g KCl, 1.15 g Na_2HPO_4, 0.2 g KH_2PO_4 in 80 mL distilled water. Make up to 100 mL with distilled water, autoclave, and store at room temperature.
4. 1 × PBS prepared from 10 × PBS with distilled water before use.
5. Phenol:chloroform:isoamyl alcohol mix (25:24:1, pH 7.9).
6. Isopropanol.
7. Sodium acetate, 3 M, pH 5.2.
8. Ethanol, 70% and 99.5% in nuclease-free water.

2.3 PCR Reagents

1. Primers for mating-type determination. Use following primers for MAT1-1-1 (AFM1), MAT1-2-1 (AFM2), or both (AFM3) to generate mating type-specific products [6].

   AFM1: CCTTGACGCGATGGGGTGG

   AFM2: CGCTCCTCATCAGAACAACTCG

   AFM3: CGGAAATCTGATGTCGCCACG

2. Primers for transposon-based typing (Rep-PCR)

   Aft1-TIR: GCTCGCTTACCGATTACGTTA

   Taf1-TIR: GGTGAGGGGGAGCCTTAGTC

   These primers bind to long terminal repeats (LTRs) of two transposons, *Aft1* and *Taf1*, in the reverse orientation. These transposons retain the terminal inverted repeats (TIRs) at both ends.

3. PCR master mix (e.g., Quick Taq™ HS DyeMix, Toyobo, Japan, #DTM-101).
4. PCR mix for mating-type PCR: 5 μL of 2 × Quick Taq™ HS DyeMix, 0.2 μL each of AFM1, AFM2, and AFM3 primers, 4.4 μL distilled water, total volume 10 μL. Primer concentration 10 μM per primer.
5. PCR mix for rep-PCR: 5 μL of 2 × Quick Taq HS DyeMix, 0.4 μL each of Aft1-TIR and Taf1-TIR primers, 3.2 μL distilled water, total volume 9 μL. Primer concentration 10 μM per primer.
6. Agarose and DNA size marker for gel electrophoresis.

2.4 Kits and Enzymes

1. Monarch™ Genomic DNA Purification Kit (e.g., New England Biolabs, #T3010): preheat gDNA Elution Buffer to 60 °C. Other components are stored at room temperature. Before using for the first time, add 140 mL of 99.5% ethanol to the washing buffer.
2. Proteinase K (20 mg/mL).
3. RNase A (100 mg/mL).

2.5 Equipment and Consumables

1. Lysing Matrix D (e.g., MP Biomedicals, #1169130-CF). Alternatively, use 1.5 mL tubes containing beads of choice.
2. 90 mm Petri dishes.
3. Sterile inoculation loops.
4. Sterile 50 mL tubes.
5. Disposable pipettes.
6. Sterile filter paper.
7. 50 mL Erlenmeyer flasks.
8. 1.5 mL microcentrifuge tubes.
9. PCR tubes.
10. Büchner funnel.
11. Shaker, microcentrifuge.
12. Counting chamber, microscope, or an automatic cell counter.

13. High-speed bead-beating homogenizer (e.g., FastPrep-24 5G, MP Biomedicals).
14. Vortex mixer.
15. PCR thermal cycler.
16. Equipment to measure concentration of nucleic acids (e.g., Qubit Fluorometer from Thermo Fisher Scientific).
17. Gel electrophoresis system.

2.6 Software Summary

The following software tools are used in this workflow. Each of these tools can be either installed via Conda or downloaded from its official repository. All software run within a Conda-managed environment (afumi_env) to ensure reproducibility and consistent tool behavior across platforms.

1. **FastQC** (v0.12.1): Used to assess raw read quality, including per-base quality scores, GC content, and sequence duplication levels [7].
2. **fastp** (v0.23.4): Used for adapter trimming, quality filtering, and base correction of sequencing reads. It also generates summary reports [8].
3. **SeqKit** (v2.10.0): Toolkit for FASTA/Q file manipulation written in Go (optimized for speed) [10].
4. **BWA-MEM2** (v2.2.1) [9]: Used for aligning paired-end short reads to the reference genome. It offers improved speed and memory efficiency over the original BWA.
5. **SAMtools** (v1.18): Used for sorting and indexing BAM files and converting between file formats as required by downstream tools [11, 12].
6. **BCFtools** (v1.17): Used for variant calling and basic manipulation of VCF files, such as filtering and format conversion [12].
7. **GATK** (v4.2.6.1): Used for advanced variant calling following the Best Practices pipeline. Key tools included in the workflow are HaplotypeCaller, BaseRecalibrator, ApplyBQSR, and MarkDuplicates [13].
8. **VCFtools** (v0.1.16): Used for filtering SNPs based on quality metrics, minor allele frequency, and completeness across samples [14].
9. **BUSCO** (v5.8.2): Used to assess genome assembly completeness using a set of conserved single-copy orthologs [15, 16].
10. **vcf2phylip** (v2.8): A Python script used to convert multi-sample VCF files into PHYLIP format for phylogenetic analysis [17].

11. **RAxML** (v8.2.13): Used to construct phylogenetic trees using maximum likelihood methods with rapid bootstrapping [18].
12. **SRA Toolkit** (v3.2.1): Used to retrieve raw sequencing data from the NCBI Sequence Read Archive (SRA), including prefetch and fasterq-dump tools [19].
13. **GetOrganelle** (v1.7.7.1): Used to assemble chloroplast, mitochondrial, or rDNA genomes from short reads [20–24].
14. **Velvet** (v1.2.10): Used to assemble the draft genome, in conjunction with VelvetOptimiser (v2.2.6), which automates the testing of different k-mer sizes and parameters for the Velvet assembler [25].
15. **tabix** (v1.11): Indexing tool for compressed genomic position-sorted files (e.g., vcf.gz). Enables efficient querying by genomic region.
16. **Pigz** (v2.8): Used to multi-threaded alternative to gzip for faster file compression.

3 Methods

3.1 Conidia Preparation

1. Inoculate PDA plates with a small number of *A. fumigatus* conidia or mycelium using a sterile inoculation loop. Incubate at 37 °C for 5 days.
2. To harvest conidia, add 5 mL Tween 80 buffer to each plate. Gently scrape the colony surface with a sterile inoculation loop to dislodge conidia.
3. Transfer the conidia suspension to a sterile 50 mL tube.
4. Count conidia under the microscope using a counting chamber under the microscope or an automatic cell counter. Use approximately 1×10^6 conidia to prepare sufficient mycelium.

3.2 Mycelia Preparation

1. Add 20 mL autoclaved PDB to a sterile 50 mL flask.
2. Inoculate with a conidia suspension containing 1×10^6 conidia (*see* **Notes 1, 2**).
3. Incubate at 37 °C with shaking at 185 × rpm for 16–18 h (*see* Fig. 1).
4. Harvest mycelia by suction filtration using a Büchner funnel lined with sterile filter paper (*see* Fig. 2).
5. Wash mycelia with distilled water to remove any residual medium. If storing the sample, freeze at −80 °C.

3.3 Genomic DNA Extraction

1. Transfer 100 mg of wet mycelium to a Lysing Matrix D or a 1.5 mL tube containing beads.

Fig. 1 *Aspergillus fumigatus* mycelia following 18 h of incubation

Fig. 2 Büchner funnel used for mycelial filtration

2. Add 500 μL of Lysis buffer and 500 μL of phenol:chloroform: isoamyl alcohol.
3. Homogenize using a high-speed bead beater. If using FastPrep-24 5G homogenizer, shake at 6.0 m/s for 30 s, and repeat twice.
4. Centrifuge for 10 min at 12,000 × g.
5. Transfer up to 720 μL of the upper aqueous phase to a new 1.5 mL tube.
6. Incubate at 70 °C for 20 min.
7. Measure the volume, and add an equal volume of phenol: chloroform:isoamyl alcohol.
8. Vortex for 5 min.
9. Centrifuge for 10 min at 12,000 × g.
10. Transfer up to 600 μL of the upper aqueous phase to a new 1.5 mL tube.
11. Measure the volume, and add 1/10 volume of sodium acetate and an equal volume of isopropanol to the solution.
12. Centrifuge for 10 min at 16,000 × g at 4 °C. Discard the supernatant.
13. Wash the pellet with 1 mL of 70% ethanol.
14. Centrifuge for 10,000 × *g*, 4 °C 5 min, discard ethanol.
15. Add 100 μL of 1 × PBS and proceed with Part 1 of the Monarch Genomic DNA Purification Kit protocol.
16. Add 10 μL of Proteinase K, incubate at 56 °C for 30 min, and vortex every 10 min.
17. Add 3 μL of RNase A, incubate at 56 °C for 30 min, and vortex every 10 min.
18. Proceed with Part 2 of the Monarch Genomic DNA Purification Kit protocol.
19. Add 400 μL of gDNA Binding Buffer (from the kit), and mix thoroughly by pulse-vortexing for 5–10 s.
20. Centrifuge for 3 min at 1000 × g to bind gDNA. Remove the flow-through.
21. Centrifuge for 1 min at 12,000 × g to clear the membrane, and discard the flow-through and collection tube.
22. Transfer column to a new collection tube and add 500 μL of gDNA Wash Buffer (from the kit).
23. Close the cap, and invert several times to ensure the wash buffer reaches the cap.
24. Centrifuge for 1 min at 12,000 × g. Discard the flow-through.
25. Reinsert the column into the collection tube. Add 500 μL gDNA Wash Buffer and close the cap.

26. Centrifuge for 1 min at 12,000 × g. Discard the collection tube and flow-through.
27. Place the gDNA Purification Column in a DNase-free 1.5 mL tube.
28. Add 100 μL preheated gDNA Elution Buffer, close the cap, and incubate 1 min at room temperature.
29. Centrifuge for 1 min at 12,000 × g.
30. Assess DNA quality by electrophoresis and a Qubit Fluorometer.

3.4 Mating-Type Determination: Colony-Direct PCR

Mating type is one of the most conserved genetic traits in the fungal kingdom. It determines mating compatibility by regulating the expression of genes involved in sexual development, such as partner recognition, formation of sex-specific structures, and meiosis. Multiple studies have demonstrated that the mating-type locus (MAT) is conserved, even in fungal species lacking known sexual stages. Thus, mating-type determination is one of the most fundamental methods for strain typing. In *A. fumigatus*, each strain has either MAT1-1 (α) or MAT1-2 (HMG) mating-type locus, and only partners harboring the opposite mating type can mate [6, 26]. A simple PCR-based method is provided to determine the mating type of *A. fumigatus*.

1. Set up the reaction mixture (*see* Table 1). Quick Taq™ HS DyeMix is used in this protocol. Equivalent commercial DNA polymerase mixes may also be used, if performance is verified.
2. Dispense the solution into PCR tubes. Using sterile toothpicks, transfer a small amount of mycelium into each tube. Perform PCR (*see* Table 2).
3. Conduct agarose gel electrophoresis to verify the band size. Use an agarose gel concentration of at least 1.5% for optimal resolution. Samples with the MAT1-1 locus produce an

Table 1
PCR reaction mixture for mating-type PCR

Components	Volume/sample
2 × Quick Taq™ HS DyeMix	5 μL
10 μM AFM1 primer	0.2 μL
10 μM AFM2 primer	0.2 μL
10 μM AFM3 primer	0.2 μL
D.W.	4.4 μL
Total	10 μL

Table 2
PCR cycling conditions for mating-type PCR

Step	Temperature	Time
Initial denaturation	94 °C	2 min
Amplification (25–30 cycles)	94 °C 55 °C 68 °C	30 s 30 s 1 min

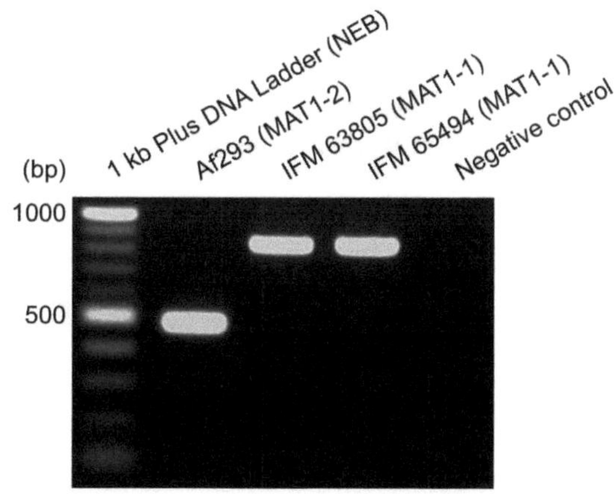

Fig. 3 Amplified products of colony-direct MAT-determining PCR (1.5% agarose gel). A sample without mycelia was used as the negative control

833-bp band, while MAT1-2 samples yield a 484-bp band (*see* Fig. 3).

3.5 Transposon-Based Typing (Rep-PCR)

Repetitive sequence-based PCR (Rep-PCR), originally developed in bacteria [27], is a method used to identify strains, verify hybridization, or detect contamination. This method has also been successfully applied to fungal species, particularly the rice blast fungus *Pyricularia oryzae* [28]. The Rep-PCR amplifies inter-transposon regions in the genome, with the primer binding in a reverse direction to the long-terminal repeats (LTRs) flanking the transposons. Unlike *P. oryzae Pot2* transposon, the genomic contents of *A. fumigatus* transposons are too low to conduct Rep-PCR with a single primer. Thus, combinations of several primers will be adopted. We describe a method of Rep-PCR using two primers targeting *Aft1* and *Taf1* transposons.

1. Set up the reaction mixture (*see* Table 3 and *see* **Note 3**). This protocol uses Quick Taq HS DyeMix; performance with other polymerases may require optimization.

Table 3
PCR reaction mixture for rep-PCR

Components	Volume/sample
2 × Quick Taq™ HS DyeMix	5 μL
10 μM Aft1-TIR primer	0.4 μL
10 μM Tafl-TIR primer	0.4 μL
D.W.	3.2 μL
Total	9 μL

Table 4
PCR cycling conditions for rep-PCR

Step	Temperature	Time
Initial denaturation	94 °C	2 min
Amplification (30 cycles)	94 °C 50 °C 68 °C	30 s 30 s 5 min

2. Dispense the solution into PCR tubes and add up to 10 ng of purified genomic DNA. PCR is performed under the conditions shown in Table 4. Conduct PCR (*see* Table 4).
3. Use agarose gel electrophoresis to compare band patterns (*see* Fig. 4). Bands up to ~5000 bp may be evaluated, although smaller bands generally provide greater reliability. The appearance of larger bands depends on DNA integrity and the fidelity of the polymerase used.

3.6 Bioinformatics Workflow Overview

This section provides a step-by-step description of the analysis workflow using the representative *A. fumigatus* isolate, IFM 63560-1 [4]. For the phylogenetic analysis, ten strains were randomly selected (Table 5).

In the command-line examples below: Lines beginning with # are **shell comments** for explanation. They are ignored by the shell when executing commands. The $ symbol denotes the start of a **shell command** and should not be typed when entering commands in the terminal.

3.6.1 Conda Environment Setup

Install and manage the necessary bioinformatics tools using Conda, a widely used package and environment management system that simplifies software installation and dependency handling. Using Conda ensures reproducibility and isolates dependencies, which is essential when working with complex bioinformatics pipelines.

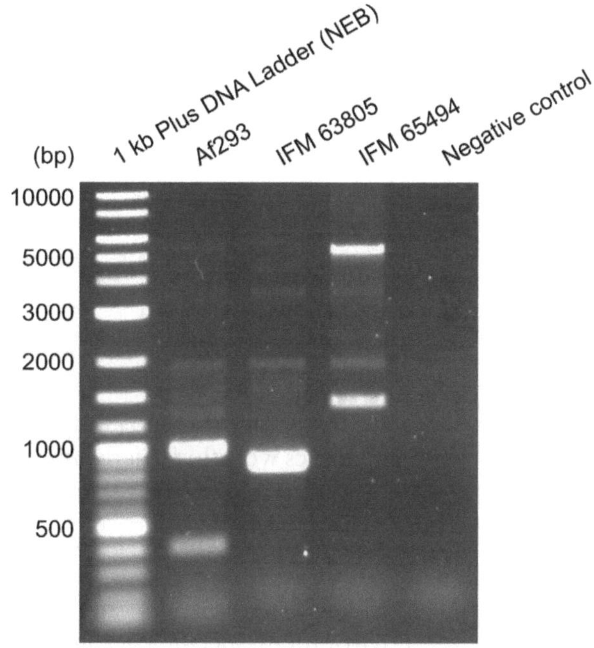

Fig. 4 Rep-PCR products electrophoresed on a 1.0% agarose gel

Table 5
Sample list used in this chapter

ID	Run ID	Country	Source	Reference
IFM 63560-1	DRR495386	Japan	Clinical	[4]
IFM 65494	DRR495423	Japan	Clinical	[4]
IFM 61578-1	DRR495432	Japan	Clinical	[4]
IFM 63805	DRR495397	Japan	Clinical	[4]
IFM 64259-1	DRR495409	Japan	Clinical	[4]
3-1-H	DRR290855	Japan	Plant	[32]
3-1-B	DRR290849	Japan	Plant	[32]
3-1-D	DRR290851	Japan	Plant	[32]
08-12-12-13	ERR769506	The Netherlands	Clinical	[2]
AFIS1415	SRR11785058	USA	Clinical	[3]

In addition to the Conda-installed tools, install GATK and vcf2phylip manually, either via Git or by downloading from their official sources.

1. Create a dedicated environment using the Miniconda distribution to include all required tools:

```
# Check operating system details
$ cat /etc/lsb-release
DISTRIB_ID=Ubuntu
DISTRIB_RELEASE=20.04
DISTRIB_CODENAME=focal
DISTRIB_DESCRIPTION="Ubuntu 20.04.5 LTS"

# Check Conda version
$ conda --version
conda 25.3.1
```

2. Create a new Conda environment specifically for this bioinformatics workflow (here: afumi_env) using Python 3.10, which is compatible with most modern bioinformatics tools.

```
$ conda create -n afumi_env python=3.10
```

This step creates an isolated directory that contains all environment-specific packages and binaries.

3. Activate the newly created environment:

```
$ conda activate afumi_env
```

The shell prompt will now indicate the active environment, implying that any software installed here will not interfere with the base system.

4. Install essential tools from the Bioconda channel:

```
$ conda install -c bioconda fastqc seqkit fastp bwa-mem2
samtools bcftools=1.10.2 sra-tools getorganelle perl-
velvetoptimiser vcftools busco=5.8.2 augustus=3.4.0 tabix
pigz
```

5. List all installed packages for documentation and troubleshooting:

```
$ conda list
```

6. Export your environment for reproducibility:

```
$ conda env export > afumi_env.yml
```

The environment can be shared and recreated on another system with:

```
$ conda env create -f afumi_env.yml
```

3.6.2 Reference Genome Preparation

The genome assembly and annotation files for *A. fumigatus* strain Af293 can be downloaded from FungiDB database. The Af293 strain is widely used as a reference in *A. fumigatus* research due to its well-characterized genome and comprehensive supporting datasets. Using this reference allows leveraging curated single-nucleotide polymorphism (SNP) information to improve the accuracy of variant detection in sequencing data from clinical or environmental isolates. These files serve as a basis for downstream analyses such as reading alignment, variant calling, and comparative genomics.

Create a directory to store the reference genome:

```
# Create a working directory for reference data.
$ mkdir ref
```

All downloaded reference data will be stored in the ref directory.

3.6.3 Raw Read Download (SRA Toolkit)

The first step in any sequencing-based bioinformatics workflow is to retrieve raw sequencing reads from a reliable source. In this protocol, publicly available datasets were accessed from the NCBI Sequence Read Archive (SRA), a comprehensive repository for high-throughput sequencing data. The SRA Toolkit is used to:

(a) Download .sra files via prefetch.

(b) Convert .sra files into standard FASTQ format using fastq-dump or fasterq-dump.

(c) Compress the FASTQ files using pigz for efficient storage and processing.

1. Create a working directory for your FASTQ data:

```
$ mkdir fastq
$ cd fastq
```

This command ensures raw files remain organized and are kept separate from processed outputs.

2. Download .sra files using prefetch. The prefetch utility downloads sequencing data directly from the SRA by accession number. Replace DRR495386 with the appropriate accession number.

```
# Download SRA files
$ prefetch DRR495386
```

This command creates a directory structure under the local SRA data repository (e.g., ~/ncbi/public/sra/ or ./fastq/DRRxxxxx/, depending on configuration).

3. Confirm the Download

```
$ tree ./
```

This command displays the directory and file structure to confirm successful downloads.

4. Convert .sra to FASTQ format. FASTQ is the standard format for raw sequencing reads, required for tools such as FastQC, fastp, and bwa-mem2.

   Option A: fastq-dump (legacy, slower).

```
$ fastq-dump DRR495386 --split-files
```

This command (--split-files) separates paired-end reads into _1.fastq and _2.fastq. Suitable for small or moderate datasets.

Option B: fasterq-dump (recommended, faster).

```
$ fasterq-dump DRR495386 --split-files -e 4
```

This method is faster and more memory-efficient, especially for large datasets.

5. Compress FASTQ files using pigz. After conversion, compress the resulting .fastq files to save storage, and improve I/O performance in downstream steps.

```
$ pigz -p 4 *fastq
```

-p 4 uses 4 CPU threads for parallel compression. Output files have the .fastq.gz extension. This significantly reduces disk usage and speeds up processing by tools that support compressed input (e.g., fastp, bwa-mem2, FastQC).

3.6.4 Quality Control and Trimming

The important step after retrieving raw FASTQ files is to assess data quality. This is done using FastQC, a widely used quality control tool that provides a comprehensive overview of raw sequencing data. It generates interactive reports to detect potential issues early in the workflow (*see* **Note 4**).

Key quality metrics evaluated by FastQC include:

(a) Per base sequence quality.

(b) GC content distribution,

(c) Sequence duplication levels.

(d) Adapter contamination.

(e) Overrepresented sequences.

Identifying these issues before read alignment or assembly enables the selection of appropriate trimming parameters and preprocessing strategies.

1. Create an output directory for FastQC reports and run FastQC on all compressed FASTQ files:

```
$ mkdir fastqc_out
$ fastqc -t 4 *gz -o fastqc_out
```

For each input FASTQ file, FastQC generates two outputs:

sample_fastqc.html—interactive quality report (open in a web browser) and

sample_fastqc.zip—compressed archive of raw metrics and plots.

2. Store all FastQC outputs in the fastqc_out/ directory for future reference.
3. Perform read preprocessing (cleaning and trimming) using fastp. fastp is an ultrafast, all-in-one FASTQ preprocessor designed for quality control and data cleaning in next-generation sequencing (NGS) workflows. Developed in C++, fastp offers high performance while integrating multiple preprocessing steps into a single tool. Example below shows the preprocessing of the paired-end sequencing dataset DRR495386.
4. Create a working directory for processed FASTQ files:

```
$ cd ../
$ mkdir fastp
$ cd fastp
```

5. Run fastp on paired-end reads:

```
$ fastp -i /data1/hi-takah/springer/fastq/DRR495386_1.fastq.
gz \
-I /data1/hi-takah/springer/fastq/DRR495386_2.fastq.gz \
-o DRR495386_1.qf.fq.gz \
-O DRR495386_2.qf.fq.gz \
-h -o DRR495386.html \
-j -o DRR495386.json \
--detect_adapter_for_pe \
-q 20
```

-i and -I: input paired-end FASTQ files.

-o and -O: output cleaned paired-end FASTQ files.

-h: HTML quality report.

-j: JSON quality report (machine-readable).

--detect_adapter_for_pe: automatically detects paired-end adapters.

-q 20: trims low-quality bases with below Q20 score.

The resulted FASTQ files (*.qf.fq.gz) are used for downstream alignment.

3.6.5 Alignment and BAM Processing

BWA-MEM2 is an optimized version of the original BWA-MEM algorithm, used to align short sequencing reads (e.g., Illumina) to a reference genome. BWA-MEM2 is fast and memory-efficient, making it suitable for high-throughput workflows.

1. Create a working directory for alignment outputs:

```
$ cd ../
$ mkdir bwa
$ cd bwa
```

2. Index the reference genome by BWA-MEM2:

```
$ bwa-mem2 index /data1/hi-takah/springer/ref/FungiDB-64_Afu-
migatusAf293_Genome.fasta
```

This command generates multiple index files required for alignment.

3. Align quality-filtered paired-end reads (*.qf.fq.gz) to the reference genome using BWA-MEM2 and output a sorted BAM file (Example: DRR495386):

```
$ bwa-mem2 mem -M -t 8 /data1/hi-takah/springer/ref/FungiDB-
64_AfumigatusAf293_Genome.fasta \
/data1/hi-takah/springer/fastp/DRR495386_1.qf.fq.gz \
/data1/hi-takah/springer/fastp/DRR495386_2.qf.fq.gz \
 | samtools sort -@ 8 -o DRR495386.sort.bam -
```

-M: marks shorter split hits as secondary (required for downstream processing).

-t 8: uses 8 CPU threads.

samtools sort: sorts BAM output for efficient access.

3.6.6 Variant Calling with GATK

This section follows GATK Best Practices for variant discovery using short-read sequencing data. It includes BAM preprocessing, quality score recalibration, and variant calling with HaplotypeCaller (*see* **Note 5**).

1. Create a software directory and install GATK:

```
#Setup GATK
$ mkdir ~/software
```

```
$ cd software
$ wget https://github.com/broadinstitute/gatk/releases/down-
load/4.2.6.1/gatk-4.2.6.1.zip
$ unzip gatk-4.2.6.1.zip
$ export PATH=$PATH:~/software/gatk-4.2.6.1
$ cd $HOME/springer
$ mkdir gatk
$ cd gatk
```

2. Clean the BAM file to remove potential formatting issues (BAM preprocessing is optional, but recommended):

```
$ gatk CleanSam /
-I /data1/hi-takah/springer/bwa/DRR495386.sort.bam \
-O DRR495386_clean.bam
```

3. Add or replace read groups information using the gatk AddOrReplaceReadGroups. It updates read group (RG) metadata in a BAM file. This is required before running tools such as MarkDuplicates, BaseRecalibrator, and HaplotypeCaller.

```
$ gatk AddOrReplaceReadGroups \
  -I DRR495386_clean.bam \
  -O DRR495386_rg.bam \
  -RGLB lib1 \
  -RGPL ILLUMINA \
  -RGPU unit1 \
  -RGSM DRR495386
```

4. Identify and mark PCR/optical duplicates using the gatk MarkDuplicates. The duplicate reads are commonly introduced during PCR amplification. These duplicates should not be used for variant calling, as they can introduce bias.

```
$ gatk --java-options "-Xmx64G" MarkDuplicates \
  -I DRR495386_rg.bam \
  -O DRR495386_rg_Markdup.bam \
  -M metrics_DRR495386.txt
```

5. Index the reference genome for GATK in multiple formats. This step is essential before running GATK tools like HaplotypeCaller, BaseRecalibrator, and others.

```
$ samtools faidx /data1/hi-takah/springer/ref/FungiDB-64_Afu-
migatusAf293_Genome.fasta
$ gatk CreateSequenceDictionary -R /data1/hi-takah/springer/
ref/FungiDB-64_AfumigatusAf293_Genome.fasta
```

6. Build the base quality recalibration model (BQSR). The BaseRecalibrator detects and models systematic errors in base quality scores produced by the sequencer. This improves the

accuracy of downstream variant calling, especially in GATK Best Practices pipelines.

```
$ gatk BaseRecalibrator --java-options "-Xmx64G" \
  -I DRR495386_rg_Markdup.bam \
  --known-sites /data1/hi-takah/springer/ref/FungiDB.vcf.gz \
  -O recal_DRR495386.table \
   -R  /data1/hi-takah/springer/ref/FungiDB-64_Afumigatu-
sAf293_Genome.fasta
```

--known-sites: specifies a VCF of known variants (e.g. dbSNP or FungiDB set) used for modeling errors.

In the example below, a VCF file was prepared containing 33,968 alleles from FungiDB and parsed it into the following format: This file can be shared upon request.

```
##fileformat=VCFv4.2
##FILTER=<ID=PASS,Description="All filters passed">
##contig=<ID=Chr1_A_fumigatus_Af293,length=4918979>
##contig=<ID=Chr2_A_fumigatus_Af293,length=4844372>
##contig=<ID=Chr3_A_fumigatus_Af293,length=4079167>
##contig=<ID=Chr4_A_fumigatus_Af293,length=3927224>
##contig=<ID=Chr5_A_fumigatus_Af293,length=3948441>
##contig=<ID=Chr6_A_fumigatus_Af293,length=3778736>
##contig=<ID=Chr7_A_fumigatus_Af293,length=2058334>
##contig=<ID=Chr8_A_fumigatus_Af293,length=1833124>
##contig=<ID=mito_A_fumigatus_Af293,length=31765>
##FORMAT=<ID=GT,Number=1,Type=String,Description="Genotype">
##bcftools_convertVersion=1.20+htslib-1.20
##bcftools_convertCommand=convert  -f  FungiDB-64_AfumigatusAf293_-
Genome.fasta  -s  FungiDB  -o  FungiDB.vcf  -c  CHROM,POS,REF,AA  --
tsv2vcf FungiDB_68.txt; Date=Tue May 27 10:15:20 2025
#CHROM    POS    ID    REF    ALT    QUAL    FILTER    INFO    FORMAT
FungiDB
Chr1_A_fumigatus_Af293    126    .    A    C    .    .    .    GT    1
Chr1_A_fumigatus_Af293    139    .    T    C    .    .    .    GT    1
Chr1_A_fumigatus_Af293    147    .    A    G    .    .    .    GT    1
Chr1_A_fumigatus_Af293    1763    .    G    T    .    .    .    GT    1
Chr1_A_fumigatus_Af293    4542    .    A    G    .    .    .    GT    1
```

7. Apply the BQSR model using ApplyBQSR. It takes the recalibration table (generated by BaseRecalibrator) and applies it to the BAM file. This produces a final, recalibrated BAM that is ready for accurate variant calling.

```
$ gatk ApplyBQSR --java-options "-Xmx64G" \
  -I DRR495386_rg_Markdup.bam \
  -bqsr recal_DRR495386.table \
  -O DRR495386_rg_Markdup_bqsr.bam
```

8. Call variants (SNPs/Indels) with HaplotypeCaller. The HaplotypeCaller is GATK's primary variant calling tool. It uses a local de novo assembly approach to detect SNPs and indels with high accuracy. In this example, GVCF mode is utilized (-ERC GVCF); it is specifically designed to support multi-sample joint genotyping in downstream analyses.

```
$ gatk HaplotypeCaller --java-options "-Xmx64G" \
   -I DRR495386_rg_Markdup_bqsr.bam \
   -O haplotypecaller_DRR495386.g.vcf.gz \
     -R /data1/hi-takah/springer/ref/FungiDB-64_Afumigatu-
sAf293_Genome.fasta \
   --dont-use-soft-clipped-bases \
   -ERC GVCF \
   -stand-call-conf 30.0 \
   --sample-ploidy 1
```

This is an example of an automation script (gatk_sample.sh):

```
<gatk_sample.sh>
#!/bin/bash
# Set sample ID
SAMPLE=DRR495386

# Set reference and known sites
REF=/data1/hi-takah/springer/ref/FungiDB-64_AfumigatusAf293_-
Genome.fasta
KNOWN_SITES=/data1/hi-takah/springer/ref/FungiDB.vcf.gz

# Step 1: Clean the BAM File
# Optional but safe to ensure compliance with format standards
gatk CleanSam \
  -I /data1/hi-takah/springer/bwa/${SAMPLE}.sort.bam \
  -O ${SAMPLE}_clean.bam

# Step 2: Add or Replace Read Groups
# Required before duplicate marking or variant calling
gatk AddOrReplaceReadGroups \
  -I ${SAMPLE}_clean.bam \
  -O ${SAMPLE}_rg.bam \
  -RGLB lib1 \
  -RGPL ILLUMINA \
  -RGPU unit1 \
  -RGSM ${SAMPLE}

# Step 3: Mark PCR/Optical Duplicates
gatk --java-options "-Xmx64G" MarkDuplicates \
  -I ${SAMPLE}_rg.bam \
```

```
  -O ${SAMPLE}_rg_Markdup.bam \
  -M metrics_${SAMPLE}.txt

# Step 5: Base Quality Score Recalibration (BQSR)
gatk BaseRecalibrator --java-options "-Xmx64G" \
  -I ${SAMPLE}_rg_Markdup.bam \
  --known-sites $KNOWN_SITES \
  -O recal_${SAMPLE}.table \
  -R $REF

# Step 6: Apply the BQSR Model
gatk ApplyBQSR --java-options "-Xmx64G" \
  -I ${SAMPLE}_rg_Markdup.bam \
  -bqsr recal_${SAMPLE}.table \
  -O ${SAMPLE}_rg_Markdup_bqsr.bam

# Step 7: Call Variants with HaplotypeCaller
gatk HaplotypeCaller --java-options "-Xmx64G" \
  -I ${SAMPLE}_rg_Markdup_bqsr.bam \
  -O haplotypecaller_${SAMPLE}.g.vcf.gz \
  -R $REF \
  --dont-use-soft-clipped-bases \
  -ERC GVCF \
  -stand-call-conf 30.0 \
  --sample-ploidy 1
```

9. Combine multiple GVCFs into a single file. When HaplotypeCaller is run in GVCF mode (-ERC GVCF) for each sample, it generates individual GVCF files that include both variant calls and reference confidence information. Before performing joint genotyping, merge these individual GVCFs into a single file using CombineGVCFs. This step is crucial for downstream analyses such as population genomics, genome-wide association studies (GWAS), and other cohort-level comparisons. In this project, a total of ten samples, including DRR495386 and nine others, are processed using the same workflow.

```
$ gatk CombineGVCFs \
   -R /data1/hi-takah/springer/ref/FungiDB-64_Afumigatu-
sAf293_Genome.fasta \
  -O all_consolidate.vcf.gz \
  -V haplotypecaller_DRR290849.g.vcf.gz \
  -V haplotypecaller_DRR290851.g.vcf.gz \
  -V haplotypecaller_DRR290855.g.vcf.gz \
  -V haplotypecaller_DRR495386.g.vcf.gz \
  -V haplotypecaller_DRR495397.g.vcf.gz \
  -V haplotypecaller_DRR495409.g.vcf.gz \
  -V haplotypecaller_DRR495423.g.vcf.gz \
  -V haplotypecaller_DRR495432.g.vcf.gz \
```

```
  -V haplotypecaller_ERR769506.g.vcf.gz \
  -V haplotypecaller_SRR11785058.g.vcf.gz
```

-R: the reference genome FASTA file (indexed with .fai and .dict).

-O: output file for the combined GVCF file.

-V: one or more input GVCF files, each generated by HaplotypeCaller in GVCF mode.

10. Joint genotyping across samples using GenotypeGVCFs. After generating individual GVCFs per sample and combining them with CombineGVCFs, the next step is joint genotyping. This step analyzes all potential variant sites across all samples simultaneously to assign genotypes (e.g., 0/0, 0/1, 1/1) and filter low-confidence calls. This is essential for population-level variant calling, comparative genomics, and association studies.

```
$ gatk GenotypeGVCFs \
   -R /data1/hi-takah/springer/ref/FungiDB-64_Afumigatu-
sAf293_Genome.fasta \
  -V all_consolidate.vcf.gz \
  -O genotype_all.vcf.gz
```

11. Select SNPs from joint genotyping output. The joint genotyping output file (genotype_all.vcf.gz) contains a mixture of variant types, including SNPs, indels, and potentially structural variants. For many downstream analyses, such as SNP-based association studies, population structure analysis, or SNP density visualization, it is often necessary to isolate only SNPs. The GATK SelectVariants tool enables this by allowing users to extract specific types of variants, thereby streamlining and optimizing the workflow for targeted analyses.

```
$ gatk SelectVariants \
  --select-type-to-include SNP \
   -R /data1/hi-takah/springer/ref/FungiDB-64_Afumigatu-
sAf293_Genome.fasta \
  -V genotype_all.vcf.gz \
  -O snps_all.vcf.gz
```

12. Hard Filtering of SNPs. The GATK recommends hard filtering for variant quality control when no machine learning-based VQSR model is available (e.g., for non-human organisms or small datasets). This step adds FILTER flags to variants that fail any user-defined expression. Variants passing all filters remain marked as PASS; others are flagged with custom names like my_filter1.

```
$ gatk VariantFiltration \
   -R  /data1/hi-takah/springer/ref/FungiDB-64_Afumigatu-
sAf293_Genome.fasta \
  -V snps_all.vcf.gz \
  -O filtered_snps_all.vcf.gz \
  --filter-name my_filter1 \
  --filter-expression "QUAL < 30.0" \
  --filter-name my_filter2 \
  --filter-expression "QD < 2.0" \
  --filter-name my_filter3 \
  --filter-expression "FS > 60.0" \
  --filter-name my_filter4 \
  --filter-expression "MQ < 40.0" \
  --filter-name my_filter5 \
  --filter-expression "MQRankSum < -12.5" \
  --filter-name my_filter6 \
  --filter-expression "ReadPosRankSum < -8.0" \
  --filter-name my_filter7 \
  --filter-expression "SOR > 3.0"
```

Filter summary:

Filter name	Expression	Description
my_filter1	QUAL <30	Low-quality variants
my_filter2	QD < 2.0	Quality score normalized by depth
my_filter3	FS > 60.0	High strand bias
my_filter4	MQ < 40.0	Low mapping quality
my_filter5	MQRankSum < −12.5	Mapping quality bias between ref./alt reads
my_filter6	ReadPosRankSum < −8.0	Read position bias
my_filter7	SOR > 3.0	Symmetric odds ratio strand bias

3.6.7 SNP Filtering with VCFtools

1. Filter SNP Variants with VCFtools. To ensure the accuracy and reliability of downstream analyses such as GWAS or phylogenetic inference, it is essential to apply stringent filtering criteria to the variant dataset. Apply stringent filtering criteria to retain only high-quality, bi-allelic SNPs with sufficient coverage across all samples. The following command demonstrates how to use VCFtools to perform such filtering on a compressed VCF file:

```
$ vcftools --gzvcf filtered_snps_all.vcf.gz  \
       --maf 0.1 \
```

```
--max-alleles 2 \
--min-alleles 1 \
--max-missing 1 \
--out snp_gwas_maf0.1_miss1 \
--recode \
--remove-filtered-all \
--remove-indels \
--not-chr mito_A_fumigatus_Af293
```

--maf 0.1: retains only SNPs with a minor allele frequency ≥ 0.1. This helps focus on common variants and removes rare or potentially spurious alleles. A threshold of 0.05 is also commonly used, depending on the study design.

--max-alleles 2: keeps only bi-allelic sites (i.e., one reference and one alternate allele). Multi-allelic variants are excluded.

--min-alleles 1: ensures at least one allele is present (helps filter out potential missing or erroneous data).

--max-missing 1: requires no missing genotypes across all samples (100% completeness). Sites with any missing data are removed.

--out snp_gwas_maf0.1_miss1: prefix for output files.

--recode: writes a new VCF with only retained variants.

--remove-filtered-all: excludes all variants that fail original filters (i.e., only variants marked PASS are kept).

--remove-indels: removes indels, retaining only SNPs.

--not-chr mito_A_fumigatus_Af293: excludes mitochondrial variants from the analysis.

2. Add Af293 reference genotype to SNP matrix. Integrate the reference strain Af293 as a homozygous reference sample into an existing filtered VCF for comparative or phylogenetic purposes.

```
# Extract SNP Positions from Filtered VCF
$ bgzip snp_gwas_maf0.1_miss1.recode.vcf
$ tabix snp_gwas_maf0.1_miss1.recode.vcf.gz
$ zcat snp_gwas_maf0.1_miss1.recode.vcf.gz | awk -F "\t"
'{print $1"\t"$2"\t"$3"\t"$4}' > snp_pos.txt
```

3. Convert TSV to VCF with reference genotypes:

```
$ bcftools convert -f /data1/hi-takah/springer/ref/FungiDB-
64_AfumigatusAf293_Genome.fasta \
  --tsv2vcf snp_pos.txt \
  --samples Af293 \
  -c CHROM,POS,ID,AA \
  -Oz -o Af293.vcf.gz
$ tabix Af293.vcf.gz
```

--tsv2vcf: convert tab-delimited file (snp_pos.txt) into VCF format.

--samples Af293: assign sample name as "Af293."

-c: specifies column order: chromosome, position, ID, reference allele.

-f: reference genome FASTA required for checking base calls.

-Oz: output compressed VCF.

4. Merge Af293 reference with existing samples.

```
$ bcftools merge Af293.vcf.gz snp_gwas_maf0.1_miss1.recode.
vcf.gz > 10strains_with_Af293.vcf
$ bgzip 10strains_with_Af293.vcf
```

This command merges the Af293 reference genotype with the filtered SNP dataset, resulting in a combined multi-sample VCF in which Af293 is represented as a homozygous reference sample.

5. Convert VCF to PHYLIP format for phylogenetic analysis.

After merging the SNP data into a single multi-sample VCF file (e.g., 10strains_with_Af293.vcf.gz), convert it into a multiple sequence alignment format suitable for phylogenetic analysis. This can be done using vcf2phylip, a Python-based script that supports PHYLIP, FASTA, and other alignment formats.

```
$ cd ~/software
$ wget https://github.com/edgardomortiz/vcf2phylip/archive/
refs/tags/v2.8.tar.gz
$ tar -xvzf v2.8.tar.gz
$ python ~/software/vcf2phylip-2.8/vcf2phylip.py \
  -i 10strains_with_Af293.vcf.gz
```

-i: input VCF file (compressed with bgzip and indexed with tabix).

-f: (optional) output alignment in FASTA format instead of PHYLIP.

This command generates the following output:

10strains_with_Af293.min4.phy: PHYLIP alignment file.

Optional log file summarizing genotype matrix statistics.

3.6.8 Phylogenetic Tree Construction with RAxML

RAxML (Randomized Axelerated Maximum Likelihood) is a powerful tool for inferring phylogenetic trees from large datasets using maximum likelihood (ML) methods. Construct a maximum likelihood phylogenetic tree with rapid bootstrapping using a PHYLIP-format SNP alignment file (RAxML Command):

```
$ raxmlHPC-PTHREADS -f a -x 12345 -p 12345 -# 1000 -m GTRCAT \
  -s 10strains_with_Af293.min4.phy \
  -n outfile \
  -T 8
```

-f a: runs a rapid bootstrap analysis followed by a best ML tree search.

-x 12345: random seed for bootstrapping (can be any integer).

-p 12345: random seed for parsimony starting tree.

-# 1000: number of bootstrap replicates (e.g., 1000).

-m GTRCAT: substitution model: GTR + CAT approximation for fast computation.

-s input.phy: input alignment file in PHYLIP format.

-n outfile: prefix for output files.

-T 8: number of threads to use (adjust based on available CPU cores).

The tree is generated using ggtree [29] (Fig. 5, *see* **Note 6**).

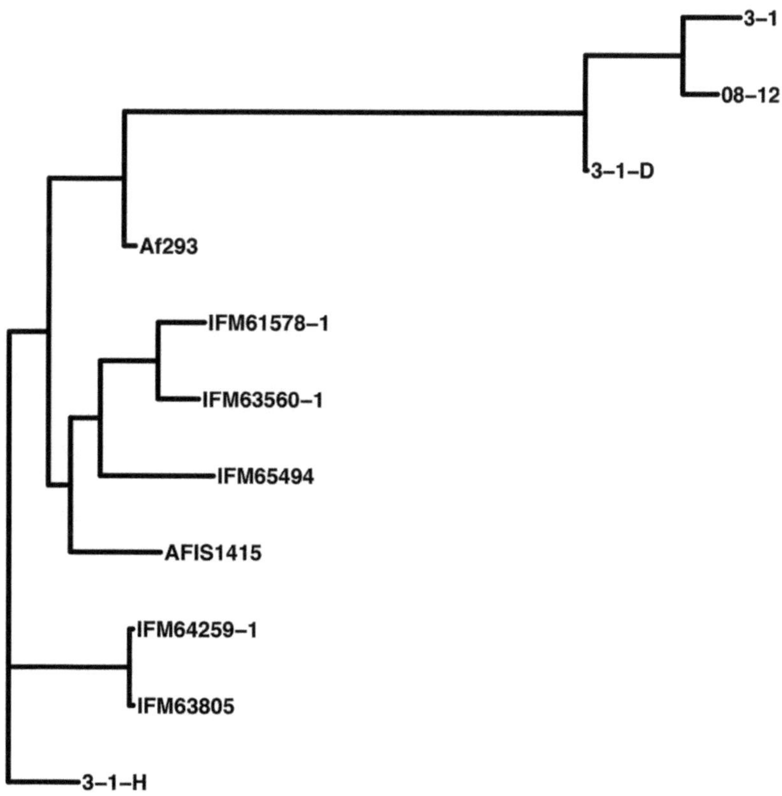

Fig. 5 Phylogenetic tree of 11 *A. fumigatus* strains, including Af293. The tree was generated using ggtree (v3.14.0) based on SNP data derived from whole-genome sequencing

3.6.9 Generate Draft Genome

This section describes the generation of a draft genome assembly using GetOrganelle and VelvetOptimiser, two widely used tools for de novo assembly of organellar and short-read genomic data (*see* **Note 7**).

Workflow Overview

Raw reads

↓

GetOrganelle → Mitochondrial genome

↓

Velvet → Draft nuclear genome

↓

BUSCO in genome mode

↓

Annotation (Beyond this protocol)

↓

BUSCO in protein mode (Beyond this protocol)

1. Set up working directory. Navigate to your project folder and create a subdirectory for assembly output:

```
$ cd ~/springer
$ mkdir -p asm/
```

2. Generate mitochondrial genome. The GetOrganelle is a specialized toolkit for assembling organelle genomes (mitochondria, plastids, etc.) from whole-genome sequencing data. It uses a seed-and-extend graph-based method and is highly effective for assembling circular mitochondrial genomes from short-read Illumina data. Set up the GetOrganelle database and run the mitochondrial genome assembly:

```
# Set up the database
$ get_organelle_config.py -a all
$ get_organelle_from_reads.py \
  -1 ../fastp/DRR495386_1.qf.fq.gz \
  -2 ../fastp/DRR495386_2.qf.fq.gz \
  -o DRR495386_getorganelle_output \
  -R 10 -k 21,45,65,85,105 \
  -F fungus_mt \
  -t 8 \
  --max-reads 5E6
```

The circular mitochondrial genome was successfully assembled, with a total length of 31,751 bp.

3. Index mitochondrial genome with BWA to enable read alignment:

```
$ bwa-mem2 index -p mito \
DRR495386_getorganelle_output/fungus_mt.K105.complete.
graph1.1.path_sequence.fasta
```

4. Map reads to the mitochondrial genome. This step involves aligning the quality-filtered sequencing reads to the assembled mitochondrial genome to identify mitochondrial-origin reads.

```
$ bwa-mem2 mem -t 8 mito \
/data1/hi-takah/springer/fastp/DRR495386_1.qf.fq.gz \
/data1/hi-takah/springer/fastp/DRR495386_2.qf.fq.gz \
 | samtools sort -@ 8 -o DRR495386.mito.sort.bam -
```

Output file contains all aligned reads.

5. Extract unmapped reads (nuclear reads)

After mapping reads to the mitochondrial genome, extract reads that did not align. These unmapped reads are presumed to be of nuclear origin.

```
$ samtools view -f 4 DRR495386.mito.sort.bam | awk -F "\t"
'{print $1}' | sort | uniq > unmapped_ids.txt
```

-f 4: filters for unmapped reads only.

6. Extract nuclear reads from FASTQ files. Retrieve sequences for the unmapped read IDs from the original FASTQ files using seqkit grep. This ensures that only nuclear reads are retained for downstream nuclear genome assembly.

```
$ seqkit grep -f unmapped_ids.txt /data1/hi-takah/springer/
fastp/DRR495386_1.qf.fq.gz | pigz -c - -p 8 > nuc_R1.fq.gz
$ seqkit grep -f unmapped_ids.txt /data1/hi-takah/springer/
fastp/DRR495386_2.qf.fq.gz | pigz -c - -p 8 > nuc_R2.fq.gz
```

Compresses output with pigz.

Output files contain cleaned nuclear paired-end reads (nuc_R1.fq.gz, nuc_R2.fq.gz).

7. Assemble nuclear genome with VelvetOptimiser

```
$ VelvetOptimiser.pl -t 4 --s 61 --e 97 \
  -f '-shortPaired -fastq.gz -separate nuc_R1.fq.gz nuc_R2.fq.
gz' \
  --o '-ins_length 350 -min_contig_lgth 500' -x 4
```

The resulting draft nuclear genome assembly comprised a total of 28,078,080 bp across 74 scaffolds (*see* Table 6).

8. Assess genome completeness with BUSCO. Perform by searching for highly conserved orthologs in genome mode (-m geno), which supports three gene prediction strategies (*see* Table 7). Several strategies can be applied; examples are below.

Table 6
Assembly statistics of the draft genome of *Aspergillus fumigatus* IFM 63560-1

# of scaffolds	Total length (bps)	Min length	N50	Ave. length	Max length
74	28,078,080	502	995,775	379,433	3,437,174

Table 7
Genome completeness assessment of *Aspergillus fumigatus* IFM 63560-1

miniprot	metaeuk	augustus
C:97.2% [S:96.9%, D:0.3%] F:0.5% M:2.4% n:4365	C:95.3% [S:95.1%, D:0.2%] F:0.9% M:3.8% n:4365	C:98.5% [S:98.3%, D:0.2%] F:0.7% M:0.8% n:4365

C: Complete BUSCOs, percentage of expected conserved genes found in full length
S: Single copy, shows how many were found once, as expected
D: Duplicated, shows how many were found more than once (can be real or assembly artifact)
F: Fragmented BUSCOs, shows genes found but only partially assembled
M: Missing BUSCOs, shows genes not detected at all
n: 4365, is a total number of BUSCO groups searched

Using miniprot (fast and efficient):

```
$ busco --download eukaryota
$ busco -c 8 -m geno \
   -l busco_downloads/lineages/eurotiales_odb12 \
   -i auto_data_97/contigs.fa -o busco_miniprot
```

Using metaeuk (sensitive, suited for metagenomes):

```
$ busco --download eukaryota
$ busco -c 8 -m geno \
   -l busco_downloads/lineages/eurotiales_odb12 \
   -i auto_data_97/contigs.fa \
   --metaeuk -o busco_metaeuk
```

Using AUGUSTUS (species-specific, customizable):

```
$ export AUGUSTUS_CONFIG_PATH=~/miniconda3/envs/afumi_env/
config/
$ busco -c 8 -m geno \
    -l busco_downloads/lineages/eurotiales_odb12 \
    -i auto_data_97/contigs.fa --augustus \
    --augustus_species aspergillus_fumigatus \
    -o busco_augustus
```

4 Notes

1. For strains with slow growth: Increase both the number of inoculated spores and the volume of medium. Avoid prolonged incubation.
2. For strains producing few or no conidia: Inoculate five or more 5 × 5 mm blocks of mycelium into PDB medium. Take care to avoid contamination from solid culture media.
3. The band patterns are affected by the DNA polymerase and the reaction conditions.
4. If read data are incomplete, contaminated, or fail quality control, consider excluding the sample from further analysis. When the contaminated cause is known, the mapping can be applied for excluding the reads.
5. Alternative SNP calling tools such as SAMtools [11, 12] and Snippy [30] (primarily designed for bacterial genomes) can also be used for variant detection.
6. The phylogenetic tree can be visualized using tools such as ggtree [29], iTOL [31], FigTree (http://tree.bio.ed.ac.uk/software/figtree/), and others.
7. SPAdes [23] is widely used for fungal genome assembly.

References

1. Denning DW (2024) Global incidence and mortality of severe fungal disease. Lancet Infect Dis 24:e428–e438. https://doi.org/10.1016/S1473-3099(23)00692-8
2. Abdolrasouli A, Rhodes J, Beale MA et al (2015) Genomic context of azole resistance mutations in aspergillus fumigatus determined using whole-genome sequencing. mBio 6: e00536. https://doi.org/10.1128/mBio.00536-15
3. Etienne KA, Berkow EL, Gade L et al (2021) Genomic diversity of azole-resistant aspergillus fumigatus in the United States. mBio 12: e0180321. https://doi.org/10.1128/mBio.01803-21
4. He X, Kusuya Y, Hagiwara D et al (2024) Genomic diversity of the pathogenic fungus Aspergillus fumigatus in Japan reveals the complex genomic basis of azole resistance. Commun Biol 7:274. https://doi.org/10.1038/s42003-024-05902-6
5. Alvarez-Jarreta J, Amos B, Aurrecoechea C et al (2024) VEuPathDB: the eukaryotic pathogen, vector and host bioinformatics resource center in 2023. Nucleic Acids Res 52:D808–D816. https://doi.org/10.1093/nar/gkad1003
6. Paoletti M, Rydholm C, Schwier EU et al (2005) Evidence for sexuality in the opportunistic fungal pathogen aspergillus fumigatus. Curr Biol 15:1242–1248. https://doi.org/10.1016/j.cub.2005.05.045
7. Andrews S (2010) FastQC: a quality control tool for high throughput sequence data. http://www.bioinformatics.babraham.ac.uk/projects/fastq. Accessed 18 Aug 2025
8. Chen S, Zhou Y, Chen Y et al (2018) Fastp: an ultra-fast all-in-one FASTQ preprocessor. Bioinformatics 34:i884–i890. https://doi.org/10.1093/bioinformatics/bty560
9. Vasimuddin M, Misra S, Li H et al (2019) Efficient architecture-aware acceleration of BWA-MEM for multicore systems. In: 2019 IEEE international parallel and distributed processing symposium (IPDPS), pp 314–324
10. Shen W, Le S, Li Y et al (2016) SeqKit: a cross-platform and ultrafast toolkit for FASTA/Q file manipulation. PLoS One 11:e0163962. https://doi.org/10.1371/journal.pone.0163962

11. Li H, Handsaker B, Wysoker A et al (2009) The sequence alignment/map format and SAMtools. Bioinformatics 25:2078–2079. https://doi.org/10.1093/bioinformatics/btp352
12. Danecek P, Bonfield JK, Liddle J et al (2021) Twelve years of SAMtools and BCFtools. Gigascience 10. https://doi.org/10.1093/gigascience/giab008
13. McKenna A, Hanna M, Banks E et al (2010) The genome analysis toolkit: a MapReduce framework for analyzing next-generation DNA sequencing data. Genome Res 20: 1297–1303. https://doi.org/10.1101/gr.107524.110
14. Danecek P, Auton A, Abecasis G et al (2011) The variant call format and VCFtools. Bioinformatics 27:2156–2158. https://doi.org/10.1093/bioinformatics/btr330
15. Manni M, Berkeley MR, Seppey M et al (2021) BUSCO update: novel and streamlined workflows along with broader and deeper phylogenetic coverage for scoring of eukaryotic, prokaryotic, and viral genomes. Mol Biol Evol 38:4647–4654. https://doi.org/10.1093/molbev/msab199
16. Manni M, Berkeley MR, Seppey M et al (2021) BUSCO: assessing genomic data quality and beyond. Curr Protoc 1(12):e323. https://doi.org/10.1002/cpz1.323
17. Ortiz EM (2019) vcf2phylip v2. 0: convert a VCF matrix into several matrix formats for phylogenetic analysis https://zenodo.org/records/2540861
18. Stamatakis A (2014) RAxML version 8: a tool for phylogenetic analysis and post-analysis of large phylogenies. Bioinformatics 30:1312–1313. https://doi.org/10.1093/bioinformatics/btu033
19. NCBI SRA-Tools (2025). https://github.com/ncbi/sra-tools. Accessed 12 Aug 2025
20. Jin J-J, Yu W-B, Yang J-B et al (2020) GetOrganelle: a fast and versatile toolkit for accurate de novo assembly of organelle genomes. Genome Biol 21:241. https://doi.org/10.1186/s13059-020-02154-5
21. Camacho C, Coulouris G, Avagyan V et al (2009) BLAST+: architecture and applications. BMC Bioinformatics 10:421. https://doi.org/10.1186/1471-2105-10-421
22. Langmead B, Salzberg SL (2012) Fast gapped-read alignment with bowtie 2. Nat Methods 9: 357–359. https://doi.org/10.1038/nmeth.1923
23. Prjibelski A, Antipov D, Meleshko D et al (2020) Using SPAdes De Novo Assembler. Curr Protoc Bioinformatics 70:e102. https://doi.org/10.1002/cpbi.102
24. Wick RR, Schultz MB, Zobel J et al (2015) Bandage: interactive visualization of de novo genome assemblies. Bioinformatics 31:3350–3352. https://doi.org/10.1093/bioinformatics/btv383
25. Zerbino DR, Birney E (2008) Velvet: algorithms for de novo short read assembly using de Bruijn graphs. Genome Res 18:821–829. https://doi.org/10.1101/gr.074492.107
26. O'Gorman CM, Fuller HT, Dyer PS (2009) Discovery of a sexual cycle in the opportunistic fungal pathogen Aspergillus fumigatus. Nature 457:471–474. https://doi.org/10.1038/nature07528
27. Versalovic J, de Bruijn FJ, Lupski JR (1998) Repetitive sequence-based PCR (rep-PCR) DNA fingerprinting of bacterial genomes. In: de Bruijn FJ, Lupski JR, Weinstock GM (eds) Bacterial genomes: physical structure and analysis. Springer Boston, pp. 437–454
28. George MLC, Nelson RJ, Zeigler RS et al (1998) Rapid population analysis of Magnaporthe grisea by using rep-PCR and endogenous repetitive DNA sequences. Phytopathology 88:223–229. https://doi.org/10.1094/PHYTO.1998.88.3.223
29. Yu G (2023) Data integration, manipulation and visualization of phylogenetic trees, First edition, Chapman & Hall/CRC computational biology series. CRC Press, Boca Raton
30. Seemann T (2015) Snippy: rapid haploid variant calling and core SNP phylogeny. GitHub. https://github.com/tseemann/snippy. Accessed 18 Aug 2025
31. Letunic I, Bork P (2024) Interactive Tree of Life (iTOL) v6: recent updates to the phylogenetic tree display and annotation tool. Nucleic Acids Res 52:W78–W82. https://doi.org/10.1093/nar/gkae268
32. Takahashi H, Oiki S, Kusuya Y et al (2021) Intimate genetic relationships and fungicide resistance in multiple strains of aspergillus fumigatus isolated from a plant bulb. Environ Microbiol 23:5621–5638. https://doi.org/10.1111/1462-2920.15724

Chapter 6

Isolation of Nucleic Acids from *Aspergillus fumigatus* for Sequencing

Sam El-Kamand and Charles Oliver Morton

Abstract

Invasive aspergillosis is a severe fungal infection that usually affects the immunocompromised, e.g., transplant patients. Sequence-based analysis has become increasingly important for understanding fungal virulence, fungal identification, and monitoring drug resistance. In this chapter, we describe methods for the isolation of fungal DNA from different sample types and quality control of sequence data.

Key words *Aspergillus fumigatus*, DNA isolation, Genome sequencing, Fungal diagnostics, Aspergillosis

1 Introduction

The fungal disease invasive aspergillosis, caused by fungi of the genus *Aspergillus*, especially *Aspergillus fumigatus*, is a major concern for immunocompromised individuals, particularly those who have undergone stem cell and solid organ transplants [1]. Two major challenges are understanding fungal virulence [2] and the development of drug resistance in the fungus [3, 4]. Analysis of genome sequences can help us to identify differences between virulent clinical isolates and colonizing isolates, which will allow identification of virulence determinants [5]. Sequencing of genomes or target genes can identify if clinical isolates are drug-resistant, allowing for improved treatment [6], and fungal genome sequencing has become increasingly important in diagnosing fungal infections [7].

The first hurdle in sequence-based analysis is obtaining fungal DNA; in medical settings, fungal samples may come in different forms, from formalin-fixed paraffin-embedded specimens to blood samples to pure isolated cultures. It is necessary to be able to isolate pure fungal DNA from the most challenging samples; the next challenge is ensuring the quality of the isolated DNA. Lower

Natalia Mach (ed.), *Aspergillus fumigatus: Methods and Protocols*, Methods in Molecular Biology, vol. 3028,
https://doi.org/10.1007/978-1-0716-5222-0_6,

quality DNA may not be suitable for genome sequencing but can be used with PCR for sequencing of target genes, e.g., *cyp51a*, involved in resistance to azole antifungal drugs [8, 9].

In this chapter, we describe how to isolate fungal DNA from clinical specimens and pure cultures and then how to ensure the quality of the DNA and genome sequences.

2 Materials

2.1 Kits for DNA Isolation

1. ISOLATE II Genomic DNA Kit for DNA isolation from serum, fungal biomass, and fungal conidia (e.g., Meridian Biosciences #BIO-52073).
2. DNeasy Blood & Tissue Kit for DNA isolation from paraffin-embedded samples, tissue biopsies, serum, fungal tissue, and fungal conidia (e.g., Qiagen #69504).
3. High Pure PCR Template Preparation Kit optimized for DNA isolation from blood; suitable for PCR-based assays (e.g., Roche #11796828001).

2.2 Beads for Bead Milling

1. Acid-washed glass beads, 710–1180 μm (e.g., Sigma-Aldrich #G1152). Use for DNA isolation from paraffin-embedded tissue and from model organisms.
2. Lysing Matrix G (e.g., MP Biomedicals #116916050). Use for isolation of high-quality DNA from fungal biomass [5].
3. MagNA Lyser Green Beads (e.g., Roche #03358941001): these are ceramic beads with low DNA retention, optimal for blood samples [10].

2.3 Reagents and Equipment

1. Ultrapure water for preparation of all solutions.
2. RNase A solution, 20 mg/mL.
3. Chloroform:isoamyl alcohol mix (24:1).
4. Red cell lysis buffer (RCLB): 10 mM Tris (pH 7.6), 5 mM $MgCl_2$, 10 mM NaCl.
5. White cell lysis buffer (WCLB): 10 mM Tris (pH 7.6), 10 mM EDTA, 50 mM NaCl, 0.2% SDS. Supplement with 200 μg/mL proteinase K.
6. Xylene.
7. Ethanol solutions: 96–100%, 80%, 50%.
8. High-speed bead-beating homogenizer (e.g., Fast Prep from MP Biomedicals #15260488).
9. Heating block compatible with 1.5 mL or 2 mL tubes, temperature range 37 °C–75 °C, best with shaker option.

10. Filtration setup including sterile filter material (e.g., Whatman filter paper, grade 1), Büchner funnel, and vacuum pump.
11. Vortex mixer.
12. Microcentrifuge.
13. Fume hood.
14. Orbital shaker incubator.
15. Sterile scalpel.

2.4 DNA Quantification and Quality Assessment

1. Benchtop fluorometer (e.g., Qubit Fluorometer #Q33238).
2. dsDNA Quantitation kit (e.g., Qubit dsDNA Broad Range Assay Kit #Q32853).
3. Tris-Borate-EDTA (TBE) buffer (10×): 0.89 M Tris base, 0.89 M boric acid, 0.02 M EDTA. Dissolve 108 g Tris base, 55 g boric acid, and 7.44 g disodium EDTA in 800 mL ultra-pure water. Adjust volume to 1 L. Store at room temperature.
4. Dye for visualizing DNA (e.g., SYBR Safe DNA gel stain from Thermo Fisher Scientific #S33102).
5. Agarose, molecular biology grade.
6. 1% agarose gels in 1× TBE buffer.
7. DNA ladder (e.g., 1 kb Plus DNA Ladder, Invitrogen).
8. UV-visible spectrophotometers (e.g. NanoDrop UV-Vis from Thermo Fisher Scientific).
9. Fluorometer for DNA quantification (e.g., Qubit, Thermo Fisher Scientific).
10. Gel electrophoresis unit and imaging/documentation equipment.
11. UV or blue-light transilluminator.

2.5 Bioinformatics Tools for DNA and Sequencing Quality Control

In the text below, highlighted font describes code examples, flags, or arguments; **bold** is used for software.

2.5.1 Software for Raw Read and Alignment Quality Assessment

1. **FastQC** (v0.12.1).
2. **Cutadapt** (v5.0).
3. **Kraken2** (v2.1.5) with PlusPF-16 database (prebuilt, available from [11, 12]).
4. **Bowtie2** (v2.5.4) or alternative short-read alignment tool (e.g., **bwa-mem2**, v2.2.1).
5. **Mosdepth** (v0.3.11).
6. **Picard** (v3.4.0).

7. **MultiQC** (v1.28).
8. md5sum or equivalent command-line checksum verifier (typically pre-installed on Unix/Linux systems).

2.5.2 Software for Assembly and Annotation Quality Control

1. **BUSCO** (v5.8.2).
2. **QUAST** (v5.3.0).

3 Methods

3.1 Contamination Control

1. Perform all DNA isolations in a class II biosafety cabinet.
2. Use PCR-grade water in all PCR reactions.
3. Optional: Prescreen DNA isolation reagents for contamination via qPCR using an ITS-targeted assay for *A. fumigatus* (*see* **Note 1**).
4. Include negative controls in all PCRs.
5. Use separate areas for master mix preparation and DNA template addition.

3.2 DNA Isolation

Four DNA extraction protocols are described in this section, depending on the sample type. Figure 1 provides a summary of protocols, showing the overall workflow for each and specific pre-lysis modifications (see Fig. 1).

3.2.1 DNA Isolation from Fungal Mycelium

This protocol is adapted for the isolation of high-quality, high-molecular-weight (HMW) DNA for genome sequencing using the Isolate II Genomic DNA Kit. It is best to prepare fungal cultures with a solution of conidia to allow for control of the inoculum. This method can also be used to isolate DNA directly from conidia.

1. Dispense 1×10^4 of *A. fumigatus* conidia into 20 mL of malt broth.
2. Incubate in orbital shaker incubator at 100 rpm, 37 °C for 4 days.
3. Recover fungal biomass by vacuum-filtering (*see* Fig. 2).
4. Transfer biomass to a 2 mL screw-cap microcentrifuge tube. Proceed immediately or store at −20 °C until required.
5. Weigh 100 mg of mycelium, and add to a 2 mL tube of Lysing Matrix G. (*see* **Note 2**).
6. Add 180 μL of lysis buffer (kit component).
7. Bead mill in a bead-beating homogenizer on max speed for 30 s (*see* **Note 3**).

	Conidia (1)	FFPE (2)	Biopsy (3)	Blood (4)
Sample Type & Preparation	• Pellet by centrifugation • Remove Supernatant	• Place sample in xylene • Wash with ethanol	• Cut biopsy into small pieces (< 1mm)	• Centrifuge to remove plasma. • Red Cell Lysis • Pellet • White Cell Lysis • Pellet
Lysis & Disruption	• Add Lysis Buffer* & Beads** • Incubate <2 h • Disrupt with vortex or bead beater***	• Add Lysis Buffer & Beads • Incubate < 2 h • Disrupt with vortex or bead beater	• Add Lysis Buffer • Incubate Overnight • Add Beads • Disrupt with vortex or bead beater.	• Add Lysis Buffer and Beads • Incubate <2 h • Disrupt with vortex mixer (30 s) or bead beater (30 s at low setting).

DNA Binding

Follow Manufacturer's Instructions, usually:

- Transfer lysed**** sample to a new 1.5 mL tube
- Add additional buffers and ethanol
- Add sample mixture to the DNA-binding column
- Use Wash Buffer and dry the column

DNA Elution

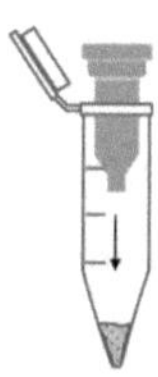

- Add Elution Buffer to the membrane in the column
- Volume should be 30 – 50 μL
- **Note:** Eluate can be added back onto the membrane and eluted again to increase DNA concentration
- **Note:** Smaller elution volumes increase DNA concentration which is desirable.

* **Note:** Lysis containing Proteinase K

** **Note:** Use acid-washed glass beads (ca. 1mm diameter) or ceramic beads (MagNA Lyser green beads (Roche) are excellent).

*****Note:** DNA shearing is a risk so calibrate the bead beating step by testing different times and intensities. Usually, 1 minute at medium setting for conidia. Shorter time periods when there are fewer fungal cells. Cycles of 1 minute bead beating followed by 1 minute on ice can prevent excessive heating of the sample, important in RNA extractions.

**** **Note:** Can add chloroform to a level just above the beads. Mix and centrifuge. Carefully remove the supernatant from the tube with a 200 μL pipette tip, be careful not to take up any chloroform. This maximises recovery from the beads, especially good for small sample volumes. Beads can be washed in PBS to improve sample recovery if chloroform is not used.

Fig. 1 Illustration of the steps in DNA isolation from: [1] *A. fumigatus* conidia, [2] formalin-fixed paraffin-embedded (FFPE) samples, [3] tissue biopsy samples, and [4] blood samples. This includes specific pre-lysis steps that are not standard and the DNA-binding and elution steps common to most DNA isolation kits

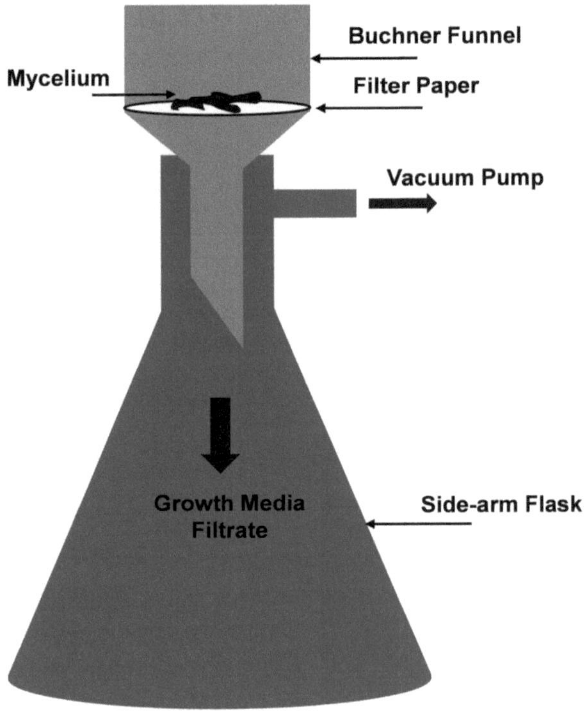

Fig. 2 Illustration of the setup for recovering fungal mycelium from a liquid culture. The culture is poured onto a disk of filter paper in the Buchner funnel. The side-arm flask is attached to a vacuum pump that provides suction to separate the growth medium from the mycelium

8. Add 1 μL of RNase A, and incubate at 37 °C for 30 min.
9. Add 25 μL of Proteinase K solution (kit component). Incubate at 56 °C for 1 h.
10. Transfer lysate from beads to a new microcentrifuge tube.
11. Add 200 μL of Buffer G3 (kit component). Incubate at 70 °C for 10 min.
12. Add 210 μL of 96–100% ethanol. Vortex briefly.
13. Place ISOLATE II Genomic DNA spin column into a collection tube (kit components).
14. Load the sample to the column, centrifuge at 11,000 × g for 1 min. Repeat this step if the volume exceeds the capacity of the column.
15. Add 500 μL of Buffer GW1 (kit component) to the column as washing step, centrifuge at 11,000 × g for 1 min. Discard the flow-through.
16. Add 500 μL of Buffer GW2 (kit component) to the column as washing step, centrifuge at 11,000 × g for 1 min. Discard the flow-through.

17. Dry column by centrifuging at 11,000 × g for 1 min.
18. Transfer column to a clean 1.5 mL tube.
19. Elute DNA with 50 μL of pre-warmed Elution Buffer G (70 °C, kit component) by adding it directly onto the silica membrane of the column. Reducing the volume of the elution buffer will increase the DNA concentration.
20. Assess DNA concentration and quality (*see* Sect. 3.3).
21. Prior to genome sequencing, confirm fungus species identity (*see* **Note 4**).

3.2.2 DNA Isolation from Tissue Samples

This protocol uses the DNeasy Blood & Tissue Kit; it is suitable for DNA extraction from tissue samples with substantial fungal growth. For downstream genome sequencing, high fungal biomass is preferred; otherwise, the protocol is suitable for PCR-based analyses of target genes. Samples like cartilage, skin, or bone require a different method.

1. Cut up to 100 mg of tissue into ≈1 mm^3 pieces using a sterile scalpel. Transfer to a 1.5 mL screw-cap microcentrifuge tube.
2. Add 180 μL of lysis Buffer ATL (kit component) containing 20 μL of the Proteinase K solution (kit component). Vortex to mix.
3. Incubate at 56 °C overnight (≥15 h). Use the thermoshaker option if available.
4. Add 200 μL of Buffer AL (kit component). Vortex to mix and incubate at 56 °C for 10 min.
5. Add acid-washed glass beads (volume ≈100 μL) (*see* **Note 2**).
6. Bead mill using bead-beating homogenizer for 1 min at medium speed. Alternatively, perform 5 cycles of 1 min vortexing (fastest setting) followed by 1 min on ice (*see* **Note 3**).
7. Add 150 μL of chloroform:isoamyl alcohol mix. Vortex 5 s, and centrifuge at 10,000 × g for 1 min.
8. Using a P200 pipette, carefully transfer the upper aqueous phase to a new 1.5 mL tube. Avoid pipetting chloroform (if some chloroform enters the pipette tip, then *see* **Note 5**).
9. Add 200 μL of96–100% ethanol to the supernatant. Vortex 5 s.
10. Load sample onto the DNA-binding column of the kit placed in a collection tube. Centrifuge at 6000 × g for 1 min and discard the flow-through. Repeat loading if there is still a sample remaining.
11. Add 500 μL of Buffer AW1 (kit component) onto the column as washing step. Centrifuge at 6000 × g for 1 min. Discard the flow-through.

12. Add 500 μL of Buffer AW2 (kit component) onto the column as washing step. Centrifuge at 20,000 × g for 3 min. Discard the flow-through.
13. Transfer column to a clean 1.5 mL tube.
14. Elute DNA by adding 50 μL of Buffer AE (kit component) to the column. Incubate at room temperature for 1 min. The low elution volume will increase DNA concentration.
15. Centrifuge at 6000 × g for 1 min to collect the eluate. Step 14 can be repeated by placing the eluate back onto the column and eluting again to further increase DNA concentration.
16. Assess DNA concentration and quality (*see* Sect. 3.3).

3.2.3 DNA Isolation from Blood Samples

This method uses the High Pure PCR Template Preparation Kit. It is based on a previously described protocol for isolating *A. fumigatus* DNA from blood samples, originally developed for PCR-based diagnostics of invasive aspergillosis [10, 13]. The protocol isolates DNA from fungal cells that have been phagocytosed by white blood cells. It has the potential to yield high-molecular-weight fungal DNA suitable for genome sequencing, provided sufficient quantities can be recovered. In contrast, extracellular DNA obtained from serum or plasma is typically more fragmented that makes it suitable for PCR but likely inadequate for genome sequencing. This method can be applied to both filamentous and unicellular fungi.

1. Collect a 3 mL EDTA blood sample in a vacutainer.
2. Transfer blood to a 15 mL Falcon tube. Add 10 mL of RCLB. Incubate for 5 min at room temperature on a shaker.
3. Centrifuge at 3000 × g for 10 min. Discard the supernatant.
4. Add 1 mL of WCLB to the pellet. Incubate at 65 °C for 30 min.
5. Centrifuge at 5000 × g for 10 min. Discard the supernatant.
6. Add 200 μL of Binding buffer (kit component) containing 20 μL of Proteinase K (kit component). Add ≈ 200 μL of MagNa lyser green ceramic beads.
7. Vortex at maximum speed for 90 s.
8. Transfer lysate to a clean 1.5 mL microcentrifuge tube.
9. Add 100 μL of isopropanol. Vortex 5 s.
10. Load sample onto a DNA-binding column (kit component) placed into a collection tube. Centrifuge at 8000 × g for 1 min. Discard flow-through.
11. Add 500 μL of Inhibitor Removal Buffer (kit component). Centrifuge at 8000 × g for 1 min. Discard flow-through.

12. Add 500 μL of Wash Buffer (kit component). Centrifuge at 8000 × g for 1 min. Discard flow-through. Repeat this step one more time.
13. Centrifuge at 13,000 × g for 10 s to dry the column. Transfer column into a new 1.5 mL microcentrifuge tube.
14. Elute DNA by adding 50 μL of pre-warmed (70 °C) elution buffer (kit component) to the column.
15. Centrifuge at 8000 × g for 1 min. The lower elution volume will increase DNA concentration.
16. Optional: Reapply eluate to column and repeat centrifugation to increase yield.
17. Assess DNA concentration and quality (*see* Sect. 3.3).

3.2.4 DNA Isolation from Paraffin-Embedded Tissue

The identity of fungi in formalin-fixed paraffin-embedded (FFPE) samples can be difficult to determine by microscopy, thus DNA isolation and PCR might be required. FFPE samples are usually small and contain limited fungal biomass, making them best suited for PCR amplification of target genes. Large FFPE samples may yield sufficient DNA for genome sequencing, though DNA quality is often sub-optimal. This protocol is adapted from the previously described method [14] and uses DNeasy Blood & Tissue Kit.

1. Start with a small block of FFPE sample. Trim excess paraffin from the sample block using a sterile scalpel.
2. Place the sample into a 2 mL screw-cap microcentrifuge tube.
3. Add 1 mL of xylene. Perform this step in a fume hood. Incubate with shaking for 15 min to dissolve paraffin.
4. Centrifuge at 16,000–20,000 × g for 2 min. Carefully discard xylene and dispose it in a designated waste container in a fume hood.
5. Add 1 mL of 96–100% ethanol. Vortex 5 s, centrifuge at maximum speed for 2 min. Carefully discard the supernatant.
6. Add 1 mL of 80% ethanol. Vortex 5 s, centrifuge at maximum speed for 2 min. Carefully discard the supernatant.
7. Add 1 mL of 50% ethanol. Vortex 5 s, centrifuge at maximum speed for 2 min. Carefully discard the supernatant.
8. Carefully remove residual ethanol without disturbing the pellet. Air-dry the sample briefly to ensure complete removal of ethanol.
9. Add 180 μL of Buffer ATL (kit component) containing 20 μL of the Proteinase K (kit component). Vortex to mix.
10. Incubate at 56 °C for at least 3 h or overnight. Using a thermoshaker is optimal for this step.

11. Add 200 μL of Buffer AL (kit component). Vortex and incubate at 56 °C for 10 min.
12. Add ≈100 μL of acid-washed glass beads (*see* **Note 2**).
13. Bead mill for 1 min at medium speed using bead-beating homogenizer. Alternatively, 5 cycles of 1 min mixing with a vortex mixer (fastest setting) and 1 min on ice can also be used (*see* **Note 3**).
14. Add 150 μL of chloroform:isoamyl alcohol mix to the tube. Vortex 5 s, centrifuge at 10,000 × g for 1 min.
15. Transfer the upper aqueous phase to a new 1.5 mL microcentrifuge tube using a P200 pipette. Avoid disturbing the chloroform interface (if some chloroform enters the pipette tip *see* **Note 5**).
16. Add 200 μL of 96–100% ethanol. Vortex 5 s.
17. Proceed with steps 10–16 from Sect. 3.2.2 (*see* **Note 6**).

3.3 Quantification and Quality Control of Extracted gDNA

The quantity and quality requirements for extracted gDNA depend on the downstream library preparation method and sequencing platform. At the time of writing, short-read paired-end sequencing on the Illumina NovaSeq 6000 is the most common strategy for *A. fumigatus* whole-genome sequencing (*see* Fig. 3a, b). However, sequencing technologies evolve rapidly (*see* Fig. 3c), and optimal platform selection depends on sample number, target coverage depth, and local availability. PCR-free protocols are generally recommended as they provide higher-quality data, and clinical isolates can be cultured to obtain the high DNA yields required. The quality control workflow described below outlines thresholds suitable for standard Illumina short-read workflows, but these should be adjusted to the specifications of the selected platform and application. Always consult the documentation for the specific library preparation kit used or seek guidance from the sequencing provider.

1. Evaluate DNA yield by fluorometry. Measure DNA concentration with, e.g., Qubit fluorometer and dsDNA BR Assay Kit, following the manufacturer's protocol. Aim for ≥2 μg of DNA, higher yields are advantageous (*see* **Notes 7–8**).
2. Evaluate purity by UV spectrophotometry. An A260/A280 ratio below 1.8 can indicate protein contamination (*see* **Note 9**). An A260/A230 ratio below 2.0 indicates contamination by organic compounds such as phenol, guanidine, or salts (*see* **Note 10**).
3. Assess fragmentation by gel electrophoresis. Run and visualize 50–100 ng of gDNA on a 1% agarose gel in 1 × TBE buffer. Electrophorese at 90 V for 45–60 min. A high-intensity, high-molecular-weight band (≥10 kb) should be visible. Smearing

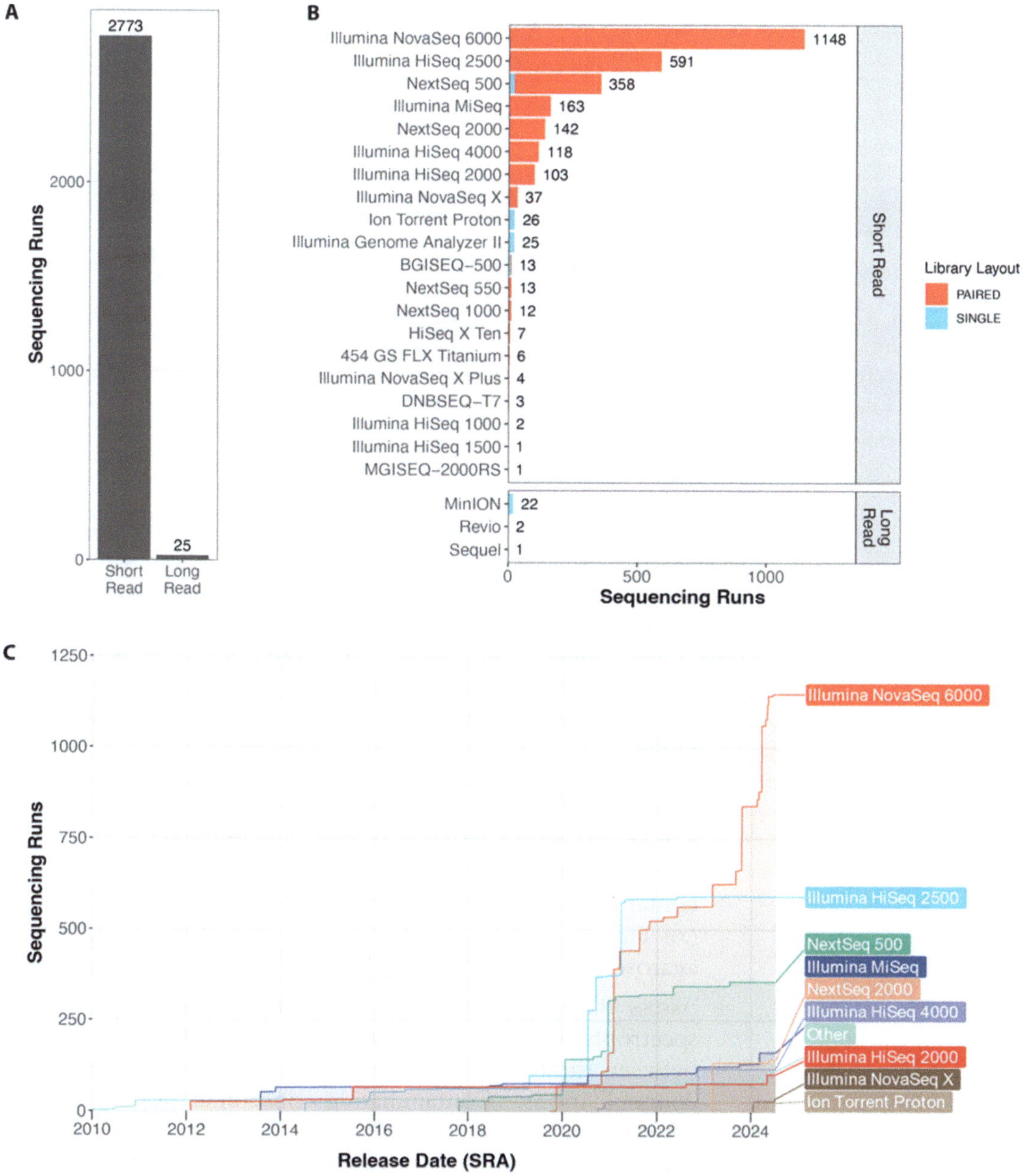

Fig. 3 Trends in whole-genome sequencing of *A. fumigatus*. (**a**) Isolates are predominantly sequenced using short-read technologies. (**b**) The most common strategy is to sequence paired-end libraries on the Illumina NovaSeq 6000 platform. The Oxford Nanopore MinION is the most widely used long-read platform. (**c**) Temporal shifts in platform usage reflect the evolving availability and cost-efficiency of sequencing technologies, as evident in data submissions to the Sequence Read Archive (SRA)

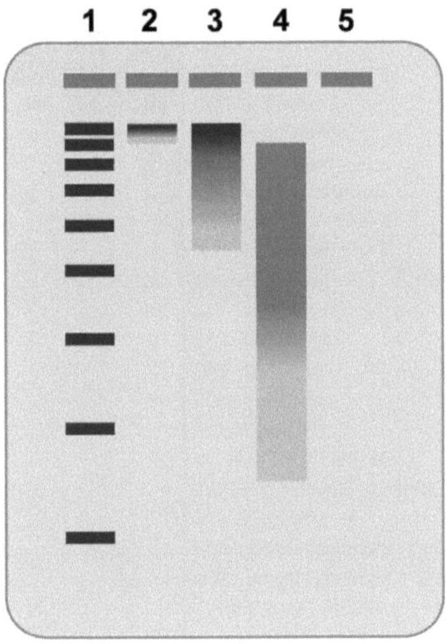

Fig. 4 Illustration of what to look for when checking the quality of genomic DNA on an agarose gel. Lane 1 shows the molecular weight ladder; HMW DNA will usually match the top band of the ladder or be slightly higher up the gel. Lane 2 is high-quality genomic DNA; it forms a tight band with minimal smearing. Lane 3 shows smearing of genomic DNA. There is still HMW DNA, and the smearing is not too severe, so this could still be used for genome sequencing. Lane 4 is fully smeared DNA with no HMW DNA. This is not suitable for genome sequencing

toward smaller sizes indicates fragmentation or RNA contamination. Mild smearing across the 2–10 kb range is acceptable for most short-read workflows (*see* **Notes 11–12**). Protein contamination may also manifest as sample retention in wells (*see* Fig. 4, an illustration of how smearing can appear on an agarose gel).

4. Assess RNA contamination. Compare fluorometry and UV spectrophotometry yield estimates. Significant discrepancies between the two may indicate RNA or single-stranded DNA contaminations. Contaminating ribosomal RNA (28S at ~4.5 kb and 18S at ~2 kb) may also be visible as distinct bands in Step 3. If downstream library preparation is sensitive to RNA contamination, treat samples with 100 μg/mL of RNase A at 37 °C for 30 min, followed by purification using a spin column or ethanol precipitation.
5. Assess suitability for library preparation and sequencing. Confirm that gDNA concentration, purity, and integrity meet or exceed thresholds stated in the library preparation documentation. For most PCR-free Illumina workflows, ≥2 μg of clean, minimally fragmented gDNA is required. For specific requirements and quality considerations for long-read sequencing platforms, *see* **Note 13**.

3.4 Quality Control of Sequencing Data

Following sequencing, data quality must be evaluated to confirm that read length, base call quality, species identity, and, most importantly, depth and breadth of coverage meet expectations. While complete quality control failure is rare for whole-genome sequencing (WGS), these assessments are essential to inform downstream processing.

1. Verify data integrity. Confirm downloaded files are intact by comparing MD5 checksums computed using the md5sum command-line utility to those provided by the sequencing facility.
2. Evaluate raw read quality. Run **FastQC** on raw ".fastq" files. Confirm that the number of bases sequenced, when divided by target genome size (29.4×10^6 bp), meets or exceeds the planned coverage (e.g., 50×). Ensure read length aligns with the specified sequencing protocol. Examine the "Per base sequence quality" module to ensure that at least the early bases in reads have good base-call confidence (*see* **Note 14**). Check the "Overrepresented sequences" and "Adapter content" modules for warnings indicative of adapter contamination. Remove contaminating adapter sequences using trimming tools such as **Cutadapt** or **Trimmomatic**. These tools can also trim reads to retain only high-confidence base calls, which may improve genome assembly quality but is not usually advisable for variant-calling applications (*see* **Note 15**).
3. Assess potential contamination and verify *A. fumigatus* identity by taxonomic classification of raw reads using **Kraken2** with the PlusPF database. Confirm that ≥90% of reads are assigned to *A. fumigatus*. A high proportion of reads classified as another species may indicate contamination, a sample swap, or misidentification of the original isolate.
4. Assess coverage metrics. Align reads to a reference *A. fumigatus* genome (e.g., GCF_000002655.1) using an aligner such as **Bowtie2** or **bwa-mem2**. Compute genome-wide depth and breadth of coverage using **Mosdepth**. Use **Picard**'s CollectAlignmentSummaryMetrics function to evaluate alignment quality. Summarize output files as an HTML report using **MultiQC**. Check median fragment size matches expectations given the sequencing library preparation protocol. Most of the genome should be covered at or near the target depth. If specific genomic regions are of interest, provide a Browser Extensible Data (BED) file to **Mosdepth** to calculate per-region mean coverage. If many regions have suboptimal but nonzero coverage, consider top-up sequencing of the same library. If a region is entirely uncovered, it may be absent from the isolate's genome, lost during library preparation (e.g., due

to over-fragmentation), or reside in a repetitive or otherwise unmappable region.

5. Perform target downstream analysis. If reference-based variant calling is the goal, the most effective quality check is to perform variant calling and examine the results. Variants should show allele frequencies close to 100%, as the sequencing derives from a haploid, pure-cultured fungus. Detected variants should include known polymorphic sites. If performing genome assembly, evaluate scaffold quality using **BUSCO** to check for the expected set of single-copy orthologs. Align reads back to the assembly and confirm that observed fragment-size distributions match library expectations. Additional assembly metrics such as NG50 and overall contiguity can be assessed using **QUAST**.

4 Notes

1. It is essential to prevent environmental contamination when working with *A. fumigatus*. The organism is ubiquitous in the environment, so strict precautions must be followed to prevent contamination. DNA isolation reagents can be prescreened for contamination via qPCR using an ITS-targeted assay for A. fumigatus as previously described [10].
2. Lysing matrix G is a mixture of 1.6 mm silicon carbide particles and 2 mm glass beads, ideal for disrupting fungal or plant tissue. Other beads can be used, e.g., glass beads (0.7–1 mm) or ceramic beads (MagNA lyser green beads). Ceramic beads are less likely to bind DNA but are more expensive.
3. Other bead milling machines can be used. Optimize the fungal lysis protocol before proceeding with sequencing. Multiple rounds of 30 s of bead milling can be used to achieve lysis; place the tubes on ice between cycles of bead milling. Test DNA quality on an agarose gel, excessive shearing is not desirable, so there should ideally be a single HMW band.
4. To confirm fungal identity the ITS regions of the ribosomal operon can be PCR amplified for sequencing. We use the ITS 4 and 5 primers to amplify the ITS 1 and ITS 2 regions, which are commonly used for fungal identification [15, 16]. The sequence obtained can be used to search the nucleotide sequence database using Blast (https://blast.ncbi.nlm.nih.gov/Blast.cgi) [17].
5. Chloroform can enter the tip when recovering the supernatant. In this case, observe the tip and allow the chloroform to sediment at the opening of the tip; it will form a layer beneath the supernatant. Once the lower chloroform layer has formed, dispense it from the tip (*see* Fig. 5).

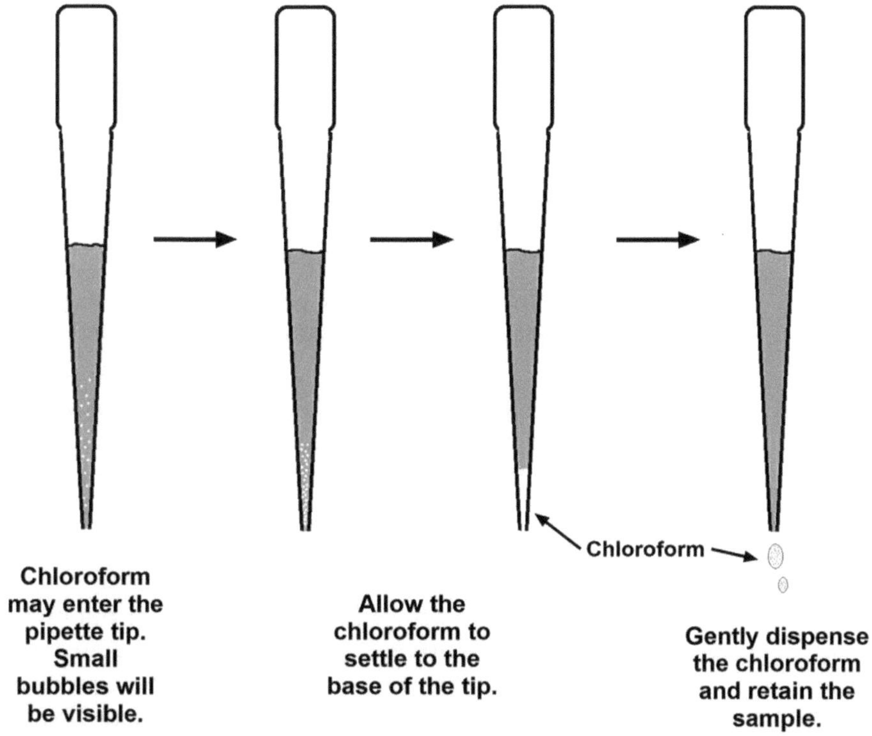

Fig. 5 An illustration of how to deal with chloroform that has been taken into a pipette tip along with the sample. Initially, the chloroform will appear as small droplets/bubbles in the sample. The chloroform is denser than the aqueous sample and will settle to the opening of the tip. Once the chloroform has settled, it can be dispensed from the tip

6. The standard DNA-binding column can be replaced by a DNA-binding column from the MinElute PCR Purification Kit (Qiagen #28004). This is important when using a small sample as it will allow for elution in a much smaller volume. When using MinElute columns, DNA can be eluted in as little as 20 μL to achieve higher concentrations.
7. Since isolates are cultured before DNA extraction, it is possible to achieve yields well above minimum input requirements. High DNA yields give sequencing facilities flexibility to dilute down to optimal concentrations or retry library preparation in case of library-preparation failure.
8. Fluorometric quantification of DNA (e.g., Qubit) is strongly preferred over UV spectrophotometry (e.g., Nanodrop), which overestimates yields in the presence of ssDNA, RNA, and degraded fragments. Fluorescent dyes bind specifically to the minor groove of double-stranded DNA, minimizing overestimation from contaminants or degraded material. We do not recommend proceeding with library preparation without fluorometric quantification. If fluorometric methods are

unavailable, increase the minimum yield threshold and assess sample integrity by gel electrophoresis to calibrate confidence in UV-based yield estimates. Additionally, request your sequencing provider confirm DNA yield by fluorometric quantification and flag any concerns before incurring library preparation or sequencing costs.

9. Increase Proteinase K concentration or extend digestion time to remove protein contamination.
10. Contamination by organic compounds such as phenol, guanidine, or salts can be resolved with an additional ethanol precipitation or column cleanup.
11. Although only severe fragmentation of gDNA will cause complete library preparation failure, library yield and coverage will decrease with the ratio of long to short fragments. This may increase the number of genomic regions where variants cannot be robustly called, or lead to larger gaps in genome assemblies.
12. Automated electrophoresis systems, such as the Agilent TapeStation, provide detailed fragment size distributions and integrity scores computed from electropherograms (e.g., DNA Integrity Number, DIN). These can guide sequencing suitability, particularly for high-throughput projects. Sequencing services may request gDNA with a minimum DIN, usually ≥5, for short-read Illumina sequencing. To approximate DIN in the absence of a TapeStation see https://www.agilent.com/cs/library/applications/5991-5258EN.pdf, which provides a map of how different DIN values appear on electrophoretic gels.
13. Similar to short-read sequencing, gDNA for long-read WGS is assessed for yield, purity, and integrity. However, requirements are more stringent. Higher yields (typically >3 μg) are required to compensate for losses during size selection of HMW (>10 kb) fragments. Long-read platforms are highly sensitive to degradation, so lower-molecular-weight smears on 1% agarose gels, tolerable for short-read sequencing; here, indicate gentler DNA extraction methods may be required. Standard extraction protocols for fungi are often inappropriate for HMW extraction, as harsh lysis steps designed to disrupt their cell walls tend to fragment DNA. Optimized protocols that minimize mechanical shear are essential. Since conventional 1% agarose electrophoresis cannot resolve large fragments, the best practice is to assess HMW fragment size distributions using pulsed-field gel electrophoresis (PFGE). In practice, however, it is more common to just ensure the average fragment length of size-filtered gDNA exceeds the maximum quantifiable size on 1% agarose gel or TapeStation Screentapes (usually 10–20 kb) and proceed with sequencing.

14. Base-call accuracy typically declines toward the end of sequencing reads. This pattern arises due to decreased signal intensity over the course of the run. Reverse reads also exhibit more pronounced degradation than forward reads.
15. Adapter sequences at the 3′ ends of short fragments may be read through during sequencing and should be removed. Adapter trimming is a standard post-processing step, and some sequencing platforms can automatically perform this during FASTQ generation. When working with public datasets or when preprocessing is uncertain, it is generally safe to apply adapter trimming again, as it only rarely affects downstream analyses. In contrast, quality-based trimming should be used cautiously. Low-confidence base calls may still be informative for variant calling, as many algorithms incorporate base-quality scores into probabilistic models. Further, a single miscalled base is unlikely to be interpreted as a true variant at a well-covered locus, particularly in haploid, clonal *A. fumigatus* isolates, where true variants are expected at 100% allele frequency. For genome assembly, moderate quality trimming may improve outcomes and is integrated into assemblers such as SKESA. However, over-trimming can remove useful sequence data and lead to fragmented assemblies. Therefore, trimming parameters should be carefully optimized.

References

1. van de Veerdonk FL, Carvalho A, Wauters J et al (2025) Aspergillus fumigatus biology, immunopathogenicity and drug resistance. Nat Rev Microbiol 23(10):652–666. https://doi.org/10.1038/s41579-025-01180-z
2. Earle K, Valero C, Conn DP et al (2023) Pathogenicity and virulence of Aspergillus fumigatus. Virulence 14(1):2172264. https://doi.org/10.1080/21505594.2023.2172264
3. Lestrade P, Meis JF, Melchers W et al (2019) Triazole resistance in Aspergillus fumigatus: recent insights and challenges for patient management. Clin Microbiol Infect 25:799–806. https://doi.org/10.1016/j.cmi.2018.11.027
4. Verweij PE, Kema GHJ, Zwaan B et al (2013) Triazole fungicides and the selection of resistance to medical triazoles in the opportunistic mould Aspergillus fumigatus. Pest Manag Sci 69:165–170. https://doi.org/10.1002/ps.3390
5. El-Kamand S, Steiner M, Ramirez C et al (2022) Assessing differences between clinical isolates of Aspergillus fumigatus from cases of proven invasive Aspergillosis and colonizing isolates with respect to phenotype (Virulence in Tenebrio molitor larvae) and genotype. Pathogens 11:428. https://doi.org/10.3390/pathogens11040428
6. Sarkar A, Rangel DE, Osherov N (2025) Leading developments in basic research on aspergillosis and mucormycosis. Fungal Biol 129:101574. https://doi.org/10.1016/j.funbio.2025.101574
7. El-Kamand S, Papanicolaou A, Morton CO (2019) The use of whole genome and next-generation sequencing in the diagnosis of invasive fungal disease. Curr Fungal Infect Rep 13:284–291. https://doi.org/10.1007/s12281-019-00363-5
8. Rabault C, Rakotoarivony Y, Houzé S et al (2024) Routine laboratory screening for cyp51a -related azole resistance in Aspergillus fumigatus using Etest: concordance with gene sequencing. Med Mycol 62(11):myae107. https://doi.org/10.1093/mmy/myae107
9. Sen P, Vijay M, Kamboj H et al (2024) cyp51A mutations, protein modeling, and efflux pump gene expression reveals multifactorial complexity towards understanding Aspergillus section Nigri azole resistance mechanism. Sci Rep 14(1):6156. https://doi.org/10.1038/s41598-024-55237-9

10. Springer J, Loeffler J, Heinz W et al (2011) Pathogen-specific DNA enrichment does not increase sensitivity of PCR for diagnosis of invasive Aspergillosis in neutropenic patients. J Clin Microbiol 49:1267–1273. https://doi.org/10.1128/JCM.01679-10
11. Wood D, Lu J, Langmead B (2025) Kraken 2, KrakenUniq and Bracken indexes. https://benlangmead.github.io/aws-indexes/k2
12. Wood DE, Lu J, Langmead B (2019) Improved metagenomic analysis with Kraken 2. Genome Biol 20:257. https://doi.org/10.1186/s13059-019-1891-0
13. Rogers TR, Morton CO, Springer J et al (2013) Combined real-time PCR and galactomannan surveillance improves diagnosis of invasive aspergillosis in high risk patients with haematological malignancies. Br J Haematol 161:517–524. https://doi.org/10.1111/bjh.12285
14. Pikor LA, Enfield KSS, Cameron H et al (2011) DNA extraction from paraffin embedded material for genetic and epigenetic analyses. J Vis Exp 49:2763. https://doi.org/10.3791/2763
15. Raja HA, Miller AN, Pearce CJ et al (2017) Fungal identification using molecular tools: a primer for the natural products research community. J Nat Prod 80:756–770. https://doi.org/10.1021/acs.jnatprod.6b01085
16. White TJ, Bruns T, Lee S et al (1990) Amplification and direct sequencing of fungal ribosomal RNA genes for phylogenetics. In: PCR Protocols. Elsevier, pp 315–322
17. Altschul SF, Gish W, Miller W et al (1990) Basic local alignment search tool. J Mol Biol 215:403–410. https://doi.org/10.1016/S0022-2836(05)80360-2

Chapter 7

Methods for Large-Scale Intraspecific Analyses of Single Nucleotide Polymorphisms in the Mitochondrial Genomes of *Aspergillus fumigatus*

Veronica Thorn and Jianping Xu

Abstract

Mitochondrial genomes (mitogenomes) are widely used in population genetics and evolutionary studies due to their high copy number per cell, conserved gene content, and informative sequence variation. In fungi, mitogenomes are particularly dynamic in their genome size and inheritance, making them valuable yet underutilized tools in genomic analyses. Here, we present a comprehensive, scalable protocol for large-scale mitochondrial population genomics using publicly available whole genome sequencing data from *Aspergillus fumigatus*, a ubiquitous mold and critical-priority human fungal pathogen. Leveraging the large collection of sequencing data from the NCBI Sequence Read Archive, this protocol enables high-throughput retrieval, processing, and analyses of mitogenome-wide single-nucleotide polymorphisms (mtSNPs) across thousands of isolates. Our workflow includes detailed steps for data acquisition, sequence read preprocessing, reference-based alignment, variant calling, and quality filtering. The data were then analyzed using workflows for phylogenetic inference, molecular and geographic population structure assessment, and recombination detection. We also introduce strategies for comparing mitochondrial to nuclear genome coverage ratios as a proxy for cellular genomic composition. The pipeline is optimized for high-performance computing environments and incorporates robust tools, such as GATK, BWA, BCFtools, and R packages, including *poppr* and *adegenet*. Though demonstrated with *A. fumigatus*, this protocol is adaptable to other eukaryotic species and provides a versatile framework for mitochondrial population genomic research across broad scales.

Key words Big data, Pathogen, Population genetics, Bioinformatics, Recombination

1 Introduction

Almost all eukaryotes have an intracellular organelle called mitochondria that contains its own genetic material, encoding genes that are essential for oxidative respiration and ATP generation. The mitochondrial genomes (mitogenomes) within each are typically much shorter than their nuclear genomes and are circular, relatively abundant, and with encoded genes well-conserved within and across groups of species [1]. Within all eukaryotic species

Natalia Mach (ed.), *Aspergillus fumigatus: Methods and Protocols*, Methods in Molecular Biology, vol. 3028,
https://doi.org/10.1007/978-1-0716-5222-0_7,

investigated so far, there are variations at the mitogenome sequence level among individuals. These features make mitogenomes a popular target for population genetic, physiological, and evolutionary studies [2, 3]. In fungi, mitogenome analysis is particularly intriguing, as, while the core gene content of fungal mitogenomes is largely conserved, their sizes can vary considerably—ranging from 20 kb to over 200 kb—both among species and among strains within the same species [1, 3, 4]. Additionally, mitochondrial inheritance in fungi often deviates from both the Mendelian inheritance pattern of the nuclear genes and genomes and the typical uniparental (usually maternal) and non-Mendelian inheritance of the mitogenomes seen in most other groups of eukaryotes, such as plants and animals [1, 5]. Such unique inheritance patterns contribute to our understanding of fungal evolution, complementing information from the analyses of nuclear genome sequences.

Although genome-wide mitochondrial single-nucleotide polymorphisms (mtSNPs) have historically been less commonly used than single-gene sequence assays, their utility for phylogenetic and population genetic analyses has been increasingly demonstrated in studies on crops, cattle, humans, and selected fungi [3, 6–10]. The higher abundance of mitochondrial DNA (mtDNA) in cells leads to higher sequence read depths, facilitating higher confidence SNP identification [11]. As the availability of high-quality whole genome sequences in NCBI's Sequence Read Archive (SRA) continues to expand, here we present big-data-focused methods for population-level analyses of the mitogenome of *Aspergillus fumigatus*, a ubiquitous mold and one of the four Critical Priority human fungal pathogens as declared by the WHO in late 2022 [12]. We would like to mention that the methods described below are applicable not only to *A. fumigatus* but also to other pathogenic fungi and eukaryotes in general.

2 Materials

In the text below highlighted font describes code examples, flags, or arguments; **bold** is used for software and ***italic bold*** font represents R package names.

2.1 Software

1. **SRA Toolkit** for downloading sequencing data from NCBI [13].
2. **GNU Parallel** for parallel execution of batch commands [14].
3. **Fastp** for read trimming and quality control [15].
4. **BWA** (Burrows-Wheeler Aligner) for sequence alignment [16].
5. **GATK** (Genome Analysis Toolkit) for variant calling [17].

6. **SAMtools** for read processing and alignment manipulations [18].
7. **Picard** for marking duplicate reads [17].
8. **BCFtools** for variant file filtering and conversion [18].
9. **vcf2phylip** for converting variant data to phylogenetic formats [19].
10. **fastreeR** for building phylogenetic trees [20].
11. **R** (including the packages: ***poppr, adegenet, vcfR, vegan, ade4, ggplot2***) for population genetic analyses and visualization [21].
12. **iTol** (Interactive Tree of Life) for tree visualization [22].

2.2 Computing Resources

1. Access to a high-performance computing (HPC) cluster or multi-core workstation.
2. Sufficient disk space and RAM to store and process large sequencing datasets.

2.3 Datasets

1. NCBI Sequence Read Archive (SRA): Raw whole genome sequencing data of *A. fumigatus.*
2. Reference genome for A. fumigatus (FASTA format).
3. Gene sequence of the nuclear single-copy gene cyp51A for estimating nuclear genome depth.
4. Publicly available metadata for SRA accessions (downloaded from NCBI).

3 Methods

Below we describe the step-by-step protocol for analyzing large-scale population mitogenome sequence data, using *A. fumigatus* as the target species.

3.1 Obtain Samples

1. First, obtain raw *A. fumigatus* whole genome sequencing data from the NCBI SRA website by using the specific search query:

```
"Aspergillus fumigatus"[orgn] AND ("biomol dna"[Properties]
AND "strategy wgs"[Properties] AND "filetype fastq"[Proper-
ties]).
```

This query filters the search results to include only whole genome sequencing (WGS) derived DNA datasets for *A. fumigatus* in FASTQ format (excluding already aligned files).

2. Click Send to in the top right corner and select Run Selector > Go. On the newly opened page, download both the Total Metadata and Accession List from the middle, "select" box.

3. Using the accession list, download the raw sequencing files remotely from NCBI using the **SRA Toolkit**. Files should be compressed in order to save space. As there are many accessions, use **GNU Parallel** in conjunction with a bash function download in parallel. The -j option should correspond to your available core count. Your input file should be the downloaded list of accessions with no header. Example bash script:

```
downloadSRA() {
    prefetch ${1}
    fasterq-dump ${1} -O Raw_Fastqs/${1}
    gzip Raw_Fastqs/${1}/*
}
export -f downloadSRA
# if your job fails or runs out of time before finishing, add
--resume before --joblog and re-run
parallel -j 32 --joblog sras.log downloadSRA {} :::: ./
accessions.txt
```

3.2 Process Samples

Sample processing steps can all be performed using **GNU parallel** or using a job array if running on a high-performance computing (HPC) cluster with job scheduling constraints.

1. Read Cleaning and Quality Control: trim adaptors and remove low-quality (Phred < 30), low-complexity reads, and those < 30bp in length using **Fastp**. Optionally, include the `--json` flag to generate an extra log file for downstream quality filtering based on estimated minimum coverage using the following formula:

$$C = \frac{L * N}{G}$$

where:

- C is the depth of coverage.
- L is the read length.
- N is the number of reads,
- G is genome length.

For *A. fumigatus*, due to the high number of SRA accessions to process (>2000), samples with not enough reads after processing can be identified for removal from further processing using **Python** and a simple bash script to improve efficiency. An example bash script:

```
passed_read_num = $(cat ${accession}.json | python -c 'import
json,sys;obj=json.load(sys.stdin);print(obj["filtering_re-
sult"]["passed_filter_reads"])')
```

```
if [[ $passed_read_num -lt ${N} ]]; then
    echo "Too Few"
    exit 0
else
    echo "Passed"
fi
```

2. Alignment: Use **BWA–MEM** [16] to align the trimmed FASTQ files to the indexed reference genome. Use the -R flag to add artificial read groups to the mapped reads as required by **GATK HaplotypeCaller** [17].

   As an example, `-R "@RG\tID:${accession}\tSM:${accession}"` where `${accession}` is your accession variable.

3. Use **Samtools** on the BWA aligned output file (${accession}_bwa_out.sam) to

   Exclude unmapped reads (`-f 4`).

   Filter by alignment length and mapping quality (`-m 30 -q 30`).

   Coordinate sort the alignment (`samtools sort`).

   Example bash script:

```
samtools view -b -h -F 4 -m 30 -q 30 -U /dev/null ${accession}
_bwa_out.sam | samtools sort - > ${accession}_sorted.bam
```

4. Mark PCR duplicates using **Picard MarkDuplicates** [17].

5. Determine average depth across the genome using **Samtools depth** [18] ensuring the `-a` flag is set to include all positions, including ones with depth of 0. Ex. `$(samtools depth -a ${accession}_sorted-md.bam | awk '{sum+=$3} END {print int((sum/NR)+0.5)}')` and record this value in a file. Optionally, accessions with too low of a depth can be removed from further processing steps for efficiency.

6. Variant Calling: Use **GATK HaplotypeCaller** to produce genomic variant call files (GVCFs), which include all genomic positions, irrespective of the presence of a variant call [17]. Having every site represented in the file allows for a joint cohort analysis in subsequent steps [17]. Ensure the `-ploidy` flag is set to 1 for a haploid genome and `-ERC GVCF` to specify GVCF output.

   Example command:

```
gatk --java-options "-Xmx8g" HaplotypeCaller \
    -ERC GVCF \
    -ploidy 1 \
    -R reference.fasta \
    -I ${accession}_sorted-md.bam \
    -O GVCFs/${accession}.g.vcf.gz
```

7. Add that accession's GVCF location to a map file for future steps.

   Example bash command:

```
echo "${accession}\tGVCFs/${accession}.g.vcf.gz\n" | tee -a
samples.map
```

3.3 Cohort Analysis

1. After generating the GVCFs, the next step is to combine these files into a single dataset for cohort-level analyses. First, **GATK GenomicsDBImport** [17] is used to create a database of all variant calls from the samples:

```
gatk --java-options "-Xmx8g" GenomicsDBImport \
     --genomicsdb-workspace-path mitoWorkspace \
     --sample-name-map samples.map
```

Here, samples.map should be a tab-delineated file with the accession and location of its associated GVCF (*see* step 6 in Sect. 3.2).

2. Next, use **GATK GenotypeGVCFs** [17] to call genotypes across all samples and produce a final VCF file containing variant information for the entire cohort:

```
gatk --java-options "-Xmx8g" GenotypeGVCFs \
     -R reference.fasta \
     -V gendb://mitoWorkspace \
     -O afMitos.vcf.gz \
     -ploidy 1 \
     -stand-call-conf 20
```

3. After the initial variant calling, filter the generated variants for quality control and frequency using **BCFtools** to include those with a quality score equal to or above 30 (`-i 'QUAL=>30'`), remove SNPs within 2bp of an indel (`-g 2`), remove sites with missing data (`-g ^miss`), and included in at least 0.1% of the population to identify robust and phylogenetically informative characters (`-q 0.001 -Q 0.999`). Example command:

```
bcftools filter -g 2 -i 'QUAL>=30' -Ou afMitos.vcf.gz |
bcftools view -g ^miss -Ou | bcftools view -q 0.001 -Q 0.999
-o filtered_afMito.vcf.gz
```

Further filtering with **BCFtools** is needed to produce a dataset including biallelic (`-m2 -M2`) SNPs (`-v snps`) for population genomic analysis:

```
    bcftools view -v snp -m2 -M2 -o biallelic_snps_afMito.vcf.
gz filtered_afMito.vcf.gz
```

3.4 Phylogeny

1. Convert your filtered VCF into a format suitable for tree construction using the program **vcf2phylip,** which will concatenate and align the filtered variants in FASTA format.
2. Construct a neighbor-joining phylogenetic tree using the ***fastreeR*** [20] package and root it with an *A. fischeri* isolate as the outgroup using the ***ape*** [21] package in **R** [23]. Use the ***fastreeR*** [20] package to calculate a distance matrix and construct a Neighbour-joining phylogenetic tree:

```
myFastaDist <- fastreeR::fasta2dist("mt_vars.fasta")
myFastaTree <- fastreeR::dist2tree(myFastaDist)
tree <- read.tree(text = myFastaTree)
```

3. Once the tree is constructed, it can be rooted using an appropriate outgroup, in this case, SRR11363404, using the ***ape*** [21] package in **R**:

```
#removes any auto-applied root
tree.unroot <- unroot(tree)

#the outgroup should be the exact accession or name of your
outgroup in the fasta
tree.rooted <- root(tree.unroot, outgroup = "\"SRR11363404\"",
resolve.root = TRUE)
#check to make sure it rooted correctly
is.rooted(tree.rooted)

write.tree(tree.rooted, "rootedTreeFasta.nwk")
```

4. Visualize the tree using **iTol**. Additionally, use the downloaded metadata for each accession to annotate the tree in **iTol** (*see* Fig. 1 as an example).

3.5 Population Genetics

1. Using the **R** package ***vcfR*** [24], import your biallelic SNP VCF file into **R** as a genind object for further processing. First import it as a ***vcfR*** object with the function `read.vcfR()`, then transform that object into genind format using the function `vcfR2genind()`, ensuring the ploidy argument is set to 1 as mitochondria are haploid.
2. Using the associated metadata, stratify the population by both content and country of origin using the **R** package ***poppr*'s** `pop()` and `strata()` functions to assign population information to each accession [25]. Add each country's latitude and longitude values as a data frame to the other category of the genind object:

```
other(mitoSnps)$latLong <- data.frame(lat = latitudes, long =
longitudes)
```

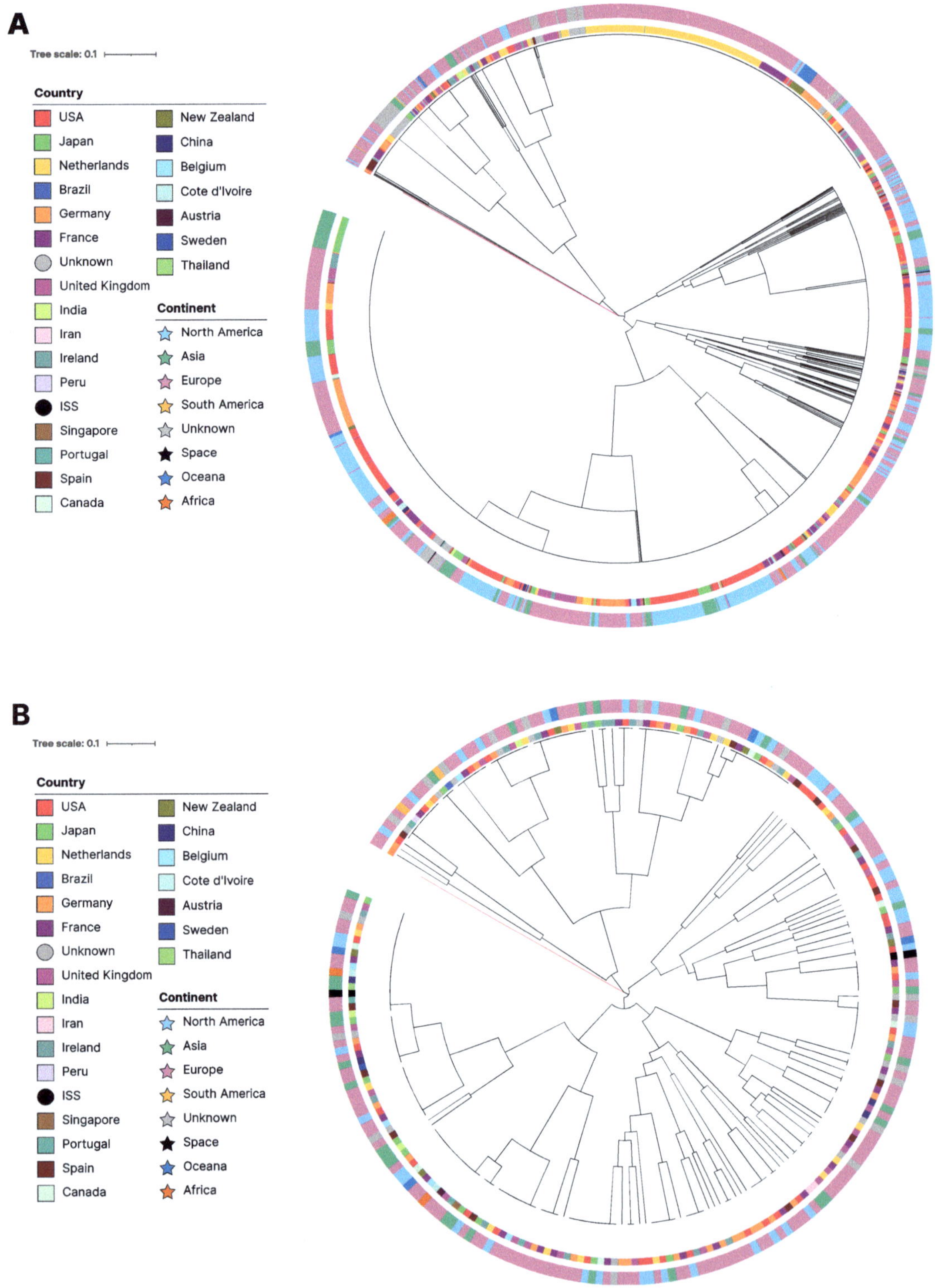

Fig. 1 (**a**). Example of neighbor-joining phylogenetic trees showing relationships among *A. fumigatus* strains based on mtSNPs. All samples included, with *A. fischeri* serving as outgroup, with its branch highlighted in red. The inner ring denotes the country of origin, and the outer ring denotes the continent of origin. In the legend, countries are denoted by squares and different colors, with circles emphasizing extraterrestrial or unknown

3. Remove the *A. fisheri* root from further analysis using ***poppr's*** `popsub()` function, setting the exclude argument to the *A. fisheri* accession [25].

3.5.1 Multi-Locus Genotype (MLG) Analysis

1. MLGs in the dataset can be found using ***poppr's*** `mlg.table()` function on the earlier genind object [25]. Then, use the `colSums()` function to find the number of countries each unique MLG appears in. Ex. `colSums(mlgstable >0)` as each column is an MLG, this will count only countries that have more than 0 samples of that MLG present in any country (row).
2. The number of unique MLGs found in each country can be found similarly with the `rowSums()` function.

3.5.2 Mantel

1. Clone correct the dataset by country using the `clonecorrect()` function in ***poppr*** [25] setting `strata = ~Country`. Convert to a genepop object using ***adegenet's*** `genind2genpop()` function, and compute the genetic distance between populations using Nei's distance via ***adegenet's*** `dist.pop()` function [26].
2. Using the ***vegan*** package's `mantel()` function [27], compare the distance matrix of the latitude and longitude between countries (found using the `dist()` function) to the Nei distance matrix.

3.5.3 AMOVA

1. Check AMOVA on both the full dataset and country-wise clone-corrected dataset (*see* Sect. 3.5.2).
2. Use `poppr.amova()` in the ***poppr*** package [25] with the argument `within = F`, avoiding within-individual variance measurements as these samples are haploid and their argument set to the highest strata level (ex. *~Continent/Country*).
3. To evaluate significance, use the **R** package ***ade4's*** `randtest()` function [28] to perform a permutation test on the above AMOVA result. Set the number of simulations to 999 for a 99.9% confidence level.

3.5.4 Population Structure

1. Apply Discriminant Analysis of Principal Components (DAPC) on both the full and country-wise clone corrected datasets in **R**.

Fig. 1 (continued) categories. Continents are represented with stars and different colors to further distinguish them from countries in the legend (from [9]). (**b**). Example of neighbor-joining phylogenetic trees showing relationships among *A. fumigatus* strains based on mtSNPs. MLGs after clone correction at the country level are shown. The inner ring denotes the country of origin, and the outer ring denotes the continent of origin. In the legend, countries are denoted by squares and different colors, with circles emphasizing extraterrestrial or unknown categories. Continents are represented with stars and different colors to further distinguish them from countries in the legend (from [9])

For both datasets, use ***adegenet*'s** `find.clusters()` function on the genind object 10 times, setting the `choose.n.clust = FALSE` and `max.n.clust` to the number of countries represented in the dataset, saving the resultant kernel (k) value statistic (Kstat) in a matrix [29].

2. These values can be plotted as boxplots with Bayesian Information Criterion (BIC) as the y-axis and k value on the x-axis. Optimal clustering values should be chosen when the BIC differences begin to decrease minimally (*see* Fig. 2a).
3. DAPC is then applied to the dataset clustered using the chosen k values using the ***adegenet*'s** `find.clusters()` and `dapc()` functions and visualized with the **R** package ***ggplot2*** [29, 30] (*see* Fig. 2). For example:

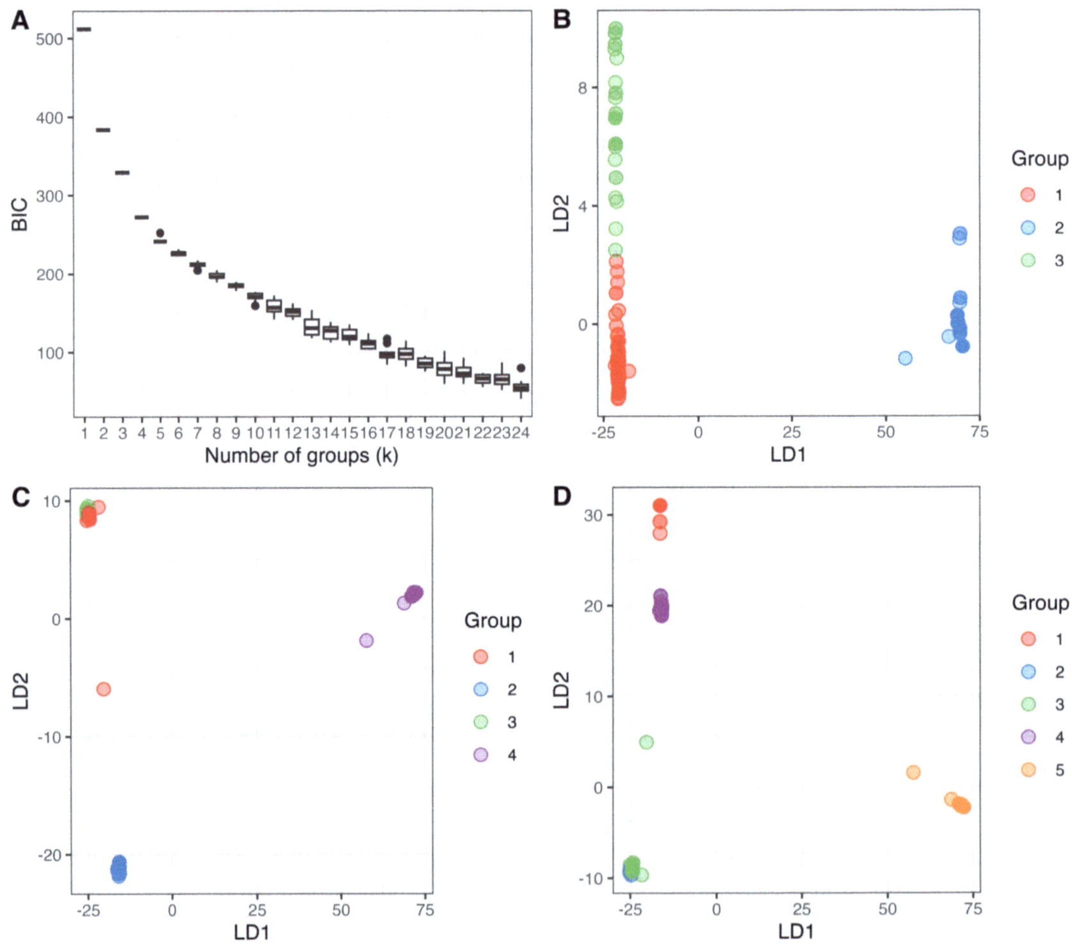

Fig. 2 Example of Bayesian Information Criterion (BIC) boxplot and Discriminant Analysis of Principal Components (DAPC) scatterplots based on mitogenome SNPs. (**a**). BIC boxplot for each putative grouping kernel (k) value for the country-wise clone corrected dataset. (**b–d**). Scatter plot of the first two DAPC loadings of the country-wise clone-corrected dataset when k = 3 (B), k = 4 (C), and k = 5 (D) (from [9])

```
clustered <- find.clusters(perCountry.cc.noRoot, n.pca =
30, n.clust = my_k)
dapc_k <- dapc(perCountry.cc.noRoot, pop = clustered$grp, n.
pca = 30, n.da = my_k)
```

4. These cluster groupings can also be used to annotate the phylogenetic tree in **iTol**.

3.6 Recombination

Three different tests can be used together to assess the presence of recombination in the mitogenome using clone-corrected genind data.

1. Multilocus Index of Association (IA) using ***poppr***'s `pair.ia()` with sample argument set to 999 for a 99.9% confidence level [25]. This test should be performed with both full and clone-corrected datasets for comparison to assess the level of clonal expansion.
2. Linkage disequilibrium (LD).
3. Four-gamete test of phylogenetic incompatibility combining ***poppr*** [25] and custom **R** functions if the dataset is small enough. If the dataset is larger than roughly 1000 unique MLGs and/or 100 loci, alternatively, use **PLINK** [31] for LD analysis and a custom script (available from https://github.com/thorn-v/FGT) for the four-gamete test.

   Example custom **R** functions and script:

```
## Functions ####
# function to check 2 loci vectors for how many combos there
are - only one allele per element
FGT <- function(l1, l2){
  # table does not include NAs by default
  cross <- table(l1, l2)
  # table makes a Punnett square basically - so each element
over 0 is a combo seen in the set then is counted and returned
  return(sum(cross > 0))
}

LD <- function(l1, l2, measure = "r2"){ # will also accept
  stopifnot(measure %in% c("r2", "D", "D.prime"))

  freqs <- table(l1, l2)/length(l1)
  p1 <- table(l1)[1]/length(l1)
  q1 <- table(l1)[2]/length(l1)
  p2 <- table(l2)[1]/length(l1)
  q2 <- table(l2)[2]/length(l1)

  D <- (freqs[1]*freqs[4]) - (freqs[2]*freqs[3])
```

```
  switch(measure,
         D = return(D),
         D.prime = return(ifelse(D < 0,
                                 D/(max(c(p1*q1, p2*q2)*-1)),
                                 D/(min(c(p1*q2, p2*q1))))),
         r2 = return(as.numeric((D^2) / (p1*p2*q1*q2)))
  )
}

### set-up ####
Pinfdf <- genind2df(noRoot.cc, sep = "")
# gets all possible combos of samples (2 in this case); the -1
is for not use country column, will not do with self so no
diagonal
loci <- combn(colnames(Pinfdf)[-1], 2)

# holder matrix for fgt pair results
results.fgt <- matrix(data = 0, nrow = length(colnames
(Pinfdf)[-1]), ncol = length(colnames(Pinfdf)[-1]))
colnames(results.fgt) <- colnames(Pinfdf)[-1]
rownames(results.fgt) <- colnames(Pinfdf)[-1]

# holder matrix for linkage disequalibrium pairs
results.ld <- matrix(data = 0, nrow = length(colnames
(Pinfdf)[-1]), ncol = length(colnames(Pinfdf)[-1]))
colnames(results.ld) <- colnames(Pinfdf)[-1]
rownames(results.ld) <- colnames(Pinfdf)[-1]

# holder for one half fgt and one half ld values
results.both <- results.ld

ld.xsq <- tibble(pair = NA, statistic = NA, p.value = NA, DF =
NA, method = NA)

## actually do the test ##
# results in an upper triangle
for(i in 1:ncol(loci)){
  x <- loci[1, i]
  y <- loci[2, i]
   results.both[x,y] <- results.fgt[x, y] <- FGT(Pinfdf[,x],
Pinfdf[,y])
    results.both[y,x] <- results.ld[x, y] <- LD(Pinfdf[,x],
Pinfdf[,y], measure = "r2")
  ld.xsq[i, 1] <- paste(x, y, sep = ":")
  ld.xsq[i, -1] <- chisq.test(table(Pinfdf[,x], Pinfdf[,y]),
                        correct = F) %>% tidy() #E defaults to
```

```
1/length(x)
}

ld.xsq <- ld.xsq %>% mutate(p.adj.bon = p.adjust(p.value,
method = "bonferroni"))
```

3.7 Mitochondrial to Nuclear Genome Ratio

1. Obtaining the average read depth of a single-copy nuclear gene, we used *cyp51A* (GeneID: 3509526), to represent the nuclear genome and compare it with the average read depth of the mitogenome.
2. Use the **BWA index** [16] on the *cyp51A* FASTA and align the trimmed SRA reads in the same manner, following the method outlined in step 3 in Sect. 3.2. Then, find the average read depth using **Samtools depth** [18] ensuring the `-a` flag is set to include all positions, including ones with depth of 0. This can be run in parallel or looped over accessions for both *cyp51A* and the whole mitochondria alignment, if not performed earlier as described in step 3 in Sect. 3.2.

   Example bash script:

```
cypDepth = $(samtools depth -a ${accession}_sorted-md.bam |
awk '{sum+=$3} END {print int((sum/NR)+0.5)}')
echo "${accession}\t${cypDepth}" >> cyp51Adepths.txt
```

3. For each sample, calculate the mitogenome to nuclear genome ratio by dividing the average mitogenome read depth by the *cyp51A* read depth.
4. Compare across strata and clusters with `kruskal.test()` and `dunn.test()` in **R** [23] for post-hoc testing. The **R** function cor() [23] can be used to determine any correlation between the ratio and average country temperature, which can be found from online resources (Ex. https://climateknowledgeportal.worldbank.org/).

References

1. Sandor S, Zhang Y, Xu J (2018) Fungal mitochondrial genomes and genetic polymorphisms. Appl Microbiol Biotechnol 102:9433–9448. https://doi.org/10.1007/s00253-018-9350-5
2. Lewin R (1987) The unmasking of mitochondrial eve. Science 238:24–26. https://doi.org/10.1126/science.3116666
3. Wang Y, Xu J (2020) Mitochondrial genome polymorphisms in the human pathogenic fungus Cryptococcus neoformans. Front Microbiol 11:706. https://doi.org/10.3389/fmicb.2020.00706
4. Song S-N, Tang P, Wei S-J et al (2016) Comparative and phylogenetic analysis of the mitochondrial genomes in basal hymenopterans. Sci Rep 6:20972. https://doi.org/10.1038/srep20972
5. Basse CW (2010) Mitochondrial inheritance in fungi. Curr Opin Microbiol 13:712–719. https://doi.org/10.1016/j.mib.2010.09.003
6. Álvarez-Iglesias V, Mosquera-Miguel A, Cerezo M et al (2009) New population and

phylogenetic features of the internal variation within mitochondrial DNA macro-Haplogroup R0. PLoS One 4:e5112. https://doi.org/10.1371/journal.pone.0005112
7. Chung H (2013) Phylogenetic analysis and characterization of mitochondrial DNA for Korean native cattle. Open J Genet 03:12–23. https://doi.org/10.4236/ojgen.2013.31003
8. Sabir JSM, Arasappan D, Bahieldin A et al (2014) Whole mitochondrial and plastid genome SNP analysis of nine date palm cultivars reveals plastid heteroplasmy and close phylogenetic relationships among cultivars. PLoS One 9:e94158. https://doi.org/10.1371/journal.pone.0094158
9. Thorn V, Xu J (2023) Mitogenome variations in a global population of Aspergillus fumigatus. JoF 9:995. https://doi.org/10.3390/jof9100995
10. Zhu Q, Gao P, Liu S et al (2016) Comparative analysis of single nucleotide polymorphisms in the nuclear, chloroplast, and mitochondrial genomes in identification of phylogenetic association among seven melon (Cucumis melo L.) cultivars. Breed Sci 66:711–719. https://doi.org/10.1270/jsbbs.16066
11. Joardar V, Abrams NF, Hostetler J et al (2012) Sequencing of mitochondrial genomes of nine Aspergillus and Penicillium species identifies mobile introns and accessory genes as main sources of genome size variability. BMC Genomics 13:698. https://doi.org/10.1186/1471-2164-13-698
12. World Health Organization (ed) (2022) WHO fungal priority pathogens list to guide research, development and public health action, 1st edn. World Health Organization, Geneva
13. Wheeler DL (2006) Database resources of the National Center for biotechnology information. Nucleic Acids Res 34:D173–D180. https://doi.org/10.1093/nar/gkj158
14. Tange O (2023) GNU Parallel 20230522 ('Charles') [stable], Zenodo. https://doi.org/10.5281/zenodo.114601. Accessed 18 Aug 2025
15. Chen S, Zhou Y, Chen Y et al (2018) fastp: an ultra-fast all-in-one FASTQ preprocessor. Bioinformatics 34:i884–i890. https://doi.org/10.1093/bioinformatics/bty560
16. Li H (2013) Aligning sequence reads, clone sequences and assembly contigs with BWA-MEM. Accessed 18 Aug 2025
17. der Auwera GV, O'Connor BD (2020) Genomics in the cloud: using Docker, GATK, and WDL in Terra, 1st edn. O'Reilly Media, Sebastopol
18. Danecek P, Bonfield JK, Liddle J et al (2021) Twelve years of SAMtools and BCFtools. Gigascience 10(2):giab008. https://doi.org/10.1093/gigascience/giab008
19. Ortiz EM (2019) vcf2phylip v2. 0: convert a VCF matrix into several matrix formats for phylogenetic analysis. https://zenodo.org/records/2540861
20. Gkanogiannis A (2023) fastreeR: phylogenetic, distance and other calculations on VCF and Fasta Files. https://github.com/gkanogiannis/fastreeR. Accessed 18 Aug 2025
21. Paradis E, Schliep K (2019) ape 5.0: an environment for modern phylogenetics and evolutionary analyses in R. Bioinformatics 35:526–528. https://doi.org/10.1093/bioinformatics/bty633
22. Letunic I, Bork P (2024) Interactive Tree of Life (iTOL) v6: recent updates to the phylogenetic tree display and annotation tool. Nucleic Acids Res 52:W78–W82. https://doi.org/10.1093/nar/gkae268
23. R Core Team (2024) R: a language and environment for statistical computing. www.R-project.org/22. Accessed 18 Aug 2025
24. Knaus BJ, Grunwald NJ (2016) vcfR: Manipulate and Visualize VCF Data. https://CRAN.R-project.org/package=vcfR. Accessed 18 Aug 2025
25. Kamvar ZN, Tabima JF, Grünwald NJ (2014) Poppr: an R package for genetic analysis of populations with clonal, partially clonal, and/or sexual reproduction. PeerJ 2:e281. https://doi.org/10.7717/peerj.281
26. Jombart T Kamvar ZN (2007) adegenet: Exploratory analysis of genetic and genomic data. https://CRAN.R-project.org/package=adegenet. Accessed 18 Aug 2025
27. Oksanen J, Simpson GL, Blanchet FG, Kindt R, Legendre P, Minchin PR, O'Hara RB, Solymos P, Stevens MHH, Szoecs E, Wagner H, Barbour M, Bedward M, Bolker B, Borcard D, Carvalho G, Chirico M, De Caceres M, Durand S, Evangelista HBA, FitzJohn R, Friendly M, Furneaux B, Hannigan G, Hill MO, Lahti L, McGlinn D, Ouellette M-H, Ribeiro Cunha E, Smith T, Stier A, Ter Braak CJF, Weedon J, Borman T (2001) vegan: community ecology package. 2. https://CRAN.R-project.org/package=vegan. Accessed 18 Aug 2025

28. Dray S, Dufour A-B, Thioulouse J (2002) ade4: Analysis of Ecological Data: Exploratory and Euclidean Methods in Environmental Sciences. https://CRAN.R-project.org/package=ade4. Accessed 18 Aug 2025
29. Jombart T, Devillard S, Balloux F (2010) Discriminant analysis of principal components: a new method for the analysis of genetically structured populations. BMC Genet 11:94. https://doi.org/10.1186/1471-2156-11-94
30. Wickham H (2016) ggplot2: elegant graphics for data analysis, 2nd edn. Use R! Springer international publishing, Cham
31. Chang CC, Chow CC, Tellier LC et al (2015) Second-generation PLINK: rising to the challenge of larger and richer datasets. Gigascience 4:7. https://doi.org/10.1186/s13742-015-0047-8

Chapter 8

Development and Validation of a Real-Time PCR Assay for the Detection of *Aspergillus fumigatus* on Panther Fusion System

Margherita Cacaci, Damiano Squitieri, Maurizio Sanguinetti, and Brunella Posteraro

Abstract

This protocol aims to provide a detailed description of the development and clinical validation of a laboratory-developed test (LDT) for the detection of *Aspergillus fumigatus* in respiratory specimens using the Panther Fusion Open Access system. This platform allows automated extraction of the pathogen genomic DNA and amplification of targets. Specifically, the assay targets the ITS1 region of fungal ribosomal DNA through a real-time PCR workflow. Specificity of primers and probe was verified in silico, and assay parameters, such as $MgCl_2$, KCl, primer/probe concentrations, and amplification protocol, were experimentally optimized for the Panther Fusion open access system. Plasmids, containing the ITS1 target, were used to assess linearity, amplification efficiency, and limit of detection (LoD), determined by probit regression.

Clinical validation was conducted using residual lower respiratory tract samples from patients with suspected pulmonary fungal infection. Diagnostic performance was evaluated by comparing the assay with conventional microbiological methods, including culture and galactomannan testing. The evaluation was conducted in accordance with the guidelines to verify and validate a clinical microbiology test.

This protocol enables laboratories to implement a reliable molecular tool for the detection of *A. fumigatus* within a fully automated platform, supporting early diagnosis of invasive pulmonary aspergillosis.

Key words Invasive-fungal infection, Pulmonary aspergillosis, Molecular diagnosis, Real-time PCR, Pulmonary infection, Panther Fusion, Laboratory-developed test (LDT)

1 Introduction

Invasive pulmonary aspergillosis (IPA) caused by *Aspergillus fumigatus* is a major cause of morbidity and mortality in immunocompromised patients, including those with hematological malignancies or underlying organ transplants [1, 2]. Early and reliable detection of *A. fumigatus* is critical for the timely initiation of a specific antifungal therapy, but traditional culture-based

Natalia Mach (ed.), *Aspergillus fumigatus: Methods and Protocols*, Methods in Molecular Biology, vol. 3028,
https://doi.org/10.1007/978-1-0716-5222-0_8,

diagnostics often have limited sensitivity and require prolonged incubation [3, 4]. Real-time polymerase chain reaction (PCR) assays have therefore become an important tool, providing faster and sensitive detection of fungal DNA in respiratory specimens [5, 6]. However, technical challenges (difficulty of efficiently extracting DNA from fungi with rigid cell walls, the presence of PCR inhibitors in clinical samples) are still reported, leading to the lack of a fully standardized PCR assay for *A. fumigatus* [7, 6]. Moreover, most of the platforms currently available that perform real-time PCR on respiratory samples require a previous DNA extraction, usually conducted using specific kits, making the time to results longer.

The Panther Fusion system (Hologic, San Diego, CA, USA) is a fully automated platform that combines extraction and amplification, reducing the hands-on time on samples and allowing the processing of multiple samples simultaneously. The Open Access functionality provided by the system allows clinical microbiology laboratories to develop laboratory-developed tests (LDTs) for custom targets or target combinations that may not be available in commercial assays, or to tailor molecular workflows according to local epidemiology and specific diagnostic needs [8]. Several LDTs on the Panther Fusion have already been successfully developed, targeting emerging viral infections, antimicrobial resistance markers, and other specialized targets, including SARS-CoV-2 subtypes, influenza A subtypes, tick-borne diseases, human mpox virus, and macrolide resistance in *Mycoplasma genitalium* [8–11]. However, its use for fungal pathogen detection remains unexplored.

Based on this evidence, we have recently reported the development of a real-time PCR assay targeting the internal transcribed spacer 1 (ITS1) region of *A. fumigatus*, specifically optimized for the Panther Fusion system [12]. This chapter describes in detail the reagents, primer and probe design, and the step-by-step protocol for implementing this assay as a Panther Fusion LDT, with the aim of supporting laboratories interested in a fully automated workflow and to expand molecular diagnostics for *A. fumigatus*. The overall workflow is illustrated in Fig. 1.

2 Materials

All solutions should be prepared using nuclease-free water. Store reagents and kits according to the manufacturer's instructions.

2.1 Biological Materials

1. *A. fumigatus* reference strain, e.g., ATCC 13093.
2. Genomic DNA extracted from reference strains obtained from ATCC.

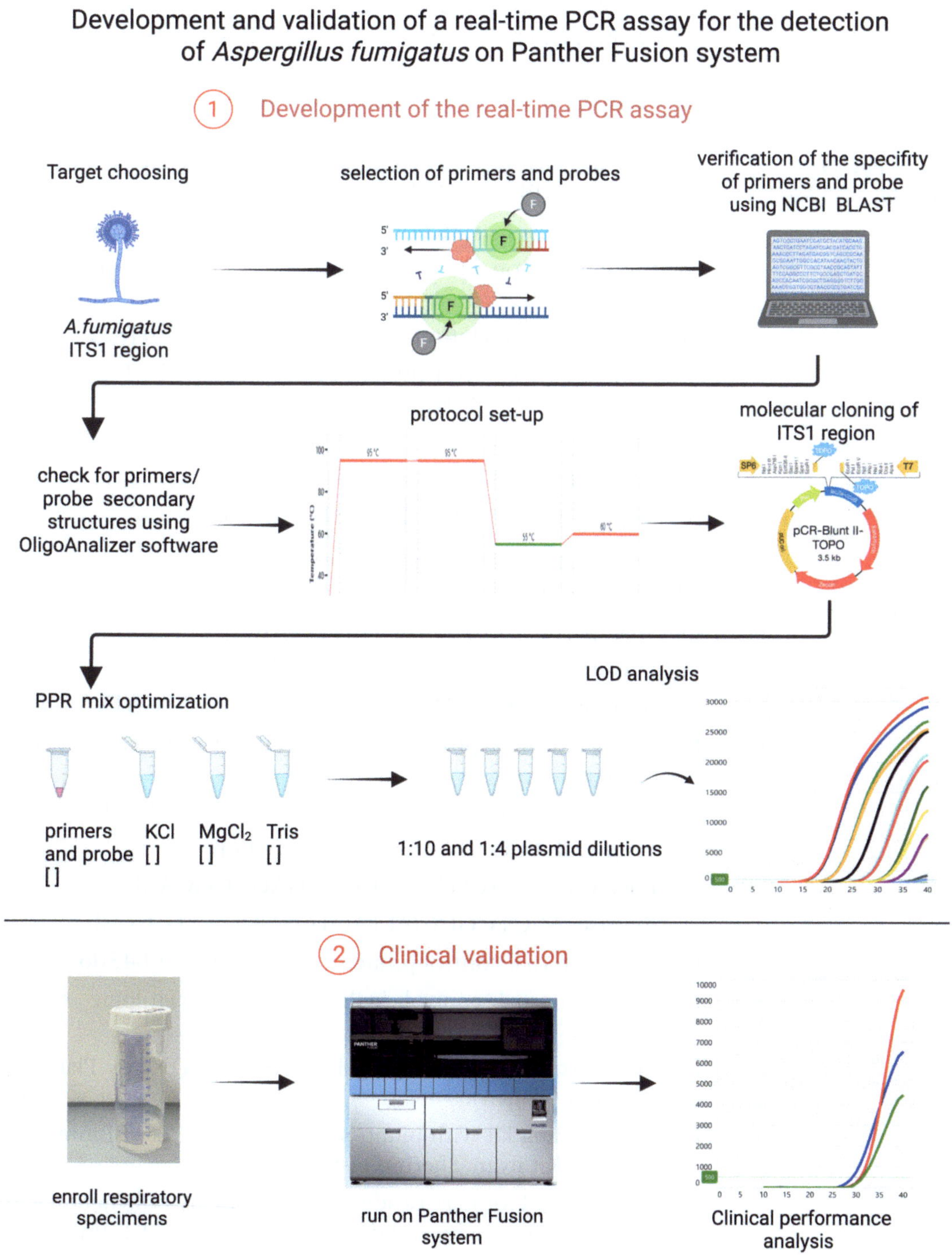

Fig. 1 Workflow of the protocol to develop and validate the real-time PCR assay for *A. fumigatus*

3. Bronchoalveolar lavage (BAL) fluid: samples previously tested negative and positive for *A. fumigatus* DNA, collected in sterile containers.
4. Competent *Escherichia coli:* e.g., *E. coli* TOP10 cells.

2.2 Reagents and Kits

1. Panther Fusion Open Access RNA/DNA enzyme cartridge (Hologic #PRD-04303).
2. Panther Fusion extraction reagents-X (Hologic #PRD-04477).
3. Panther Fusion internal control-X (IC-X) (Hologic #PRD-04476).
4. Panther Fusion elution buffer (Hologic #PRD-04334).
5. Panther Fusion reconstitution buffers I and II (Hologic #PRD-04333 and #PRD-04804).
6. Panther Fusion oil (Hologic #PRD-04335).
7. Tris buffer: 1 M stock.
8. Magnesium chloride ($MgCl_2$): 0.8 M stock.
9. Potassium chloride (KCl): 1 M stock.
10. Nuclease-free water.
11. ITS1 primers for Panther Fusion assay:

    Forward: 5'-GCCCGCCGTTTCGAC-3'

    Reverse: 5'-CCGTTGTTGAAAGTTTTAACTGATTAC-3'
12. ITS1 Hydrolysis probe for Panther Fusion assay (e.g., TaqMan probe):

    5'-HEX-CCCGCCGAAGACCCCAACATG-BHQ1-3'
13. ITS1 PCR primers for plasmid generation:

    Forward: 5'-AACGACTCCCCAGAGCCGGA-3'

    Reverse: 5'-CTCGCCGTTACTGAGGCAATCCC-3'
14. DNA Internal control primers (Hologic #PRD-04306).
15. DNA Internal control hydrolysis probe (Hologic #PRD-04308).
16. DNA extraction kit for fungal tissues (e.g., DNeasy Plant Pro Kit from Qiagen).
17. PCR cloning kit with blunt-end vector (e.g., Zero Blunt PCR Cloning Kit from Thermo Fisher Scientific).
18. KAPA HiFi HotStart ReadyMix: high-fidelity polymerase with proofreading activity and dNTPs (e.g., Roche #KK2601).
19. T4 DNA Ligase.
20. Plasmid miniprep kit (e.g., QIAprep Spin Miniprep Kit from Qiagen).

21. PCR purification kit (e.g., QIAquick PCR Purification Kit from Qiagen).
22. LB agar plates.
23. LB broth.
24. Kanamycin sulfate: 50 μg/mL stock solution.
25. Molecular biology grade agarose to prepare 1.5% agarose gels, DNA stain, and DNA ladder for electrophoresis gels.

2.3 Consumables

1. Multi-tube units (MTU) (Hologic #104772-02).
2. Panther Fusion tube trays (Hologic #PRD-04000).
3. Disposable conductive liquid sensing filter tips: 1000 μL (e.g., from Tecan).
4. Nuclease-free certified disposable filter tips: 10, 20, 50, 100, 200, and 1000 μL.
5. Panther waste bag kit (Hologic #902731).
6. Panther waste bin cover (Hologic #504405).
7. PPR tubes: Open Access Primer/Probe tubes and caps: 2.0 mL (supplied by Hologic to be used with the Panther Fusion system).
8. Panther Fusion Open Access Pack (Hologic #PRD-04305).
9. Panther Fusion Specimen Lysis Tubes (SLT) (Hologic #PRD-04339).

2.4 Instrumentation

1. Panther Fusion System with Open Access module from Hologic.
2. Biological safety cabinets (Class II) for reagent preparation and sample handling.
3. Vortex mixer.
4. Microcentrifuge.
5. Orbital shaker incubator.
6. Calibrated pipettes.
7. Microvolume spectrophotometer (e.g., Nanodrop from Thermo Fisher Scientific).
8. PCR Thermal cycler.
9. Gel electrophoresis setup with gel documentation system.
10. USB stick/device for data storage and transfer.

2.5 Software and Tools

1. **myAccess** v2.1.2.1 or newer versions (Panther Fusion protocol creation and analysis, Hologic),
2. Basic Local Alignment Search Tool (**BLAST**) program from the National Center for Biotechnology Information (available at https://blast.ncbi.nlm.nih.gov).

3. Statistics program (e.g., **GraphPad Prism**).
4. **RStudio package** including **GLM package**, version 2023.12.1.
5. **OligoAnalyzer** (available from https://www.idtdna.com).

3 Methods

3.1 Primer and Probe Selection

1. Download ITS1 sequence using GenBank accession number GU319985. Select ITS1 primers and ITS1 Hydrolysis probe targeting the ITS1 region of *A. fumigatus*. In this protocol, the selection is based on a previously published assay [13]. The oligonucleotide sequences are listed in Sect. 2.2 (steps 13–14).
2. Verify specificity of the selected sequences by performing a **BLAST** search to confirm the absence of cross-reactivity with non-target organisms.
3. Check for secondary structures such as hairpins or self-dimers using **OligoAnalyzer**.
4. Synthesize the oligonucleotides through a commercial supplier and reconstitute them according to the manufacturer's instructions.
5. Store aliquots at −20 °C until use.

3.2 Molecular Cloning of A. fumigatus *ITS1 Region*

1. Extract genomic DNA from *A. fumigatus* strain ATCC 13093 using fungal DNA extraction kit.
2. Design ITS1 PCR primers for plasmid generation targeting the ITS1 region based on GenBank accession number GU319985.1. These primers are distinct from the primers used in Sect. 3.1. Verify their specificity using **BLAST**.
3. Amplify the ITS1 region using KAPA HiFi HotStart ReadyMix.
4. Visualize the PCR product by agarose gel electrophoresis and confirm the expected amplicon size.
5. Purify the PCR amplicon using a PCR purification kit.
6. Ligate the purified product into the Zero Blunt PCR Cloning kit using T4 DNA ligase enzyme.
7. Transform the ligation product into TOP10 *E. coli* using standard heat-shock transformation.
8. Plate the transformants on LB agar supplemented with 50 μg/mL kanamycin.
9. Screen single colonies by PCR using both vector-specific primers (provided in the Zero Blunt PCR Cloning kit) and the ITS1 primers used in step 2 to confirm the presence of the insert.

10. Inoculate positive clones into LB broth supplemented with 50 μg/mL kanamycin and incubate with 220 rpm shaking at 37 °C overnight (12–16 h).
11. Extract plasmid DNA using a miniprep kit.
12. Quantify DNA concentration using a microvolume spectrophotometer.
13. Calculate plasmid copy number per μL using the standard molecular weight and length formula:

    Copy number = (plasmid amount in ng $\times$ 6.022 $\times$ 10^{23}) $\div$ (plasmid length in bp $\times$ 1 $\times$ 10^{9} $\times$ 660).

3.3 Open Access Protocol Setup and Upload on Panther Fusion

In the following section **bold** font is used for software and *italic* is used for commands and biological terms.

Custom laboratory-developed tests (LDT) can be implemented on the Panther Fusion system using Open Access protocols created with the **MyAccess** software. Key steps are described below.

3.3.1 Protocol Creation in MyAccess

1. Open **MyAccess** and create a new protocol. Assign a unique name (automatically prefixed with "LDT").
2. Select protocol type, use *liquid level* (low) and *target type* (DNA).
3. Enter the protocol ID, author name, and description.
4. Set fluorescence channel *HEX* and assign the analyte name (e.g. *Aspergillus fumigatus*). Enable internal control (*channel 5* by default).
5. Configure the thermal profile for *A. fumigatus* detection. The protocol is listed in Table 1 (*see* **Note 1**).
6. Evaluate amplification performance using BAL fluid samples negative for fungal DNA and artificially spiked with plasmid containing the target region, and BAL fluid samples previously tested positive for *A. fumigatus* by conventional microbiological methods.

Table 1
Final amplification protocol for *A. fumigatus* LDT on Panther Fusion system

Step	Temperature	Duration	Cycles
Initial denaturation	95 °C	2 min	1
Denaturation	95 °C	5 s	
Annealing	55 °C	13 s	
Extension + detection	60 °C	19 s	40

7. Set the cycle threshold (Ct) cutoff for positivity at 37.
8. Export protocol in a USB stick/device by selectin *Export protocol* command.

3.3.2 Protocol Upload to Panther Fusion

1. Log in to the Panther Fusion system as an administrator.
2. Insert the USB stick/device containing the exported protocol files.
3. In the software interface, navigate to *Admin* > *More* > *Manage Open Access Protocols.*
4. Click *Import*, select the protocol file, and confirm the upload.
5. Restart the instrument to finalize installation. Once imported, the protocol is ready to use.

3.4 Reagent and Sample Preparation

1. Prepare the Primer/Probe Reconstitution (PPR) mix by combining the ITS1 primers for Panther Fusion assay, ITS1 Hydrolysis probe for Panther Fusion assay, $MgCl_2$, KCl, Tris-HCl buffer, and DNA internal control primers and DNA internal control hydrolysis probe. The PPR mix is used to rehydrate and activate the lyophilized enzyme reagent supplied in the Panther Fusion cartridge (*see* **Note 2**). Final concentrations of each component must be optimized to achieve best assay performance. Use only 2.0 mL PPR tubes, as required by the system.
2. Perform PPR optimization experiments using negative BAL fluid spiked with plasmid containing the target region at a final concentration of 4.9 $\log_{10}$ copies/reaction. Transfer the mixture into an SLT tube and load it onto the instrument. Conduct experiments in triplicate (*see* **Note 3**).
3. Optimize $MgCl_2$ concentration by testing concentrations 0–4 mM in PPR mix (*see* Table 2). In the protocol described here, the optimal concentration is 4 mM.
4. Optimize KCl concentration by testing concentrations 50 mM and 65 mM in PPR mix (*see* Table 3). In the protocol described here, the optimal concentration is 50 mM.
5. Optimize primer concentration by testing concentrations 0.2 mM and 0.4 mM in PPR mix (*see* Table 4). In the protocol described here, the optimal concentration is 0.2 mM.
6. Optimize probe concentration by testing concentrations 0.1 mM and 0.2 mM in PPR mix (*see* Table 5). In the protocol described here, the optimal concentration is 0.1 mM (*see* Table 6 for the final optimized PPR mix used in this protocol).
7. Add 250–400 μL of Panther Fusion oil on top of the PPR mix to prevent evaporation and maintain onboard stability. The exact oil volume should be determined according to Hologic

Table 2
Reagent concentrations and final reaction volumes calculated for 24 tests at $MgCl_2$ concentrations of 0, 2, 3, and 4 mM

Reagents	Final PCR concentrations	Final volume for 24 tests (μL) 0/2/3/4 mM
$MgCl_2$	0 mM/2 mM/3 mM/4 mM	0/2.13/3.19/4.25
KCl	50 mM	53.13
Tris	12.50 mM	10.63
Forward primer	0.2 μM	21.25
Reverse primer	0.2 μM	21.25
Probe	0.1 μM	10.63
DNA IC primers	100×	10.63
DNA IC probe	100×	10.63
Water	–	711.85/709.72/708.66/707.60

Table 3
Reagent concentrations and final reaction volumes calculated for 24 tests at KCl concentrations of 50 and 65 mM

Reagents	Final PCR concentrations	Final volume for 24 tests (μL) 50/65 mM
$MgCl_2$	4 mM	4.25
KCl	50 mM/65 mM	53.13/69.06
Tris	12.50 mM	10.63
Forward primer	0.2 μM	21.25
Reverse primer	0.2 μM	21.25
Probe	0.1 μM	10.63
DNA IC primers	100×	10.63
DNA IC probe	100×	10.63
Water	–	707.60/691.67

Open Access instructions provided with the system (*see* **Note 4**).

8. Load the PPR tubes into the reusable Open Access tube holders (up to 4 tubes per holder) and insert them into the designated positions within the Universal Fluids drawer compartment. Using the Panther Fusion software interface, assign each PPR tube to the corresponding Open Access protocol.

Table 4
Reagent concentrations and final reaction volumes calculated for 24 tests at primer concentrations of 0.2 μM and 0.4 μM

Reagents	Final concentrations	Final volume for 24 tests (μL) 0.2/0.4 μM
$MgCl_2$	4 mM	4.25
KCl	50 mM	53.13
Tris	12.5 mM	10.63
Forward primer	0.2 μM/0.4 μM	21.25/42.50
Reverse primer	0.2 μM/0.4 μM	21.25/42.50
Probe	0.1 μM	10.63
DNA IC primers	100×	10.63
DNA IC probe	100×	10.63
Water	–	707.60/665.1

Table 5
Reagents concentrations and final reaction volumes calculated for 24 tests at probe concentrations of 0.1 μM and 0.2 μM

Reagents	Final concentrations	Final volume for 24 tests (μL) 0.1/0.2 μM
$MgCl_2$	4 mM	4.25
KCl	50 mM	53.13
Tris	12.50 mM	10.63
Forward primer	0.2 μM	26.56
Reverse primer	0.2 μM	26.56
Probe	0.1 μM	10.63/21.25
DNA IC primers	100×	10.63
DNA IC probe	100×	10.63
Water	–	696.98/686.36

9. Load the Panther Fusion Open Access RNA/DNA enzyme cartridges via *Load Reagents > Load Cartridges* in the software. Up to 14 cartridges can be loaded at once. The instrument transfers them automatically to onboard refrigerated storage (4 °C), with capacity for 28 cartridges in total.

Table 6
Final PPR mix used in the study

Reagent	Concentration
$MgCl_2$, mM	4
KCl, mM	50
Tris, mM	12.5
Forward primer, μM	0.2
Reverse primer, μM	0.2
Probe, μM	0.1
DNA IC primers	100×
DNA IC probe	100×

10. Load samples into sample racks and insert them into the Sample Compartment of the instrument. Assign the Open Access protocol to each sample via the *Default Sample Values* screen. Each sample can be assigned up to five PCR reactions with the same sample ID. If more than five are required, the system will prompt the generation of a new sample ID from the same tube.
11. Start the run. The system automatically begins test processing when the following conditions are met:
    - System is in *Ready* state,
    - All reagents (cartridges, PPR tubes) and consumables are loaded,
    - All drawers/doors are closed,
    - Internal mixing is complete.

3.5 Result Management

1. After assay completion, export results generated by the locked Open Access protocol to a USB stick (in .GPR format).
2. Review and analyze amplification curves and data using the *Analyze* function in **MyAccess** software.

3.6 Determination of the Limit of Detection, Linearity, and Amplification Efficiency of the Assay

1. Prepare a stock solution of the plasmid at a known concentration in nuclease-free water (e.g., 1×10^5 copies/reaction). Use the plasmid containing the cloned *A. fumigatus* ITS1 region (*see* Sect. 3.2) as the standard for all dilution series.
2. Perform 1:10 and 1:4 serial dilutions in negative BAL fluid previously processed in Panther Fusion lysis buffer (*see* **Note 5**) to cover a dynamic range (e.g., from 10^5 to <10 copies/reaction).

Table 7
Limit of detection analyses (LoD)

Concentration (log_{10} copies/reaction)	No. of replicates detected/no. of replicates tested	Mean Ct	Standard deviation	Coefficient of variation (%)
5.90	5/5	18.9	0.30	1.60
4.90	5/5	21.8	0.35	1.62
3.90	5/5	24.9	0.21	0.86
2.90	5/5	28.8	0.32	1.13
1.90	5/5	31.1	0.23	0.75
1.30	5/5	34.5	0.55	1.60
0.70	0/5	37.1	0.52	1.42

3. Use the optimized PPR mix (*see* Sect. 3.4).
4. Load diluted plasmid samples into PPR tubes on the Panther Fusion system.
5. Test each dilution in five replicates to assess precision and reproducibility.
6. Include a negative control (BAL without plasmid DNA) in each run.
7. Run the assay and record Ct values or detection results (positive/negative) for each replicate.
8. Calculate the proportion of positive results for each dilution (*see* Table 7).
9. Import data into **RStudio** and perform Probit regression analysis using the **GLM package** to determine the limit of detection (LoD), defined as the concentration corresponding to a 95% detection probability (*see* Fig. 2).
10. Assess linearity by performing linear regression in **GraphPad** and calculate the correlation coefficient (R [2]). Acceptable R^2 is ≥ 0.98 (*see* Fig. 3).
11. Calculate amplification efficiency using the equation E = 10^ (−1/slope), based on the slope of the linear portion of the log-transformed standard curve. "E" represents the theoretical efficiency, and "slope" is derived from a plot of the logarithm of the initial DNA concentration (x-axis) against the Ct value (y-axis). Acceptable amplification efficiency range: 95–110%.

3.7 Verification of the Analytical Specificity of the Assay

1. To assess the specificity of the *A. fumigatus* assay, spike negative BAL fluid specimens with genomic DNA extracted from a panel of purchased reference strains.

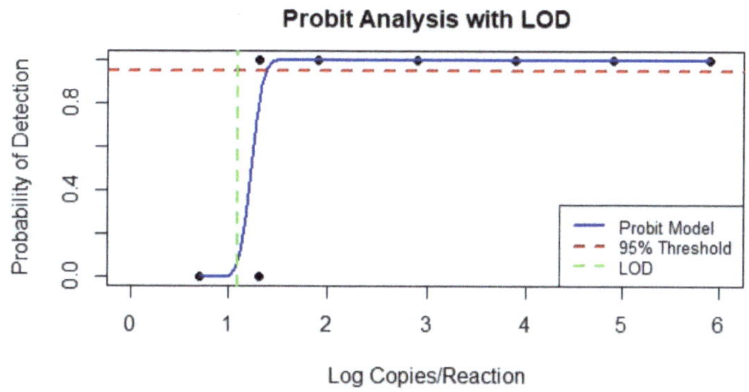

Fig. 2 Probit regression: estimation of the limit of detection (LoD) of the LDT for *A. fumigatus* assay in copies/reaction. The green dashed lines represent the C95 value (LoD). The solid blue line and dashed red line represent the best fit of the probit regression (theoretical frequency) and its 95% CI, respectively. The circles represent the observed frequency

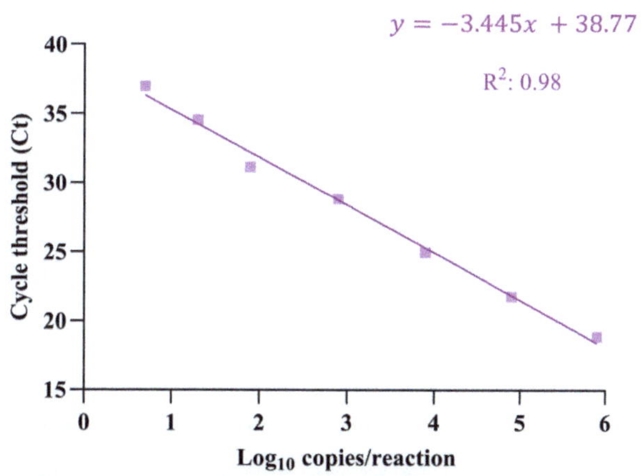

Fig. 3 Linear regression analysis of *A. fumigatus* target on the Panther Fusion. Each dilution of plasmid DNA spiked into BAL was tested 5 times, and the mean Ct value at each concentration was plotted. The formulas represent the linear regression equation along with the correlation coefficients for the LDT

2. The panel should include *A. fumigatus* as a positive control and other clinically relevant fungal, bacterial, and viral pathogens of the lower respiratory tract to evaluate potential cross-reactivity (*see* Table 8).

3.8 Clinical Validation of the Laboratory-Developed Test

1. Use clinical samples from patients submitted for mycological diagnostics. Eligible samples include lower respiratory tract samples such as BAL fluid and tracheal aspirates. Include *A. fumigatus*-positive and -negative samples, based on routine diagnostics.

Table 8
Analytical specificity panel

Organism	Source	Concentration
Acinetobacter baumannii	ATCC 19606	10^5 genome copies/rx
Citrobacter koseri	ATCC 25408	10^5 genome copies/rx
Enterobacter cloacae	ATCC 13047	10^5 genome copies/rx
Enterococcus faecium	ATCC 7000221	10^5 genome copies/rx
Enterococcus faecalis	ATCC 19433	10^5 genome copies/rx
Escherichia coli	ATCC 25922	10^6 genome copies/rx
Haemophilus influenzae	ATCC 33391	10^6 genome copies/rx
Klebsiella oxytoca	ATCC 13182	10^5 genome copies/rx
Klebsiella pneumoniae	ATCC 43816	10^6 genome copies/rx
Legionella pneumophila	ATCC 33152	10^5 genome copies/rx
Moraxella catarrhalis	ATCC 25238	10^6 genome copies/rx
Morganella morganii	ATCC 25830	10^5 genome copies/rx
Proteus mirabilis	ATCC 29906	10^4 genome copies/rx
Proteus vulgaris	ATCC 29905	10^4 genome copies/rx
Pseudomonas aeruginosa	ATCC 27853	10^6 genome copies/rx
Serratia marcescens	ATCC 13880	10^4 genome copies/rx
Staphylococcus aureus	ATCC 25923	10^5 genome copies/rx
Streptococcus pneumoniae	ATCC 49619	10^5 genome copies/rx
Streptococcus pyogenes	ATCC 19615	10^6 genome copies/rx
Aspergillus fumigatus	ATCC 13073	10^6 genome copies/rx
Aspergillus flavus	ATCC 12693	10^5 genome copies/rx
Aspergillus nidulans	ATCC 38163	10^5 genome copies/rx
Aspergillus Niger	ATCC 16404	10^6 genome copies/rx
Aspergillus oryzae	ATCC 11494	10^5 genome copies/rx
Aspergillus terreus	ATCC 20542	10^5 genome copies/rx
Fusarium solani	ATCC 201839	10^5 genome copies/rx
Rhizopus oryzae	ATCC 20344	10^5 genome copies/rx
Candida albicans	ATCC 76615	10^6 genome copies/rx
Candida parapsilosis	ATCC 22019	10^5 genome copies/rx
Pneumocystis jirovecii	MYA-5006SD[a]	10^5 genome copies/rx

[a]Synthetic DNA

2. Load 500 μl of the sample on Panther Fusion platform and assign the LDT protocol (*see* Sect. 3.3) (*see* **Note 6**).
3. Include the Panther Fusion internal control-X (IC-X) with the Panther Fusion extraction reagents-X in each run to monitor extraction and amplification performance.
4. Analyze Ct values and amplification curves using the *Analyze* function in **MyAccess**.
5. Compare results to standard-of-care diagnostics (e.g., fungal culture, galactomannan, β-D-glucan) using the following metrics:
   - Positive percent agreement (PPA) and negative percent agreement (NPA) with 95% confidence intervals, based on comparison with the reference method.
   - Clinical agreement: defined as ≥90% concordance with reference method (IVD standards [14]).
   - Precision: defined as ≥90% agreement within ±3 Ct values across technical replicates (*see* **Note 6**).
   - Concordance analysis: perform Cohen's kappa coefficient calculation to assess agreement strength between LDT and standard methods.

4 Notes

1. Initial amplification settings followed the standard two-step protocol recommended by Hologic for Open Access assays on the Panther Fusion system. While satisfactory amplification was observed in spiked samples, suboptimal amplification efficiency and poor Ct reproducibility were noted in clinical samples. A modified three-step protocol, introducing a distinct annealing step at 55 °C, resulted in improved reproducibility and sensitivity with BAL fluid specimens. The modified protocol is provided (*see* Table 1).
2. When preparing the PPR mix, ensure thorough homogenization of primers and probe. Add primers and probe first, followed by nuclease-free water. Mix well before adding the remaining components.
3. The oil added at the end of the PPR mix preparation must form an interface with the mix. Add the oil slowly, tilting the test tube.
4. During optimization experiments and limit-of-detection (LoD) analysis, plasmid DNA dilutions may result in sample volumes that exceed the capacity or alter the composition of

the Panther Fusion Specimen Lysis Tubes (SLT). To avoid adding extra lysis buffer volume, use Aptima specimen aliquot tubes (Hologic #503762) fitted with penetrable caps (Hologic #105668). These tubes are compatible with the Panther Fusion system, which aspirates the sample directly through the cap.

5. To optimize sample homogenization and facilitate nucleic acid extraction, if the respiratory specimens appear mucoid, a volume-to-volume pretreatment with Sputasol is performed at 37 °C for 30 min. If the specimens appear bloody, use Cytolit (e.g., from Hologic) under the same conditions. If both features are present, combine the two reagents and incubate at 37 °C for 30 min.
6. For precision analysis, test eight *A. fumigatus*-positive clinical samples in triplicate.

References

1. El-Baba F, Gao Y, Soubani AO (2020) Pulmonary Aspergillosis: what the generalist needs to know. Am J Med 133:668–674. https://doi.org/10.1016/j.amjmed.2020.02.025
2. Bassetti M, Giacobbe DR, Agvald-Ohman C et al (2024) Invasive fungal diseases in adult patients in intensive care unit (FUNDICU): 2024 consensus definitions from ESGCIP, EFISG, ESICM, ECMM, MSGERC, ISAC, and ISHAM. Intensive Care Med 50:502–515. https://doi.org/10.1007/s00134-024-07341-7
3. Estella Á, Martín-Loeches I, Núñez MR et al (2023) Microbiological diagnosis of pulmonary invasive aspergillosis in critically ill patients with severe SARS-CoV-2 pneumonia: a bronchoalveolar study. Ann Clin Microbiol Antimicrob 22(1):90. https://doi.org/10.1186/s12941-023-00626-7
4. Badiee P, Ghasemi F, Jafarian H (2022) Role of biomarkers in the diagnosis of invasive aspergillosis in immunocompromised patients. Ann Clin Microbiol Antimicrob 21:44. https://doi.org/10.1186/s12941-022-00539-x
5. Wickes BL, Wiederhold NP (2018) Molecular diagnostics in medical mycology. Nat Commun 9:5135. https://doi.org/10.1038/s41467-018-07556-5
6. Kidd SE, Chen SC-A, Meyer W et al (2019) A new age in molecular diagnostics for invasive fungal disease: are we ready? Front Microbiol 10:2903. https://doi.org/10.3389/fmicb.2019.02903
7. Mengoli C, Cruciani M, Barnes RA et al (2009) Use of PCR for diagnosis of invasive aspergillosis: systematic review and meta-analysis. Lancet Infect Dis 9:89–96. https://doi.org/10.1016/S1473-3099(09)70019-2
8. Stellrecht KA, Cimino JL, Maceira VP (2020) The panther fusion system with open access functionality for laboratory-developed tests for influenza a virus subtyping. J Clin Microbiol 58(6):e00188-20. https://doi.org/10.1128/JCM.00188-20
9. Grant J, Atapattu N, Dilcher M et al (2023) Development of real-time RT-PCR assays to detect measles virus on the Hologic Panther Fusion® System. J Clin Virol 159:105355. https://doi.org/10.1016/j.jcv.2022.105355
10. Villa RD, Pentella MA, Benfer JL (2024) A laboratory-developed assay for clade II human mpox virus on the panther fusion open access system. J Infect Dis 229:S132–S136
11. Lindroth Y, Hansson L, Forslund O (2025) An automated commercial open access assay for detection of mycoplasma genitalium macrolide resistance. APMIS 133:e13477. https://doi.org/10.1111/apm.13477
12. Cacaci M, Talamonti D, Menchinelli G et al (2025) A laboratory-developed assay for the simultaneous detection of aspergillus fumigatus and pneumocystis jirovecii pulmonary pathogens. J Fungi (Basel) 11(4):280. https://doi.org/10.3390/jof11040280

13. Walsh TJ, Wissel MC, Grantham KJ et al (2011) Molecular detection and species-specific identification of medically important Aspergillus species by real-time PCR in experimental invasive pulmonary aspergillosis. J Clin Microbiol 49:4150–4157. https://doi.org/10.1128/JCM.00570-11

14. Yusuf E, Schijffelen MJ, Leeflang M (2024) How to verify and validate a clinical microbiology test before it can be used in routine diagnostics: a practical guide. Clin Microbiol Infect 30:1261–1269. https://doi.org/10.1016/j.cmi.2024.06.028

Chapter 9

RAPD-PCR for Typing and Identification of *Aspergillus fumigatus*

María del Rocío Reyes-Montes, Esperanza Duarte-Escalante, and María Guadalupe Frías-De-León

Abstract

RAPD is a PCR variant that uses a single primer and is performed based on the principle that, as the annealing temperature decreases, the primer finds multiple sequences in the template DNA with which it can hybridize when these sites are close to each other and in opposite orientations, resulting in amplification of the DNA sequence between the sites to produce a template DNA band. The RAPD method has been used for the typing and identification of *Aspergillus* species, characterization of environmental sources, and the detection of intraspecific and interspecific genetic variability. For this purpose, a wide variety of primers have been used with varying results. Primer OPF-01 has proven helpful in typing *Aspergillus* spp. and identifying *A. fumigatus*, as it generates species-specific polymorphic patterns for this fungus. This chapter describes the RAPD method for identifying *A. fumigatus* with primer OPF-01.

Key words RAPD, Typing, Identification, *Aspergillus fumigatus*, OPF-01, Species-specific patterns

1 Introduction

RAPD (random amplified polymorphic DNA) is a PCR variant that uses a single primer and is performed based on the principle that, as the annealing temperature decreases, the primer finds multiple sequences in the template DNA with which it can hybridize when these sites are close to each other and in opposite orientations, resulting in amplification of the DNA sequence between the sites to produce a template DNA band. Since hybridization occurs at different sites in the genomic DNA, multiple bands of various sizes are obtained from the same genomic DNA and constitute a "fingerprint" of that genome [1].

The RAPD method was developed independently by two research groups [1, 2]. Williams et al. [1] note that in the RAPD methodology, unlike PCR, knowledge of the target DNA sequence is not required, and for amplification of genomic DNA, only a

Natalia Mach (ed.), *Aspergillus fumigatus: Methods and Protocols*, Methods in Molecular Biology, vol. 3028, https://doi.org/10.1007/978-1-0716-5222-0_9,

primer with an arbitrary sequence of 10 nucleotides hybridizing to loci randomly distributed throughout the genome is used. A modification of this procedure is to use two primers in each reaction, which results in the amplification of more and smaller fragments than the standard RAPD technique. These new bands originate from other genomic regions that were not amplified with individual primers. The advantage of two-primer RAPDs is that they enable more reactions to be performed with a limited number of primers, thereby generating a greater number of markers [3]. Polymorphisms between individuals result from sequence differences in one or both primers binding sites, originating from point mutations (insertions and deletions), which prevent or avoid stable association with the primer, produce changes that alter the size of the amplified fragment, or can prevent amplification of the target DNA. Hence, these polymorphisms are visible as the presence or absence of a specific RAPD band. On the other hand, these polymorphisms, detected simply as DNA segments that are amplified from one parent but not the other, are inherited in a simple Mendelian manner, and only dominant inheritance can be detected [4]. The fragments obtained by the RAPD method are separated by electrophoresis in agarose gels and revealed by staining with a DNA dye. It is pertinent to mention that there are variants of RAPD, such as AP-PCR (amplified arbitrary primers polymerase chain reaction) [2], DAF (DNA amplification fingerprinting) [5], and RAM (random amplified polymorphic DNA) [6].

Many fungi have been investigated using the RAPD method since its initial application, including *Aspergillus*, which causes aspergillosis—a spectrum of diseases dependent on host factors and the immune response. Noninvasive forms include allergic bronchopulmonary aspergillosis and allergic fungal rhinosinusitis, while more invasive diseases include chronic pulmonary aspergillosis and invasive pulmonary aspergillosis (IA) [7, 8]. Pathogenic *Aspergillus* species are the leading cause of infections and deaths. As a result, many cryptic species have recently been discovered at the conventional section or species complex level of *Aspergillus* through multi-locus DNA sequence analysis [9]. Among the clinical forms, IA is predominantly caused by *A. fumigatus*, mainly in immunosuppressed individuals. Its incidence is higher in patients undergoing hematopoietic stem cell transplantation, solid organ transplantation, or intensive chemotherapy for hematological malignancies, particularly those with prolonged neutropenia [10]. Therefore, the RAPD method has been utilized for the typing and identification of *Aspergillus* species, characterization of environmental sources, and the detection of intraspecific and interspecific genetic variability [11].

In particular, the OPF-01 primer has been used in RAPD tests [12, 13] and can therefore be utilized as a tool for identifying this species. This method represents an alternative approach for the identification of *Aspergillus* species because it discriminates three species, *A. fumigatus*, *A. tubingensis*, and *A. ustus*, as well as other closely related species [12, 13].

2 Materials

2.1 Reagents and Buffers

All solutions must be prepared with ultrapure water (e.g., Milli-Q, Merck Millipore). Chemicals should be of molecular biology grade. Use nuclease-free water for PCR reagents (*see* **Notes 1** and **2**)

1. Potato dextrose agar (PDA) plates: 200 g potato infusion, 20 g dextrose, 20 g agar, make up to 1 L with distilled water, autoclave. Pour 10 mL per plate, cool, and store at 4 °C.
2. YEPG medium: 1% yeast extract, 2% peptone, 2% glucose. Make up to 1 L with distilled water, mix, autoclave, and store at 4 °C.
3. DNeasy Plant Mini Kit for DNA extraction (e.g., Qiagen, or equivalent).
4. PCR amplification buffer (10×): supplied with DNA polymerase.
5. $MgCl_2$ stock solution: 10 mM in nuclease-free water.
6. dNTP mix: 2 mM each of dATP, dCTP, dGTP, dTTP.
7. Taq DNA polymerase, 5 U/μL.
8. Primer OPF-01 stock, 100 μM: 5′-ACGGATTCTG-3′.
9. Template DNA from *Aspergillus* isolates, 10 ng/μL working stock.
10. 10× TBE buffer: 0.89 M Tris base, 0.89 M boric acid, 20 mM EDTA, pH 8.0 in ultrapure water.
11. Agarose, molecular biology grade.
12. SYBR Safe DNA gel stain (or equivalent DNA stain).
13. Phage λ DNA as a reference for quantification (10, 30, 50 ng/μL).
14. DNA size marker: 100 bp ladder.

2.2 Equipment and Consumables

1. Orbital shaker incubator, 37 °C.
2. Filtration setup including sterile filter material, Büchner funnel, and vacuum flask.
3. Gel electrophoresis unit and imaging/documentation system (e.g., Synoptics photodocumenter, Syngene).
4. UV spectrophotometer (e.g., NanoDrop), PCR thermocycler.

3 Methods

3.1 DNA Extraction

1. Grow *Aspergillus* spp. on PDA plates until sporulation is visible.
2. Prepare a conidial suspension and inoculate into 50 mL YEPG medium. Incubate at 37 °C on an orbital shaker for 3 days until mycelial growth is observed.
3. Collect mycelial biomass by filtration, freeze at −20 °C until use.
4. Extract genomic DNA using the DNeasy Plant Mini Kit.
5. Assess DNA integrity by electrophoresis on 1% agarose gel in TBE buffer, stained with SYBR Safe DNA stain. Compare band intensity with reference phage λ DNA loaded at 10, 30, and 50 ng/μL concentration. Additionally, measure DNA concentration by spectrophotometry.

3.2 RAPD-PCR Setup

1. Prepare the RAPD-PCR reaction in a 25 μL volume containing: 2.5 μL of 10× PCR amplification buffer (final concentration 1×), 3.0 μL of $MgCl_2$ stock solution (final 2.5 mM), 2.5 μL dNTP mix (final 200 μM each), 1.0 μL primer OPF-01 stock (100 pmol per reaction), 0.2 μL 5 U of Taq DNA polymerase (final 1 U), and 2.0 μL of template DNA (final 20 ng). Make up to 25 μL with nuclease-free water.
2. Run PCR with the following cycling conditions: 1 cycle of 7 min at 94 °C, followed by 45 cycles of 1 min at 92 °C, 1 min at 35 °C, 1 min at 72 °C, and a final extension of 5 min at 72 °C.
3. Separate PCR products on a 1.5% agarose gel stained with SYBR Safe DNA gel stain. Visualize and document the gels using an imaging/documentation system (*see* Fig. 1).

3.3 Optimization of RAPD-PCR Conditions

Since one of the limitations of RAPD is its reproducibility, standardization of the test is recommended in each laboratory. Factors that must be optimized in the RAPD reaction include enzyme concentration, magnesium concentration (*see* **Note 3**), primer annealing temperature (*see* **Note 4**), DNA concentration (*see* **Note 5**), and dNTPs concentration (*see* **Note 6**). All these parameters must be strictly controlled to obtain reproducible results.

1. Prepare 25 μL RAPD-PCR reactions containing 1× PCR amplification buffer and test the following parameters independently.
2. Test the dNTP mix at the following concentrations: 100, 150, and 200 μM.

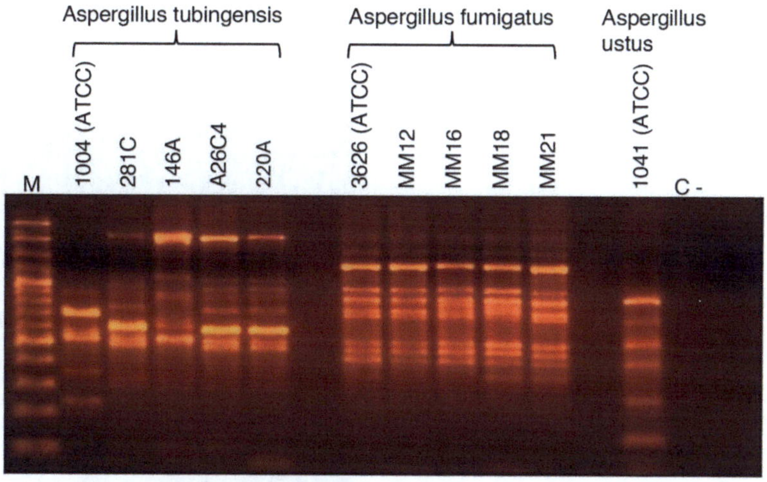

Fig. 1 Polymorphic patterns generated by RAPD with primer OPF-01 for *A. tubingensis*, *A. fumigatus*, and *A. ustus*. It can be seen that the pattern of *A. fumigatus* is species-specific. M: molecular size marker. C-: negative control

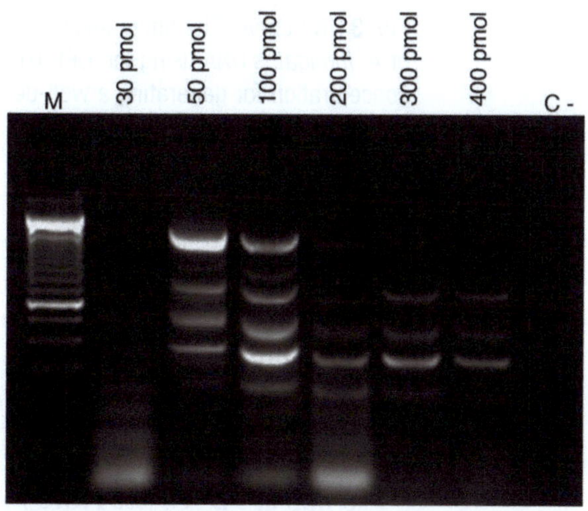

Fig. 2 Oligonucleotide curve. The figure shows the amplification of *A. fumigatus* DNA using different concentrations of the OPF-01 primer. As can be seen, the optimal primer concentration for generating a well-defined polymorphic pattern was 100 pmol. M: molecular size marker. C-: negative control

3. Test the Primer OPF-01 at the following concentrations: 30, 50, 100, 200, and 400 pmol/μL (*see* Fig. 2).
4. Test the $MgCl_2$ at the following concentrations: 1.5, 2.0, 2.5, 3.0, and 3.5 mM.
5. Test the template DNA at the following concentrations: 5, 10, 15, and 20 ng per reaction (*see* Fig. 3).

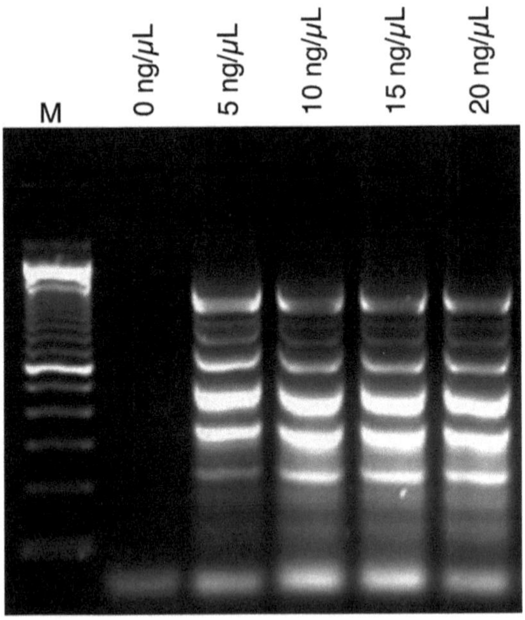

Fig. 3 DNA curve. The figure shows the amplification of different concentrations of *A. fumigatus* DNA using the OPF-01 primer. As can be seen, the optimal DNA concentration for generating a well-defined polymorphic pattern was 20 ng/μL. M: molecular size marker

6. Test the Taq DNA polymerase at the following concentrations: 0.5, 1.0, 1.5, and 2.0 U.
7. Include a negative control in each PCR assay containing all reagents except *A. fumigatus* template DNA, to exclude contamination.
8. Run PCR with the following cycling conditions: 1 cycle of 7 min at 94 °C, followed by 45 cycles of 1 min at 92 °C, 1 min at 35 °C, 1 min at 72 °C, and a final extension of 5 min at 72 °C. (*see* **Note 7**).
9. Separate PCR products on a 1.5% agarose gel stained with SYBR Safe DNA gel stain (*see* **Note 8**). Visualize and document the gels using imaging/documentation system.
10. Repeat optimized RAPD-PCR reactions on three separate days to verify reproducibility.

4 Notes

1. Primers containing 40% to 60% G + C should be selected to avoid the formation of internal secondary structures that lead to erroneous results.

2. During RAPD assays, we recommend using reagents from a single commercial brand, particularly *Taq* polymerase, as we have observed that using a *Taq* polymerase with different activity and specificity can lead to variations in polymorphic patterns. It could be argued that both the activity and specificity of the different polymerases depend on slightly different temperatures and reaction preferences, which affect the outcome of potential competitive reactions between the amplified products in the first and most critical cycles [14].
3. It has also been essential for us to use the optimal concentration of $MgCl_2$ in the RAPD tests, which is obtained by performing $MgCl_2$ concentration curves, since the magnesium ion (Mg^{2+}) acts as a cofactor for DNA polymerase activity, allowing the incorporation of dNTPs during polymerization. Therefore, it is essential to determine the optimal magnesium concentration for optimal PCR performance, as low Mg^{2+} concentrations result in little or no PCR product due to reduced polymerase activity. On the other hand, high Mg^{2+} concentrations can produce nonspecific PCR products due to increased stability of primer-template complexes.
4. In addition to amplifying at the primer's melting temperature (Tm), we suggest testing different annealing temperatures (5 °C below the primer's Tm), as this influences primer specificity. Too low a temperature can lead to nonspecific amplifications, while too high a temperature can reduce yield.
5. We recommend that DNA concentration curves be performed to determine the optimal amount of DNA, as higher DNA concentrations result in DNA extensions without defined bands, while at lower concentrations, unreproducible patterns are observed [15]. In our experience, this step is crucial to consistently use the same concentration of DNA in all experiments to ensure the reproducibility of the polymorphic patterns generated by RAPD. This is because different concentrations of DNA result in different polymorphic patterns.
6. It has been suggested that the recommended final concentration of each dNTP is generally 0.2 mM, since higher concentrations may be helpful in some cases, especially in the presence of high levels of Mg^{2+}, since Mg^{2+} binds to dNTPs and reduces their availability for incorporation. However, dNTP concentrations that exceed the optimal can inhibit PCR. The amount of magnesium ions and the annealing temperature affect the intensity of the amplified bands. Increasing the annealing temperature results in a decrease in the profiles due to the increased specificity of the primer binding site, causing the low-intensity bands to disappear [16]. In our experience, the suggested

concentration of 0.2 mM dNTPs has always given us good results, so we recommend using this concentration for RAPD assays, since high primer concentrations produce DNA extensions that prevent precise detection of the amplification bands, and low primer concentrations make their detection difficult, as do low nucleotide triphosphate concentrations [15, 17] due to incorrect dNTP incorporation.

7. To ensure optimal results, it is crucial that RAPD assays are always performed in the same thermal cycler, since when using different thermal cyclers, we have observed variation in banding patterns.
8. The visualization of polymorphic patterns can vary significantly depending on the DNA dye used. Therefore, it is recommended not to modify the dye brand or concentration during RAPD assays to avoid affecting the clarity and sharpness of the bands observed in agarose gels. If all the suggestions mentioned above are considered, the reproducibility of the results is guaranteed, since the main disadvantage of this method is the reproducibility of RAPDs [18].

References

1. Williams JG, Kubelik AR, Livak KJ et al (1990) DNA polymorphisms amplified by arbitrary primers are useful as genetic markers. Nucleic Acids Res 18:6531–6535. https://doi.org/10.1093/nar/18.22.6531
2. Welsh J, McClelland M (1990) Fingerprinting genomes using PCR with arbitrary primers. Nucleic Acids Res 18:7213–7218. https://doi.org/10.1093/nar/18.24.7213
3. Hu J, van Eysden J, Quiros CF (1995) Generation of DNA-based markers in specific genome regions by two-primer RAPD reactions. PCR Methods Appl 4:346–351. https://doi.org/10.1101/gr.4.6.346
4. Halldén C, Hansen M, Nilsson NO et al (1996) Competition as a source of errors in RAPD analysis. Theor Appl Genet 93:1185–1192. https://doi.org/10.1007/BF00223449
5. Caetano-Anollés G, Bassam BJ, Gresshoff PM (1991) DNA amplification fingerprinting using very short arbitrary oligonucleotide primers. Biotechnology (N Y) 9:553–557. https://doi.org/10.1038/nbt0691-553
6. Zietkiewicz E, Rafalski A, Labuda D (1994) Genome fingerprinting by simple sequence repeat (SSR)-anchored polymerase chain reaction amplification. Genomics 20:176–183. https://doi.org/10.1006/geno.1994.1151
7. Walsh TJ, Anaissie EJ, Denning DW et al (2008) Treatment of Aspergillosis: clinical practice guidelines of the Infectious Diseases Society of America. Clin Infect Dis 46:327–360. https://doi.org/10.1086/525258
8. Pagano L, Akova M, Dimopoulos G et al (2011) Risk assessment and prognostic factors for mould-related diseases in immunocompromised patients. J Antimicrob Chemother 66 Suppl 1:i5–i14. https://doi.org/10.1093/jac/dkq437
9. Houbraken J, Kocsubé S, Visagie CM et al (2020) Classification of *Aspergillus, Penicillium, Talaromyces* and related genera (Eurotiales): an overview of families, genera, subgenera, sections, series and species. Stud Mycol 95:5–169. https://doi.org/10.1016/j.simyco.2020.05.002
10. Dagenais TRT, Keller NP (2009) Pathogenesis of *Aspergillus fumigatus* in invasive Aspergillosis. Clin Microbiol Rev 22:447–465. https://doi.org/10.1128/CMR.00055-08
11. Shams-Ghahfarokhi M, Aghaei-Gharehbolagh S, Aslani N et al (2014) Investigation on distribution of airborne fungi in outdoor environment in Tehran, Iran. J Environ Health Sci Eng 12:54. https://doi.org/10.1186/2052-336X-12-54
12. Valencia-Ledezma OE, Castro-Fuentes CA, Duarte-Escalante E et al (2022) Selection of

polymorphic patterns obtained by RAPD-PCR through qualitative and quantitative analyses to differentiate *Aspergillus fumigatus*. J Fungi (Basel) 8. https://doi.org/10.3390/jof8030296

13. Castro-Fuentes CA, Frías-De-León MG, Del González-Villaseñor MC et al (2024) Evaluation of primers OPF-01, P54, and 1253 to identify *A. fumigatus, A. flavus, and A. niger* from polymorphic patterns obtained by RAPD-PCR. Pathogens 13. https://doi.org/10.3390/pathogens13070574
14. Schierwater B, Ender A (1993) Different thermostable DNA polymerases may amplify different RAPD products. Nucleic Acids Res 21: 4647–4648. https://doi.org/10.1093/nar/21.19.4647
15. Williams JG, Hanafey MK, Rafalski JA et al (1993) Genetic analysis using random amplified polymorphic DNA markers. Methods Enzymol 218:704–740. https://doi.org/10.1016/0076-6879(93)18053-f
16. Caetano-Anollés G, Bassam BJ (1993) DNA amplification fingerprinting using arbitrary oligonucleotide primers. Appl Biochem Biotechnol 42:189–200. https://doi.org/10.1007/BF02788052
17. Munthali M, Ford-Lloyd BV, Newbury HJ (1992) The random amplification of polymorphic DNA for fingerprinting plants. PCR Methods Appl 1:274–276. https://doi.org/10.1101/gr.1.4.274
18. Beebe SE, Ochoa I, Skroch P et al (1995) Genetic diversity among common bean breeding lines developed for Central America. Crop Sci 35:1178–1183. https://doi.org/10.2135/cropsci1995.0011183X003500040045x

Part III

Methods for Studying Gene Regulation

Chapter 10

Electrophoretic Mobility Shift Assay in *Aspergillus fumigatus*

Vijendra Arya, Sanjoy Paul, and W. Scott Moye-Rowley

Abstract

One of the classic techniques to allow detection of direct DNA-binding by sequence-specific transcription factors is the electrophoretic mobility shift assay (EMSA). EMSA takes advantage of the reduced mobility of a protein:DNA complex during gel electrophoresis compared to the naked DNA alone. This rapid and straightforward method has been used to identify DNA target elements in organisms ranging from bacteria to humans. Here we describe its use with a focus on the AtrR transcription factor from the filamentous fungus *Aspergillus fumigatus*.

Key words *Aspergillus fumigatus*, Electrophoretic mobility shift assay, EMSA

1 Introduction

The electrophoretic mobility shift assay (EMSA) is a highly effective and powerful technique for studying DNA-protein interactions [1]. It is primarily employed to identify and characterize protein-DNA interactions, such as the recognition of specific DNA sequences by transcription factors. DNA for EMSA can be labeled with radioisotopes like ^{32}P or with other nonradioactive labels like biotin [2]. The general protocol for EMSA is mixing labeled nucleic acid probe with protein, separating by non-denaturing polyacrylamide or agarose gel electrophoresis, and then visualizing mobility shifts by a detection method appropriate for the label. The principle of EMSA is based on the differential migration of DNA-protein complexes in a gel matrix under the influence of an electric field. DNA-binding proteins slow down the migration of DNA, resulting in a shift of the DNA band in the gel compared to free DNA. It can provide both qualitative and quantitative data for DNA-protein binding [3]. The mobility shift assay offers simplicity, sensitivity, and versatility, working with small concentrations of various nucleic acid sizes and structures while detecting different binding

Natalia Mach (ed.), *Aspergillus fumigatus: Methods and Protocols*, Methods in Molecular Biology, vol. 3028,
https://doi.org/10.1007/978-1-0716-5222-0_10,

conditions. It accommodates proteins ranging from small oligopeptides to large transcription complexes, functioning effectively with both purified proteins and crude cell extracts. These advantages explain the assay's enduring popularity among researchers.

Aspergillus fumigatus is a ubiquitous filamentous fungus that represents one of the most significant fungal threats to human health worldwide (Reviewed in [4, 5]). This opportunistic pathogen causes invasive aspergillosis—a devastating infection with high mortality rates even with treatment and predominantly targeting immunocompromised individuals, including transplant recipients, cancer patients undergoing chemotherapy, and those with chronic respiratory conditions. This fungus possesses remarkable environmental resilience, sophisticated virulence mechanisms, and increasing resistance to available antifungal drugs, thereby creating an urgent need for scientific investigation (discussed in [6]).

By identifying binding sites for key transcription factors, EMSA has enhanced our understanding of the complex networks controlling fungal growth, pathogenicity, and azole resistance in *Aspergillus fumigatus* (recently reviewed in [7]). Our lab has been investigating the roles of key transcription factors such as AtrR [8, 9] and SrbA [10], as well as other DNA-binding proteins involved in azole resistance and virulence in *Aspergillus fumigatus.* EMSA experiments have been part of the approach we use to identify DNA regions bound by these transcription factors. We will illustrate the utility of EMSA analysis of transcription factor DNA-binding using the AtrR transcriptional regulator as a model protein.

2 Materials

2.1 Biological Materials

1. *Escherichia coli* expression strains for recombinant protein production.
2. Expression plasmids (e.g., pET vectors with 6 × His or GST tags).
3. Codon-optimized synthetic DNA fragments corresponding to DNA-binding domains of transcription factors of interest, designed for protein expression in *E. coli.* The AtrR DNA-binding domain is used in this protocol as an example.
4. Recombinant purified protein expressed in *E. coli* for EMSA (AtrR protein of *A. fumigatus* is used here as an example).
5. Single-stranded DNA oligonucleotides for radiolabeling, corresponding to promoter fragments of interest. AtrR response elements (ATREs) sequences from the *cyp51A* and *abcG1* promoters of *A. fumigatus* are used in this protocol as examples to demonstrate AtrR binding.

6. Biotinylated oligonucleotides for short probes: custom-synthesized complementary oligos with 5′ biotin modifications (e.g., 34 bp). The ATRE repeat from the *cyp51A* promoter is used here as an example to demonstrate AtrR binding.
7. Biotinylated PCR primers for long probes (optional to **step 6**): Two PCR primers of 20 nucleotides with 5′ biotin modifications for amplification of promoter fragments (100–400 bp) by PCR.
8. Unlabeled (cold) DNA probes correspond to the EMSA probe sequences. ATRE fragments from *cyp51A* or *abcG1* are used in this protocol as an example.
9. Salmon sperm DNA: 5 mg/mL stock solution, used as a non-specific competitor.

2.2 Reagents and Kits

1. EMSA Reaction buffer: 4% glycerol, 20 mM Tris-HCl (pH 8.0), 60 mM potassium chloride, 5 mM magnesium chloride, 100 μg/mL BSA, 1 mM DTT, 20 μg/mL poly(dI-dC) nonspecific competitor.
2. PBS: 137 mM NaCl, 10 mM phosphate, 2.7 mM KCl, pH 7.4.
3. PBST: PBS supplemented with 0.1% Tween 20.
4. 10× binding buffer for biotinylated probe detection: 250 mM Tris pH 7.5, 400 mM NaCl, 350 mM KCl, 10 mM DTT. For transcription factors containing zinc finger motifs, add 100 mM $ZnSO_4$.
5. 10× TBE buffer: 0.9 M Tris base, 0.9 M boric acid, 2 mM EDTA, pH 8.0.
6. 10× DNA loading dye: 10 mM Tris, 30% glycerol (v/v), 0.25% (w/v) bromophenol blue.
7. T4 Polynucleotide Kinase enzyme (e.g., New England Biolabs #M0201S).
8. 10× T4 Polynucleotide Kinase Reaction buffer (e.g., New England Biolabs #B0201S).
9. [γ-^{32}P] ATP (radioactive ATP).
10. Molecular biology grade water.
11. 50% glycerol.
12. 50 mM $MgCl_2$.
13. DTT: 1 M stock solution.
14. Poly(dI-dC): 1 mg/mL stock solution as nonspecific competitor.
15. PCR purification kit (spin-column based).
16. Polyacrylamide solution (29:1 acrylamide:bisacrylamide) for 4–6% native gels.

17. Nylon membranes for electrotransfer.
18. Streptavidin IRDye 800 nm (Li-Cor Biosciences).
19. Intercept blocking buffer (e.g., Li-Cor Biosciences).
20. SDS: 1% solution.
21. NP-40: 1% solution.
22. BSA: 10 mg/mL stock solution.
23. Scintillation cocktail for use with scintillation counter (commercial, ready-to-use).

2.3 Consumables

1. PCR tubes and microcentrifuge tubes (1.5 and 2 mL).
2. Scintillation vials.
3. Autoradiography film or phosphor screen.
4. Whatman 3 mm filter paper.
5. Plastic wrap (Saran wrap).
6. Disposable gloves and protective gear for handling radioisotopes.

2.4 Instrumentation

1. Electrophoresis apparatus for polyacrylamide gels.
2. Gel dryer (80 °C).
3. UV crosslinker.
4. Autoradiography equipment: phosphorimager or X-ray film developer.
5. Scintillation counter.
6. Rocker or orbital shaker.
7. Infrared imaging system with application software (e.g., Li-Cor Odyssey).
8. Image Studio Lite software (Li-Cor, version 4.0).
9. Gel documentation system.
10. Electrotransfer apparatus for DNA transfer to nylon membranes.
11. Microcentrifuge.
12. Vortex mixer.

3 Methods

Recombinant proteins are expressed in *E. coli* and purified using affinity chromatography (*see* **Note 1**). Yields are typically 0.5–5 mg/mL of purified protein.

3.1 EMSA Using Radiolabeled DNA

1. Prepare single-stranded DNA oligonucleotides corresponding to promoter fragments of interest (e.g., ATREs from the *cyp51A* or *abcG1* promoters of *A. fumigatus*). Label up to 50 pmol of 5′ ends in a 50 μL reaction volume: mix 5 μL of 10× T4 Polynucleotide Kinase Reaction buffer, 50 pmol of [γ-32P] ATP, 2 μL of T4 Polynucleotide Kinase (20 U), and molecular biology grade water to 50 μL total volume. Incubate at 37 °C for 30 min. Stop reaction by heating at 70 °C for 10 min.
2. Purify the labeled DNA fragments using a PCR purification kit. In the final step, elute DNA with 30 μL of elution buffer provided in the kit.
3. Quantify probe radioactivity using scintillation counting. Use 20,000–100,000 CPM of labeled probe with recombinant DNA-binding protein (e.g., AtrR protein, 0.5–5 μg) in each assay.
4. Prepare binding reactions in the EMSA Reaction buffer. For each reaction, include 20,000 CPM of radiolabeled probe and adjust the volume to 40 μL with water. Add recombinant protein in increasing concentrations (0.1–2 μg) and adjust the volume to 10 μL with EMSA Reaction buffer. Prepare a negative control reaction containing radiolabeled probe without recombinant protein (*see* **Note 2**).
5. Incubate the reaction mixture at 30 °C for 45 min.
6. Separate protein-DNA complexes from free probe by electrophoresis on a 4–5% native nondenaturing gel made with 1 × TBE buffer. Run the gel at 160 V at room temperature until the bromophenol blue tracking dye reaches three quarters of the gel length.
7. Place the gel on 3 mm Whatman paper, cover with Saran wrap on top of gel:paper. Dry using gel dryer at 80 °C for 30 min.
8. Expose the dried gel to radiographic film or phosphor screen at −70 °C for 12–16 h.
9. Develop images by autoradiography or with a phosphorimager using a PMT setting of 250–750 and an exposure time of 5–20 min.
10. Determine the protein concentration required to shift approximately 50% of the probe (*see* **Note 3**). Repeat the EMSA with the same protein concentration but include a 40–50-fold molar excess of unlabeled cold probe prior to adding the labeled probe. A specific EMSA reaction is sensitive to the levels of this cold competitor probe. Binding of recombinant AtrR to ATRE fragments from the *cyp51A* and *abcG1* promoters is shown in Fig. 1.

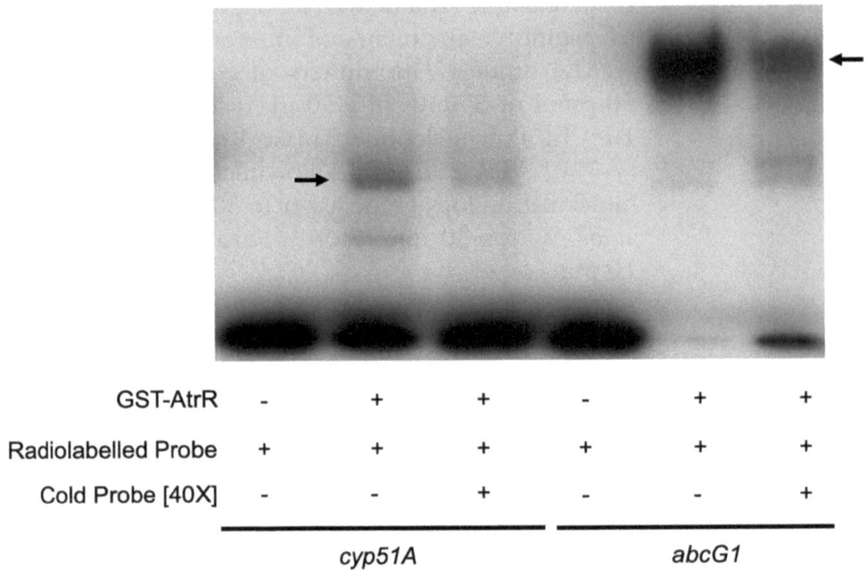

Fig. 1 EMSA analysis of AtrR binding to two different target promoters. GST-AtrR DNA-binding domain purified from *E. coli* was used in an EMSA reaction with a ^{32}P-labeled 200 bp DNA probe prepared by PCR from either the *cyp51A* promoter region (with a single ATRE (AtrR response element)) or *abcG1* promoter region (with two ATREs). Specificity of the probe was confirmed by depletion of the shifted band upon addition of the cold probe in 40 molar excess. Locations of the AtrR:DNA complexes are indicated by the arrows

3.2 EMSA Using Biotinylated DNA

1. Prepare biotinylated probes. For short probes, anneal biotinylated oligonucleotides to form a duplex by mixing equimolar amounts of the two single-stranded oligos in annealing buffer, heating it to 95 °C for 5 min, and slowly cooling down to room temperature. This approach we have used for the 34 bp ATRE repeat from the *A. fumigatus cyp51A* promoter [9]. Alternatively, use two biotinylated PCR primers to prepare a particular promoter fragment of 100–400 bp by PCR (*see* **Note 4**).
2. Set up the EMSA reaction in a final volume of 20 μL. For each reaction, mix 2 μL of 10× binding buffer, 4 μL of 50% glycerol, 2 μL of $MgCl_2$, 1 μL of salmon sperm DNA, 0.1–2 μg of recombinant protein (e.g., AtrR here), and 10–20 fmol of biotinylated probe. Adjust volume with molecular biology grade water. For competitive EMSA, add a 30–40-fold molar excess of unlabeled cold probe prior to adding the labeled probe (*see* **Note 5**).
3. Determine the amount of protein required to shift roughly 50% of the probe by testing several concentrations of recombinant protein. Always add a negative control reaction without protein.
4. Incubate binding reactions at room temperature for 30 min.

5. Prepare 4–6% native non-denaturing polyacrylamide gel in 1 × TBE buffer. Pre-run the gel for 30–60 min at 100 V to remove contaminants and stabilize the temperature while the binding reaction takes place.
6. Load each EMSA reaction onto the pre-electrophoresed gel. Include one lane with 20 μL of EMSA Reaction buffer containing bromophenol blue. Run the gel at 100 V until the bromophenol blue dye has migrated two-thirds to three-quarters of the gel length.
7. Transfer DNA from the gel to a nylon membrane by electroblotting in prechilled 1 × TBE buffer at 100 V for 45–60 min.
8. Cross-link the transferred DNA to the nylon membrane using a UV cross-linker (DNA side facing the UV light) for 5 min.
9. Block the membrane with 1% SDS in 4 mL Li-Cor Intercept blocking buffer on a rocker shaker at room temperature for 30 min.
10. Replace the blocking solution with 1% SDS in 4 mL Li-Cor Intercept blocking buffer containing Streptavidin IRDye 800 nm (1:1000 dilution). Incubate on a rocker shaker at room temperature for 30–40 min. Minimize exposure of the membrane to light during this step and subsequent washing steps.
11. Wash the membrane three times for 5 min with 5 mL PBST buffer and once for 5 min with 5 mL PBS.
12. Scan the membrane using the Infrared imaging system and application software. Quantify the bands using Image Studio Lite software.

    An example of this assay to analyze AtrR binding to the *cyp51A* ATRE repeat is shown in Fig. 2.

4 Notes

1. To produce recombinant *A. fumigatus* transcription factors in *E. coli*, we use several modifications of otherwise standard protein purification strategies. First, we typically try to produce only the DNA-binding domain of the factor in order to minimize the amount of heterologous protein that must be expressed in the *E. coli* host. Second, the region encoding this segment of a factor of interest is obtained as a codon-optimized DNA region by chemical synthesis. Finally, we often use a purification moiety (like the hexa-histidine region from pET vectors [11]) that is attached to the C-terminus of the DNA clone expressing the DNA-binding domain of interest. In this way, we enhance the recovery of full-length protein, as only this expressed fragment will be purified using an appropriate

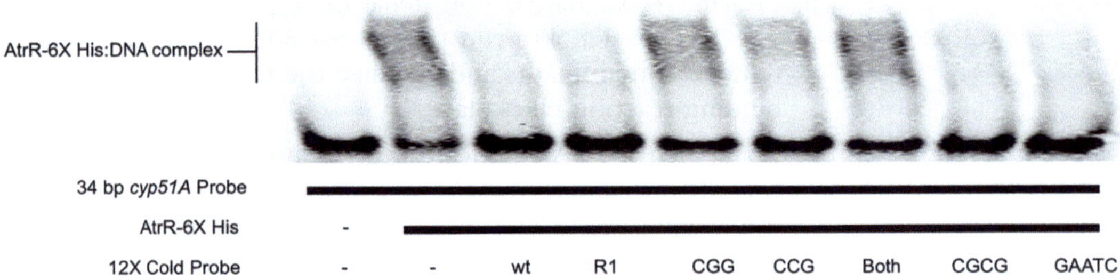

Wild Type (wt)	GAATCACGCGGTCCGGATGTGTGCTGAGCCGAAT
EcoR1 (R1)	GAATCACGCGGTCCGGATGTGGAATTCGCCGAAT
CGG	GAATCACGCGGTCATTATGTGTGCTGAGCCGAAT
CCG	GAATCACGCGGTCCGGATGTGTGCTGAGATTAAT
Both	GAATCACGCGGTATTGATGTGTGCTGAGTATAAT
CGCG	GAATCAattaGTCCGGATGTGTGCTGAGCCGAAT
GAATC	tAATaACGCGGTCCGGATGTGTGCTGAGCCGAAT

Fig. 2 EMSA using a biotin-labeled 34 bp probe from the *cyp51A* promoter. AtrR DNA-binding domain-6× His purified from *E. coli* was used in an EMSA reaction with a 34 bp DNA oligonucleotide probe corresponding to a region of the *cyp51A* promoter. Location of the AtrR:DNA complex is indicated at the left. The 34 bp region contains an ATRE (AtrR response element) as well as an SRE (SrbA response element). The ATRE is the DNA highlighted in yellow, while the SRE is represented by the blue font. Red font corresponds to the bases that were changed from the original sequence. The sequence of the unlabeled probes used as competitors to establish specificity in this EMSA is listed in the table below. Oligonucleotides lacking biotin were included at a 12-fold molar excess relative to the biotinylated 34 bp probe from binding to AtrR-6 × His. Note that oligonucleotides with changes in the ATRE (CGG, CCG, or both) did not compete effectively. The other mutant forms of the 34 bp element still reduced the formation of the AtrR:DNA complex

column. We have also successfully used N-terminal fusions, such as glutathione S-transferase (GST) [12], to produce GST-AtrR. GST-AtrR was purified effectively using a glutathione-agarose column, and we use this fusion protein below in the EMSA employing radioactive DNA. For many proteins, efficient purification requires testing multiple different affinity tags. GST and hexa-histidine represent two good starting points for the production of recombinant factors from *E. coli*.

2. To prepare the proper master reaction mixture for biotinylated EMSA, multiply each value given by the desired number of reactions plus one. For five reactions, multiply each value by 6. For example, mix 12 μL of 10× binding buffer, 24 μL of 50% glycerol, 12 μL of 50 mM $MgCl_2$, 6 μL of 5 mg/mL salmon sperm DNA, and 36 μL of water. Then, each individual reaction receives 15 μL of this mixture before the addition of recombinant protein and biotinylated probe.

3. The sensitivity of the EMSA reaction makes determining the minimum amount of protein to observe a particular protein: DNA complex important. The addition of a basic protein (as many transcription factors are) to an EMSA reaction can lead to altered probe mobility even without specific binding.
4. When preparing biotinylated probes by PCR, order oligonucleotides that lead to the production of a desired DNA fragment, both in standard and 5′-biotinylated forms. This will allow the biotinylated probe to be prepared and then the unlabeled competitor using the standard oligonucleotides.
5. Competition EMSA reactions are important to establish the specific nature of a given protein:DNA complex. Competition can be enhanced with the addition of the unlabeled competitor fragment to the protein of interest prior to the addition of the labeled probe.

Acknowledgments

We thank Dr. Bao Vu (Department of Microbiology and Immunology, University of Oklahoma Health Sciences Center) for helpful discussions during the early stages of this work.

References

1. Hellman LM, Fried MG (2007) Electrophoretic mobility shift assay (EMSA) for detecting protein-nucleic acid interactions. Nat Protoc 2: 1849–1861. https://doi.org/10.1038/nprot.2007.249
2. Rodgers JT, Patel P, Hennes JL et al (2000) Use of biotin-labeled nucleic acids for protein purification and agarose-based chemiluminescent electromobility shift assays. Anal Biochem 277:254–259. https://doi.org/10.1006/abio.1999.4394
3. Fried MG (1989) Measurement of protein-DNA interaction parameters by electrophoresis mobility shift assay. Electrophoresis 10:366–376. https://doi.org/10.1002/elps.1150100515
4. Fisher MC, Denning DW (2023) The WHO fungal priority pathogens list as a game-changer. Nat Rev Microbiol 21:211–212. https://doi.org/10.1038/s41579-023-00861-x
5. Casalini G, Giacomelli A, Antinori S (2024) The WHO fungal priority pathogens list: a crucial reappraisal to review the prioritisation. Lancet Microbe 5:717–724. https://doi.org/10.1016/S2666-5247(24)00042-9
6. van de Veerdonk FL, Carvalho A, Wauters J et al (2025) Aspergillus fumigatus biology, immunopathogenicity and drug resistance. Nat Rev Microbiol. https://doi.org/10.1038/s41579-025-01180-z
7. Kordana N, Johnson A, Quinn K et al (2025) Recent developments in Aspergillus fumigatus research: diversity, drugs, and disease. Microbiol Mol Biol Rev 89:e0001123. https://doi.org/10.1128/mmbr.00011-23
8. Hagiwara D, Miura D, Shimizu K et al (2017) A novel Zn2-Cys6 transcription factor AtrR plays a key role in an azole resistance mechanism of Aspergillus fumigatus by co-regulating cyp51A and cdr1B expressions. PLoS Pathog 13:e1006096. https://doi.org/10.1371/journal.ppat.1006096
9. Paul S, Stamnes M, Thomas GH et al (2019) AtrR is an essential determinant of azole resistance in Aspergillus fumigatus. mBio 10. https://doi.org/10.1128/mBio.02563-18
10. Willger SD, Puttikamonkul S, Kim K-H et al (2008) A sterol-regulatory element binding protein is required for cell polarity, hypoxia adaptation, azole drug resistance, and virulence in Aspergillus fumigatus. PLoS Pathog 4:

e1000200. https://doi.org/10.1371/journal.ppat.1000200

11. Shilling PJ, Mirzadeh K, Cumming AJ et al (2020) Improved designs for pET expression plasmids increase protein production yield in Escherichia coli. Commun Biol 3:214. https://doi.org/10.1038/s42003-020-0939-8
12. Guan KL, Dixon JE (1991) Eukaryotic proteins expressed in Escherichia coli: an improved thrombin cleavage and purification procedure of fusion proteins with glutathione S-transferase. Anal Biochem 192:262–267. https://doi.org/10.1016/0003-2697(91)90534-z

Chapter 11

Chromatin Immunoprecipitation in *Aspergillus fumigatus*

Sanjoy Paul, Mark A. Stamnes, and W. Scott Moye-Rowley

Abstract

Aspergillus fumigatus is the major human filamentous fungal pathogen. While its importance as a pathogen is well-appreciated, our understanding of its biology is still at an early stage. One valuable area of study has been the analysis of gene expression in A. *fumigatus*. Dissection of the transcriptional circuits that control drug resistance and virulence in this fungus has been facilitated by the use of chromatin immunoprecipitation (ChIP), an approach pioneered in the mid-1980s. The use of ChIP coupled with next-generation sequencing (ChIP-seq) allows the determination of the genomic binding sites of a transcription factor of interest. Here, we describe the application of ChIP in *A. fumigatus* and detail different approaches to appropriately shear chromatin from this organism. This sheared chromatin can be used for analysis of factor association with single genes by qPCR or for downstream ChIP-seq applications.

Key words *Aspergillus fumigatus*, Chromatin immunoprecipitation, ChIP

1 Introduction

Chromatin immunoprecipitation (ChIP) is a powerful experimental technique used to study the interaction between proteins and DNA within cells [1]. It allows identification of whether specific DNA-binding proteins, such as transcription factors, are associated with particular genomic regions. ChIP coupled with high-throughput sequencing (ChIP-seq) allows for genome-wide mapping of protein binding sites and provides unique insight into how DNA-binding proteins like transcription factors mediate their function (reviewed in [2, 3]). By pinpointing where transcription factors bind across the genome, ChIP-seq reveals how these proteins might influence gene activation or repression, shedding light on regulatory networks. ChIP-seq has also allowed researchers to identify specific DNA modifications (*see* [4] for a review), such as histone acetylation or methylation, which play a critical role in determining whether genes are active or silenced.

Performing ChIP-seq typically involves cross-linking DNA and proteins in living cells, fragmenting the chromatin, and using

Natalia Mach (ed.), *Aspergillus fumigatus: Methods and Protocols*, Methods in Molecular Biology, vol. 3028,
https://doi.org/10.1007/978-1-0716-5222-0_11,

antibodies specific to the DNA-binding protein under study to selectively immunoprecipitate the protein-DNA complexes. The associated DNA fragments are then purified and analyzed to determine their sequence and identity. This technique has significantly advanced the field of epigenomics and gene regulation studies, contributing to a better understanding of the functional elements of the genome, such as enhancers, silencers, and promoters.

The utility of ChIP-seq to study protein-DNA interactions has expanded the use of this technique to filamentous fungi, including *Aspergillus fumigatus*, the pathogen causing invasive aspergillosis affecting over two million cases annually, with a high mortality rate in immunocompromised patients [5, 6]. Given the growing concern of drug resistance in *A. fumigatus*, particularly to azole antifungal drugs [7], our lab has been focused on understanding the action of transcription factors such as AtrR [8, 9] and SrbA [10] that play key roles in antifungal drug resistance. We have elucidated the network of genes they regulate to confer azole drug resistance in this fungal pathogen, primarily utilizing single-gene ChIP or ChIP-seq techniques. Given below is a detailed protocol for ChIP that can be adapted for use with *A. fumigatus*.

2 Materials

2.1 Strains

Aspergillus fumigatus strain of choice (e.g., AfS35 and Af293). If a suitable ChIP antibody for the protein of interest is not available, the experiment requires a strain in which the protein is expressed with an epitope tag (e.g., HA- or FLAG) (*see* **Note 1**).

2.2 Cultivation Media, Buffers, and Solutions

1. Sabouraud Dextrose Media (SDM.): 0.5% peptone, 0.5% tryptone, 4% glucose, pH 5.6 ± 0.2 at 25 °C, sterilize by autoclaving.
2. Resuspension buffer: 0.4 M sucrose, 10 mM Tris-HCl, pH 7.5, 1 mM EDTA. Filter-sterilize and store at room temperature. Add 1 mM PMSF and 1% formaldehyde solution just before use.
3. Glycine solution: 2 M glycine in distilled water. Filter-sterilize and store at room temperature.
4. ChIP Lysis buffer: 50 mM Hepes-KOH, pH 7.5, 150 mM NaCl, 1 mM EDTA, 1% Triton X-100, 0.1% deoxycholate (e.g., Sigma #D6750), 0.1% SDS. Add 1 mM PMSF and 1 × fungal protease inhibitor cocktail (e.g., Sigma #P8215: supplied as 100×) just before use.
5. Blocking buffer: ChIP Lysis buffer supplemented with 10 μL BSA (10 mg/mL) and 2 μL yeast tRNA (100 μg/μL) (e.g., Invitrogen #AM7119).

6. Washing buffer: 0.25 M LiCl, 1% NP40 (e.g., Nonidet P40 or Fluka #74385), 1% deoxycholate, 1 mM EDTA.
7. Elution buffer: 1% SDS, 0.1 M $NaHCO_3$, 0.2 mg/mL proteinase K, 1 mM DTT. Prepare fresh before use.
8. 1 × TE buffer: 10 mM Tris-HCl, pH 8.0, 1 mM EDTA, pH 8.0. Store at room temperature.

2.3 Reagents

1. PMSF: 100 mM stock in isopropanol; add fresh to buffers before use.
2. Protease inhibitor cocktail: supplied as 100× (e.g., Sigma #P8215), dilute freshly to 1× before use.
3. Formaldehyde: 37% aqueous stock solution (e.g., Sigma #F8775), dilute freshly to 1% before use.
4. Antibody specific to the target protein. Examples used in this protocol include monoclonal HA antibodies (Invitrogen #2-2.2.14 and Covance #16B12), monoclonal FLAG antibodies (Sigma #F1804 and Proteintech #20543-1-AP), and custom rabbit polyclonal antibodies (anti-AtrR, anti-FfmA).
5. Control IgG antibody: for negative control immunoprecipitation (e.g., Abcam #ab46540).
6. RNase A: DNase- and protease-free (e.g., Thermo Scientific #EN0531).
7. PCR purification kit (e.g., QIAquick PCR Purification Kit from Qiagen #28104).
8. PCR Primers: custom-synthesized oligonucleotides specific to the genomic regions of interest (*see* **Note 2**).
9. Reagents for PCR including Taq DNA polymerase, PCR reaction buffer, and dNTP mix.
10. iTaq Universal SYBR Green Supermix (Bio-Rad #1725120).
11. Nuclease-free water.

2.4 Equipment and Consumables

1. Erlenmeyer flasks (125 mL, 500 mL).
2. Vacuum filtration system with sterile filters.
3. Orbital shaker incubator.
4. Liquid nitrogen.
5. Mortar and pestle.
6. Centrifuges (benchtop and refrigerated, capable of 12,000× *g*).
7. Covaris E220 focused-ultrasonicator with AFA Fiber Pre-Slit Snap-Cap microtubes (Covatis #500239 and #520077).
8. Diagenode Bioruptor sonicator with water bath cooling system.

9. Protein A or Protein G magnetic beads (Dynabeads, e.g., Invitrogen #10001D and #10003D).
10. Magnetic stand for microtubes.
11. Fluorometer for DNA quantification (e.g., Qubit, Thermo Fisher Scientific).
12. Agarose gel electrophoresis setup.
13. Bioanalyzer (e.g., Agilent 2100) used with High Sensitivity DNA kit (Agilent, #5067-4626).
14. PCR thermocycler.
15. Real-time PCR instrument (e.g., Bio-Rad CFX Opus 96).

3 Methods

3.1 Growing and Fixing Cells

1. Inoculate 1 × 10^6 *A. fumigatus* conidia/mL into 100 mL of SDM in a 500 mL Erlenmeyer flask. Use fresh conidia that are not older than 1 week. Incubate at 37 °C with 200 rpm shaking for 16 h. Harvest mycelia by vacuum filtration.
2. Start cross-linking by resuspending mycelia in 50 mL of Resuspension buffer in a 125 mL flask. Incubate with 100 rpm shaking for 15 min at room temperature (*see* **Note 3**). Stop cross-linking by adding 2.5 mL (1/20 volume) of 2 M glycine to reach a final concentration of 0.1 M. Incubate for 5 min.
3. Collect the mycelia by vacuum filtration and rinse with sterile water. Dry the mycelium gently with paper towels, and transfer to a 15 mL screw cap tube. Freeze immediately in liquid nitrogen. Proceed directly to tissue lysis or store samples at − 70 °C until ready for mycelial lysis.
4. Grind 500–1000 mg of frozen mycelia in liquid nitrogen to a fine powder using a mortar and pestle. Transfer the amount of powder that fills up to the 1 mL mark of a 15 mL screw cap tube. Weigh ground mycelia, add 2–5 mL of ChIP Lysis buffer (1 mL per 200 mg of ground mycelia), vortex thoroughly. Incubate at room temperature for 10 min followed by 10 min on ice (*see* **Note 4**).

3.2 DNA Fragmentation Using a Focused Ultrasonicator

1. Centrifugate the suspension (*see* Subheading 3.1, **step 4**) at 5000× *g* at 4 °C for 5 min.
2. Transfer the supernatant to an AFA Fiber Pre-Slit Snap-Cap microtube (maximum 130 μL per tube).
3. Shear DNA using E220 focused ultrasonicator with following parameters: *peak incident power* (W), 175; *duty factor*, 10%; *cycles per burst*, 200; *treatment time* (s), 600–900 (optimize as

needed); *temperature* (°C), 8; *sample volume* (μL), 130; *E220—intensifier* (pn500141), yes. (*see* **Note 5**).

4. Collect samples into 1.5 mL microcentrifuge tubes and centrifuge at 12,000× *g* at 4 °C for 5 min.
5. Transfer supernatant to a fresh 1.5 mL tube on ice. Either proceed directly to immunoprecipitation or store at −70 °C. Reserve 25 μL of supernatant to provide an input control after reverse cross-linking.

3.3 DNA Fragmentation Using Bioruptor for Sonication

1. Sonicate samples (*see* Subheading 3.1, **step 4**) on ice using Bioruptor and following setup: 30 s on and 30 s off at power level H for a total of 30 min. Replace the ice every 10 min. Use cooled water to fill the water bath in Bioruptor (maximize efficiency at 4 °C). The maximum recommended sample volume is 300 μL.
2. Centrifugate the suspension at 10,000× *g* at 4 °C for 5 min.
3. Transfer approximately 250 μL of the supernatant into a new microcentrifuge tube. Either proceed directly to immunoprecipitation or store at −70 °C. Reserve 25 μL of supernatant to provide an input control after reverse crosslinking.

3.4 Checking DNA for Input Control

1. Add 25 μL of the supernatant collected after sonication to 75 μL of the Elution buffer. Incubate overnight at 65 °C.
2. Add 1 μL of RNase A (1/10 dilution of 10 mg/mL stock) and incubate at room temperature for 30 min.
3. Purify DNA using a PCR purification kit. As the last step, elute with 50 μL elution buffer provided with the kit.
4. Assess DNA shearing efficiency using standard gel electrophoresis with 1% agarose gel (expect maximal intensity around 200–500 bp).
5. Alternatively, to **step 4**, assess DNA shearing efficiency using Agilent 2100 Bioanalyzer and High Sensitivity DNA kit: Run 3 μL of input DNA diluted to 3 ng/μL to determine the DNA size distribution accurately.

3.5 Immunoprecipitation

1. Add the specific antibody to 500 μL of sheared chromatin after sonication (*see* **Note 6**). Adjust to a final dilution of 1:100 to 1: 500, depending on the antibody's affinity and specificity. Incubate at 4 °C overnight (12–16 h).
2. Pipet 50 μL of well-mixed Protein A or G magnetic Dynabeads, depending on the origin of the primary antibody, into a 1.5 mL tube for each sample. Wash the beads twice with 500 μL of ChIP Lysis buffer by inversion, using a magnetic stand to remove the supernatant. Do not allow beads to dry.

3. Add 100 μL of blocking buffer to the washed beads, incubate with rotation at 4 °C for 1 h. Remove the blocking buffer from the beads using a magnetic stand.
4. Add the antigen-antibody complexes from **step 1** to the prepared beads. Incubate with rotation at 4 °C for 2–3 h.
5. Add 0.5 mL of Lysis Buffer, mix by inverting 6 times, place on a magnetic stand, and discard the supernatant.
6. Add 0.5 mL of washing buffer, mix by inverting 6 times, place on a magnetic stand, and discard the supernatant. Repeat this wash once more.
7. Add 0.5 mL of 1× TE buffer, mix by inverting 6 times, place on a magnetic stand, and discard the supernatant. Repeat this wash once more.
8. Elute DNA with 50 μL of freshly prepared Elution buffer and incubate at 65 °C for 10 min.
9. Place on a magnetic stand and transfer the supernatant to a new tube. Repeat the elution with an additional 50 μL. The final elution volume is 100 μL. In parallel, mix 25 μL of the input fraction with 75 μL of the elution buffer for the input control.
10. Incubate eluates and input control at 65 °C overnight (12–16 h).
11. Add 1 μL of RNase A prepared as a 1:10 dilution of 25 mg/mL stock and incubate at room temperature for 30 min.
12. Recover DNA using a PCR purification kit. In the last step, elute in 50 μL of elution buffer from the kit.
13. Measure DNA concentration using a microvolume fluorometer with 5 μL of sample (*see* **Note 7**).

3.6 Nonquantitative PCR

1. Perform PCR amplification using regular Taq DNA polymerase. Prepare reactions with the following samples: input control, DNA pulled down without specific antibody (negative control), and DNA pulled down with antibody specific to the protein of interest. Use 1 μL of each sample for 20 μL PCR reactions.
2. Carry out PCR cycling under the following conditions: initial denaturation at 95 °C for 3 min, followed by 29 cycles of denaturation at 95 °C for 30 s, annealing at 60 °C for 30 s, and extension at 72 °C for 30 s. Final extension is at 72 °C for 5 min.
3. Analyze PCR products by gel electrophoresis. A strong band is expected for the input control sample, a faint or no band for the negative control sample, and strong bands in samples immunoprecipitated with the specific antibody. *See* Fig. 1 for an example.

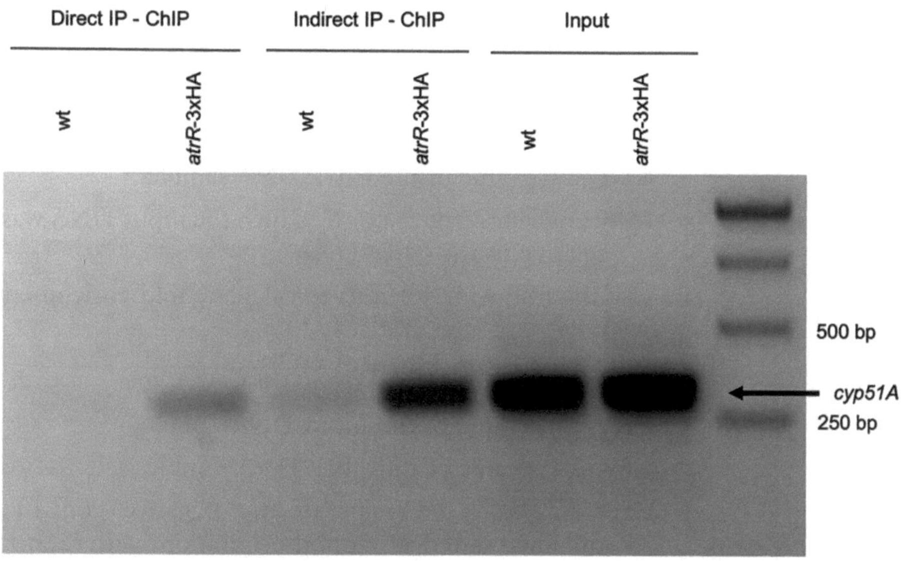

Fig. 1 Enrichment analysis of the *cyp51A* promoter by PCR with ChIP-ed DNA. Immunoprecipitation (IP) with an HA antibody was performed in wild-type (AfS35) and *atrR*-3 × HA-tagged *A. fumigatus* strains. In the direct IP approach, the HA antibody was incubated with protein G Dynabeads at 4 °C for 4 h, followed by incubation with the sheared chromatin cell lysate (prepared using a Covaris focused-ultrasonicator) at 4 °C for 12 h. In the indirect IP approach, the antibody was incubated with chromatin prior to capture with Protein G Dynabeads. The input sample was obtained from DNA purified from the sheared chromatin cell lysate before IP. PCR was then carried out with primers targeting the *cyp51A* promoter region (-69 to -415 relative to the *cyp51A* gene), which includes the 34 bp element that is duplicated to form the TR34 repeat commonly found in azole-resistant *A. fumigatus* clinical isolates [12]

3.7 Quantitative PCR

1. Use the same primers as in Subheading 3.6. Prepare 20 μL qPCR reaction: 1 μL of ChIP-ed DNA or 1 μL of 10× diluted input control, 0.5 μM of each primer, 10 μL of iTaq Universal SYBR Green Supermix, make up to volume with nuclease-free water.
2. Perform quantitative PCR on a real-time thermocycler. If using a Bio-Rad CFX Opus 96 qPCR instrument, use the following conditions: initial denaturation at 95 °C for 2 min, followed by 39 cycles of denaturation at 95 °C for 15 s with annealing at 60 °C for 30 s.

3.8 Quantification of DNA Enrichment by ChIP-qPCR

ChIP-qPCR data must be normalized for sources of variability, including the amount of chromatin, the efficiency of immunoprecipitation, and DNA recovery (*see* **Note 8**). Quantification involves several steps to ensure accurate measurement of DNA enrichment at specific genomic regions. Two commonly used methods are described below. We prefer analyzing ChIP-qPCR data using the percentage input calculation, as this includes normalization for both background levels and input chromatin.

1. Use the following formula to calculate the percentage input:

$$\text{Percentage Input} = \frac{2^{(\text{CtInput}-\text{CtChIP})}}{\text{dilution factor}} \times 100$$

Ct_{Input}: Ct value of the input DNA sample.

Ct_{ChIP}: Ct value of the ChIP DNA sample.

Dilution factor: the factor by which the input DNA was diluted relative to the ChIP DNA.

2. Use the following formula to calculate fold enrichment:

$$\text{Fold Enrichment} = \frac{2^{(\text{CtInput}-\text{CtChIP})}}{2^{(\text{CtInput}-\text{CtNegativeControl})}}$$

Ct_{Input}: Ct value of the Input DNA sample.

Ct_{ChIP}: Ct value of the ChIP DNA sample.

$Ct_{NegativeControl}$: Ct value of the negative control sample, obtained after immunoprecipitation with a nonspecific antibody (e.g., IgG), or a knock-out/untagged strain for the gene under study.

3. After validation of enrichment by ChIP-qPCR, the same ChIP DNA can also be used for library construction if genome-wide analysis by ChIP-seq is desired (*see* **Note 9**).

4 Notes

1. If a suitable ChIP-grade antibody for the protein of interest is not available, it may be necessary to generate or obtain an epitope-tagged *A. fumigatus* strain (e.g., HA- or FLAG-tagged allele). We successfully used AfS35 or Af293 backgrounds.
2. In our hands, optimal primers produced amplicons of 100–250 bp size, melting temperature 58–62 °C, GC content 40–60%. Establish a melting curve for the primer pair using quantitative real-time PCR to ensure no multiple peaks are formed. If multiple peaks are observed, design a new set of primers to ensure a single peak and a single amplicon. For the melting curve analysis, we use the following parameters: dwell time 5 s, set point 60 °C, temperature change 0.5 °C, and end temperature 95 °C.
3. The efficiency of formaldehyde cross-linking can vary. Adjust the concentration of formaldehyde if necessary to achieve efficient cross-linking of chromatin.
4. Filamentous fungi such as *A. fumigatus* form mycelia with a tough cell wall structure; therefore, it is important to efficiently lyse the mycelia in order to release chromatin and allow subsequent chromatin shearing, enabling a successful ChIP

reaction. We have found the best method to efficiently lyse mycelia to expose fixed chromatin for further processing is to grind mycelia to a fine powder by using a mortar and pestle in the presence of liquid nitrogen.

5. Another important consideration during ChIP is the process of chromatin shearing. We have found that the use of traditional sonication for chromatin shearing in *A. fumigatus* can be inconsistent between different strains and growth conditions, even though it can be optimized by varying the duration of each burst and the number of cycles used. Excessive sonication can cause denaturation of proteins and may not be suitable for ChIP. Instead, we have more successfully relied on using an ultrasonicator from Covaris based on adaptive focused acoustics (AFA) for getting sheared DNA fragments of consistent lengths. The focused-ultrasonicator provides rapid and complete chromatin shearing, ensuring uniform fragment sizes. The controlled bursts of high-frequency acoustic energy ensure consistent results across multiple samples, reducing variability and enhancing reproducibility. AFA operates in an isothermal, noncontact environment, maintaining sample integrity by preventing heat-induced damage. More recently, we have also used the Diagenode Bioruptor with good results. We describe the use of both these means of chromatin shearing.

6. One of the biggest challenges in using ChIP is the availability of ChIP-grade antibodies that can efficiently immunoprecipitate the protein of interest. Although many vendors provide information about the applications for which an antibody is tested, it is always prudent to validate the antibody through immunoprecipitation (IP) and immunoblotting before performing ChIP. Many commercially available antibodies fail to work in immunoprecipitation assays, likely due to their inability to recognize native proteins or insufficient specificity. For successful ChIP, the protein of interest in the immunoprecipitated should constitute at least 50% of the signal when assessed by immunoblotting in a typical IP/western assay. We have used commercially available antibodies, including monoclonal HA from Invitrogen #2-2.2.14 and Covance #16B12; monoclonal FLAG from Sigma #F1804 and Proteintech #20543-1-AP, as well as custom-made rabbit polyclonal antibodies (anti-AtrR, anti-FfmA), successfully in both single-gene ChIP as well as ChIP-seq reactions [9, 11]. It is essential to use high-quality antibodies to ensure specific binding to the target protein, reducing nonspecific interactions that can lead to background noise and false positives, and thereby enhancing the enrichment of the target protein-DNA complexes, improving the signal-to-noise ratio and the overall quality of the ChIP data. Reliable antibodies are critical to provide consistent results

across different experiments and samples, which is essential for validating findings and drawing accurate conclusions from ChIP. In our experience, the use of indirect immunoprecipitation for ChIP is more effective in *A. fumigatus*. Indirect immunoprecipitation consists of initial incubation of the sheared chromatin with the primary antibody, followed by capture of antigen-antibody complexes with protein A or G Dynabeads (Invitrogen). Improving the efficiency of the immunoprecipitation step can significantly enhance the enrichment of target protein-DNA interactions, particularly when using low-affinity antibodies or when the target protein is expressed at limited levels.

7. ChIP often produces low DNA yields that may not be visible on gels. Fluorometric quantification (e.g., Qubit) is more reliable than absorbance-based spectrophotometry.
8. Normalization to reference gene: To control for variations in PCR efficiency and sample preparation, use a reference gene that is not expected to be enriched in the ChIP sample. Normalize the ChIP signal to the reference gene signal. We have used promoter regions for constitutively expressed genes such as tef1, tubA, and actA. PCR efficiency: To correct for differences in PCR efficiency between target and reference genes, calculate PCR efficiency for each primer set using a standard curve. Adjust the quantification accordingly.
9. The same DNA can be used for library preparation if genome-wide ChIP-seq analysis is desired. Library preparation may be performed in-house or outsourced to commercial providers. We have used commercial sources to make libraries and carry out next-generation sequencing (e.g., Admera Health Biopharma Services, https://www.admerahealth.com/genomic) with good results.

Acknowledgments

The ChIP protocol was established in our lab with invaluable input from the Robert Cramer laboratory at Dartmouth College.

References

1. Johnson KD, Bresnick EH (2002) Dissecting long-range transcriptional mechanisms by chromatin immunoprecipitation. Methods 26: 27–36. https://doi.org/10.1016/S1046-2023(02)00005-1
2. Robertson G, Hirst M, Bainbridge M et al (2007) Genome-wide profiles of STAT1 DNA association using chromatin immunoprecipitation and massively parallel sequencing. Nat Methods 4:651–657. https://doi.org/10.1038/nmeth1068
3. Wold B, Myers RM (2008) Sequence census methods for functional genomics. Nat Methods 5:19–21. https://doi.org/10.1038/nmeth1157

4. Furey TS (2012) ChIP-seq and beyond: new and improved methodologies to detect and characterize protein-DNA interactions. Nat Rev Genet 13:840–852. https://doi.org/10.1038/nrg3306
5. Fisher MC, Denning DW (2023) The WHO fungal priority pathogens list as a game-changer. Nat Rev Microbiol 21:211–212. https://doi.org/10.1038/s41579-023-00861-x
6. Morrissey CO, Kim HY, Duong T-MN et al (2024) Aspergillus fumigatus-a systematic review to inform the World Health Organization priority list of fungal pathogens. Med Mycol 62. https://doi.org/10.1093/mmy/myad129
7. Lockhart SR, Chowdhary A, Gold JAW (2023) The rapid emergence of antifungal-resistant human-pathogenic fungi. Nat Rev Microbiol 21:818–832. https://doi.org/10.1038/s41579-023-00960-9
8. Hagiwara D, Miura D, Shimizu K et al (2017) A novel Zn2-Cys6 transcription factor AtrR plays a key role in an azole resistance mechanism of Aspergillus fumigatus by co-regulating cyp51A and cdr1B expressions. PLoS Pathog 13:e1006096. https://doi.org/10.1371/journal.ppat.1006096
9. Paul S, Stamnes M, Thomas GH et al (2019) AtrR is an essential determinant of azole resistance in Aspergillus fumigatus. mBio 10. https://doi.org/10.1128/mBio.02563-18
10. Blatzer M, Barker BM, Willger SD et al (2011) SREBP coordinates iron and ergosterol homeostasis to mediate triazole drug and hypoxia responses in the human fungal pathogen Aspergillus fumigatus. PLoS Genet 7:e1002374. https://doi.org/10.1371/journal.pgen.1002374
11. Paul S, Stamnes MA, Moye-Rowley WS (2023) Interactions between the transcription factors FfmA and AtrR are required to properly regulate gene expression in the fungus Aspergillus fumigatus. G3 (Bethesda) 13. https://doi.org/10.1093/g3journal/jkad173
12. Snelders E, van der Lee HAL, Kuijpers J et al (2008) Emergence of azole resistance in Aspergillus fumigatus and spread of a single resistance mechanism. PLoS Med 5:e219. https://doi.org/10.1371/journal.pmed.0050219

Part IV

Methods for Modeling Host-Pathogen Interactions

Chapter 12

Micropatterning Substrates Using Deep UV Lithography: A Model to Study Host-Pathogen Interactions of *A. fumigatus* with Adherent Cells

Julien Polleux, Lea Heinrich, Susanne Perkhofer, and Natalia Mach

Abstract

Coculturing fungal pathogens with epithelial cell monolayers offers a powerful in vitro model for investigating host-pathogen interactions. Recent advances in bioengineering—particularly the use of micropatterned substrates—allow for the spatial confinement of alveolar epithelial cells into densely packed islands that closely resemble the structural organization of the flattened alveolus. This chapter outlines a comprehensive protocol that includes: (1) deep UV lithography for the fabrication of micropatterned glass coverslips, (2) culture of alveolar epithelial cells to generate uniform arrays of cell monolayer islands, and (3) infection with *Aspergillus* conidia to enable reliable quantification of host-pathogen dynamics.

Key words Micropatterning, Alveolar cells, Cell monolayer, *Aspergillus fumigatus*, Host-pathogen interactions

1 Introduction

Microfabrication techniques have a well-established history in life sciences [1]. Over the past two decades, the rapid expansion of biological applications—including cell biology, tissue engineering, coculture systems, bioassays, and biosensors—has led to a significant increase in interdisciplinary publications integrating modern microfabrication methods. For instance, these developments have adapted technologies such as microlithography for use with a variety of substrates, such as glass, plastics, hydrogels, and elastomers, across a range of molecules and cell types, in both two-dimensional (2D) and three-dimensional (3D) formats [2]. However, for researchers in biology labs without access to specialized microfabrication facilities, ease of implementation is a crucial factor [3, 4].

In their natural setting within organs and tissues, cells exist in a highly structured microenvironment. This environment, comprising the extracellular matrix (ECM) and neighboring cells, imposes

Natalia Mach (ed.), *Aspergillus fumigatus: Methods and Protocols*, Methods in Molecular Biology, vol. 3028,
https://doi.org/10.1007/978-1-0716-5222-0_12,

specific boundary conditions that influence cell shape, mechanics, and function [5]. The biochemical and mechanical properties of the microenvironment determine which molecules are available for cell adhesion, thereby influencing intracellular signaling pathways. These signaling cascades, in turn, affect the organization and dynamics of cytoskeletal networks, which accurately regulate critical cellular functions [6].

Despite this, conventional cell culture conditions do not replicate these critical aspects of in vivo microenvironment. On a Petri dish, cells are exposed to a flat, rigid, and homogeneous surface that bears little resemblance to their native surroundings. While such conditions have long been used to support cell survival and proliferation outside the body, they present an artificial context that lacks key environmental cues. Micropatterning techniques provide a way to partially reconstitute in vivo conditions during in vitro cell culture [7]. These methods involve creating culture substrates with microscale features that define specific patterns of cell adhesion, which shape the cellular response to defined environmental signals. Innovative studies show how microenvironmental cues, from cell adhesion to tissue architecture, direct cell behavior, including polarity, growth, adhesion, migration, cytoskeletal organization, differentiation, intercellular coordination, and interactions with pathogens, among others [7, 8].

Understanding host-pathogen interactions is crucial for developing effective treatments against infectious diseases [9]. In the case of bacterial or fungal infections through the respiratory system, in vitro research often relies on coculturing pathogens with epithelial cell monolayers, which serve as a model to study fungal adhesion, invasion, and host immune responses [10]. However, cellular monolayers are difficult to obtain on a large scale and lack the structural as well as biochemical complexity of native tissues, potentially limiting the physiological relevance of experimental findings.

To address this limitation, our group reported the micropatterning of alveolar epithelial cellular monolayers to investigate the internalization of fungal spores. The multicellular islands can be considered analogous to a flattened alveolus, where cells are packed together in a confined, dense arrangement. This organization has been shown to accelerate the internalization of fungal conidia and allow for more reliable quantification, as reported by Schiefermeier-Mach et al. [11]. Such improvements make micropatterned cellular monolayers an excellent tool for studying *Aspergillus* infections under conditions that more closely resemble in vivo environments (*see* Fig. 1).

Due to the significance and the ease of use of this method, this chapter describes how to (1) generate micropatterned islands of fibronectin via deep UV lithography, (2) culture alveolar cells to obtain cellular monolayer islands, and (3) infect them with *Aspergillus fumigatus* (*A. fumigatus*) conidia. This experimental setup

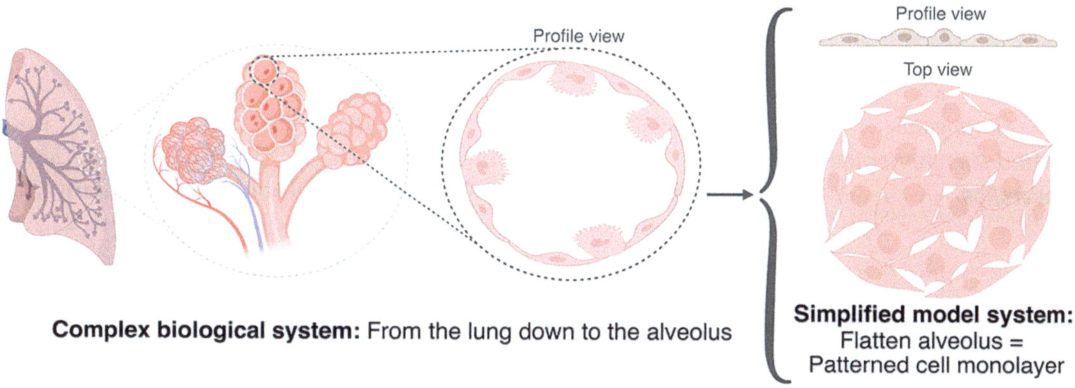

Fig. 1 From lung alveoli to micropatterned cell monolayers. Schematics describing the concept of an in vitro alveolar-like model to study the internalization of fungal spores by alveolar epithelial cells

enables more reliable quantification of pathogen internalization, providing a robust and reproducible system for studying host-pathogen interactions and facilitating high-throughput screening of antifungal strategies.

2 Materials

2.1 Polyethylene Glycol (PEG)-Coated Glass Coverslips

1. 18 mm square glass coverslips (#1, thickness 0.13–0.16 mm).
2. Home-made Teflon holder to stack 10 coverslips of 18 mm (*see* **Note 1**).
3. Pyrex beaker (150 mL).
4. 12-well plate for tissue culture, non-treated.
5. Cylindrical round bottom reaction kettle flask (500 mL) with a lid and a clamp.
6. Steel tweezers.
7. Ultrasonic bath.
8. Heated oil bath or drying oven able to heat to 80 °C.
9. Nitrogen gas cylinder equipped with a pressure reducer and a spray gun.
10. Argon gas cylinder equipped with a pressure reducer and a simple thin tubing.
11. Deionized and double-distilled Milli-Q water (Milli-Q water).
12. Piranha solution: 75 mL sulfuric acid (95–98%), 25 mL hydrogen peroxide (30%).
13. Linear PEG-silane powder (Rapp Polymere—122000-71, Germany).
14. Anhydrous toluene, reagent grade >99%.

15. Isopropanol, reagent grade >98%.
16. Methanol, reagent grade >98%.

2.2 Fibronectin-Coated Micropatterned Coverslips

1. Home-made deep UV ozone oven built with a bulb (low-pressure mercury lamp NIQ 60/35 XL long-life lamp, quartz tube, 60 W from Heraeus Noblelight, Germany), an electronic ballast (EVG 65–80 W), and a fan (*see* **Note 2**).
2. Custom-designed square chromium-coated quartz photomask of 25 mm (Compugraphics, Jena) (*see* **Note 3**).
3. Laboratory film (e.g., Parafilm, PM-996).
4. 10 cm circular glass petri dish.
5. Dental glue eco-sil (Picodent): two-compound adhesive elastomer.
6. Glass lid of a staining dish.
7. 1× Dulbecco's PBS (PBS) without Ca^{2+} and Mg^{2+}.
8. Fibronectin solution: dilute to 10 μg/mL with PBS.

2.3 Growing Cells on Fibronectin-Coated Micropatterned Coverslips

1. CO_2 incubator for cell cultures.
2. Inverted light microscope.
3. Manual or automated cell counter.
4. Adherent cell line, A549 cells, in this protocol.
5. Centrifuge capable of 1000 × rpm.
6. Trypsin EDTA 0.5%.
7. Cell culture growth medium: RPMI-1640, 10% FBS, 100 μg/mL penicillin/streptomycin, 0.2 mg/mL L-glutamine.
8. Seeding medium: RPMI-1640, 1% FBS, 100 μg/mL penicillin/streptomycin, 0.2 mg/mL L-glutamine.

2.4 Infection of Cells Grown on Fibronectin-Coated Micropatterned Coverslips with A. fumigatus Conidia

1. Incubator for fungal cultures, 37 °C (*see* **Note 4**).
2. CO_2 incubator.
3. Rocker shaker capable of at least 30 × rpm.
4. Neubauer counting chamber.
5. Inverted light microscope.
6. Centrifuge for 50 mL tubes capable of 4000 × rpm.
7. Sterile inoculation loop.
8. Fibronectin-coated micropatterned coverslips with cells.
9. A. *fumigatus* strain, for example DAL (=CBS144-89) [12, 13].
10. Sabouraud 4% glucose agar (peptone 10.0 g/L, D(+)-glucose 40.0 g/L, agar 15 g/L): Dissolve 65 g of Sabouraud 4% glucose agar in 1 L Milli-Q water, autoclave at 121 °C for 15 min (*see* **Note 5**).

11. Spore buffer: Dissolve 0.9% (w/v) NaCl and 0.1% (v/v) Tween 20 in Milli-Q water. Sterilize the solution in an autoclave at 121 °C for 15 min. Cool to and store at room temperature.
12. Infection medium: RPMI-1640 medium, 10% FBS, 0.2 mg/mL L-glutamine.

3 Methods

3.1 PEG-Coated Coverslips

1. Stack 10 square glass coverslips on a Teflon holder.
2. Place it in a Pyrex beaker filled with Piranha solution.
3. Incubate at least 4 h.
4. Discard the Piranha solution and extensively wash coverslips with deionized water (*see* **Note 6**).
5. Sonicate the coverslips 4 times for 5 min in Milli-Q water.
6. Carefully dry the coverslips with nitrogen.
7. Place the Teflon holder with clean glass coverslips in a 500 mL cylindrical round-bottom reaction kettle flask. Up to three Teflon holders can be placed.
8. Add 50 mg of linear PEG-silane.
9. Add 250 mL of anhydrous toluene to completely immerse the Teflon holders.
10. Partially close the flask with the lid and flow argon inside to remove air and humidity. Do the same with the toluene bottle and PEG-silane vial before closing and seal them with Parafilm (*see* **Note 7**).
11. Seal the lid of the flask with a clamp and heat at 80 °C, either in a heated incubator or in an oil bath, overnight to covalently graft a PEG monolayer on glass slides.
12. Take the flask out of the oven and let it cool down for 30 min.
13. Remove the coverslips from toluene, place them in a crystallizing dish filled with isopropanol, sonicate for 5 s, and incubate for 5 min.
14. Discard isopropanol, refill the dish with fresh one, sonicate for 5 s. and incubate for 5 min. Repeat this procedure one more time.
15. Discard isopropanol, fill the dish with methanol, sonicate for 5 s, and incubate for 5 min. Repeat this procedure one more time.
16. Dry the coverslips one after the other with nitrogen. Twelve-well culture plates are used to conveniently house individual coverslips in separate wells.

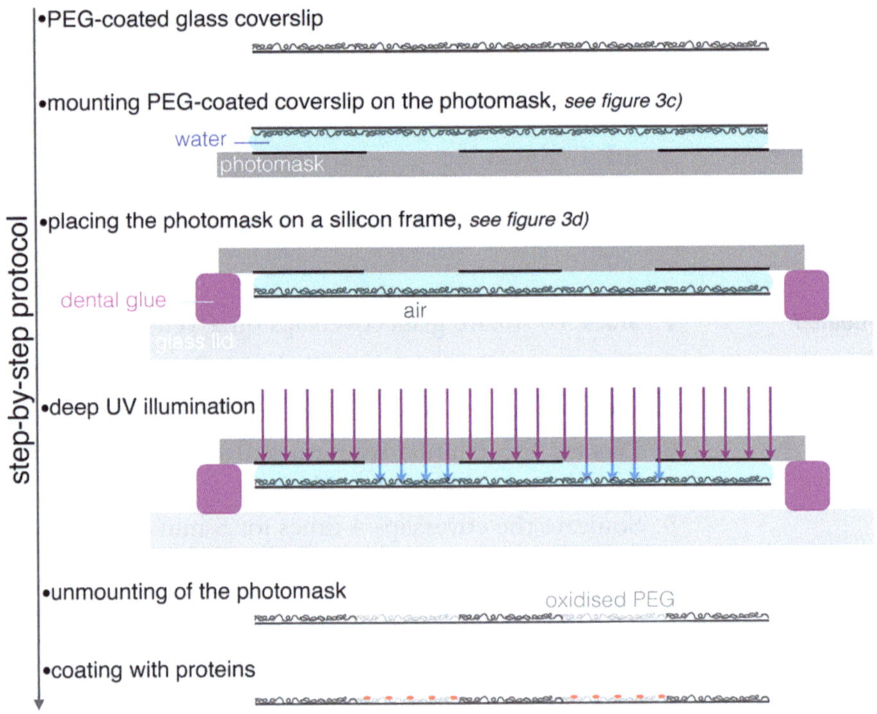

Fig. 2 Deep UV lithography. Experimental step-by-step protocol illustrating each step to perform deep UV lithography

17. Keep PEG-coated coverslips at 4 °C, if not used for the next step. Coverslips can be stored at this temperature for up to 3 weeks without loss of functionality.

3.2 Micropatterning Fibronectin on PEG-Coated Coverslips

1. Mix the two silicone-based compounds in a mass ratio of 1:1 to prepare the dental glue.
2. Draw a 25 mm silicon frame of dental glue with a spatula on the glass lid of a staining dish and let it polymerize for at least 15 min (*see* Fig. 2 and 3a).
3. Place the modified glass lid at −20 °C.
4. Place a 2.5 μL drop of Milli-Q water on a quartz photomask (*see* Fig. 2 and 3b, *see* **Note 8**).
5. Slowly flatten the drop with a PEG-coated coverslip to avoid forming air bubbles. The distance between the photomask and the PEG-coated slides is around 7.5 μm (*see* Fig. 2 and 3c, *see* **Notes 9** and **10**).
6. Take out the modified glass lid from the freezer and place the photomask upside down on the silicone frame to slow down undesirable water evaporation (*see* Fig. 2 and 3d, *see* **Note 11**).

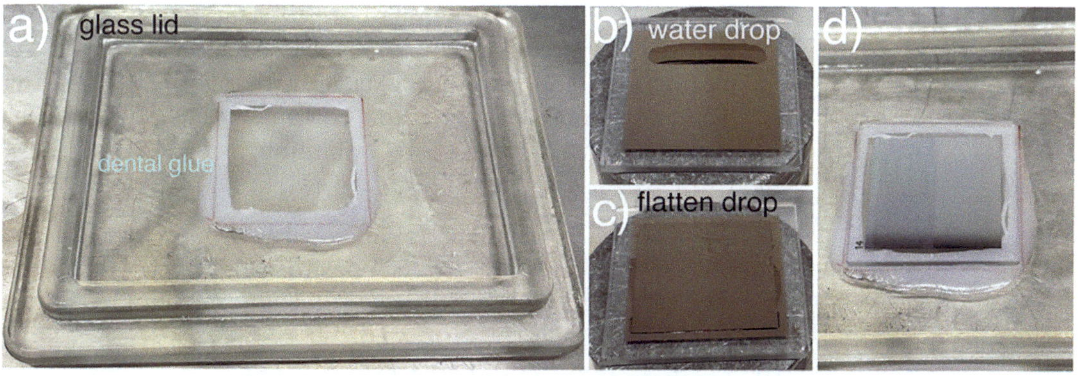

Fig. 3 (**a**) Glass lid of a staining dish with a frame of dental glue. (**b**) 2.5 μL drop of Milli-Q water deposited on a quartz photomask and (**c**) flattened with a PEG-coated coverslip. (**d**) Photomask placed upside down on the silicon frame

7. Place the sandwich system under the deep UV bulb and illuminate at 5 cm distance for 5 min in a fume hood. Deep UV light with a wavelength of 185 nm oxidizes PEG and micro-locally impairs its antifouling properties (*see* **Note 12**).
8. Stop the illumination and immediately transfer the sandwich system to a 10 cm glass petri dish filled with Milli-Q water to detach the patterned PEG-coated coverslip from the photomask (*see* **Note 13**).
9. Collect the glass coverslip with the metal tweezers and dry it with nitrogen (*see* **Notes 14** and **15**).
10. After mixing the two compounds to prepare the dental glue, deposit a small drop of this mixture close to the periphery of a 3 cm culture dish. Place one corner of the patterned coverslip on it. Leave it to polymerize for at least 10 min. This will immobilize the coverslip on the dish.
11. Place a 150 μL drop of fibronectin diluted in PBS on the glued coverslip, flatten it with 18 × 18 mm piece of Parafilm, and incubate the system at 4 °C overnight.
12. Wash once with PBS before seeding cells.

3.3 Growing Cells on Fibronectin-Coated Micropatterned Coverslips

1. Culture adherent cells. In this protocol, A549 cells are grown in cell culture growth medium at 37 °C and 5% CO_2.
2. Preheat PBS, trypsin EDTA, and seeding medium to 37 °C.
3. Aspirate medium from the flask and wash cells 3× with PBS.
4. Detach cells by adding 500 μL of trypsin EDTA.
5. Incubate at 37 °C for 3 min and control with a microscope for cell detachment.
6. Resuspend cells in fresh seeding medium and centrifuge at 1000 × rpm for 5 min. Resuspend the pellet in 1 mL seeding medium.

7. Count cell number. Depending on the cell type and micropattern size, different cell numbers are used. Prepare a total cell number of 1×10^6 for one coverslip with micropatterns with a diameter of 60 μm and 85 μm. For one coverslip with 28 μm micropatterns, prepare a total cell number of 1×10^5.
8. Incubate for 2 h at 37 °C and 5% CO_2 in seeding medium.
9. Control cell attachment to micropatterns and formation of cellular islands. Wash coverslips with cells very carefully one time with PBS and one time with seeding medium to remove nonattached cells (*see* **Notes 16** and **17**).
10. Incubate at 37 °C and 5% CO_2 overnight.

3.4 Infection of Micropatterned Cellular Islands with A. fumigatus Conidia

1. Grow *A. fumigatus* on Sabouraud 4% glucose agar in cell culture filter-cap flasks at 37 °C for 10 days. To maintain humidity, place a beaker filled with sterile Milli-Q water inside the incubator.
2. Carefully add 7 mL of spore buffer to flasks, and incubate flasks on a rocker shaker at 30 × rpm for 10 min.
3. Transfer the conidia suspension to a 50 mL Falcon tube, and centrifuge at 4000 × rpm for 5 min at room temperature.
4. Discard the supernatant and resuspend the conidial pellet in 5 mL of fresh spore buffer.
5. Store the conidial suspension at 4 °C for up to 3 weeks.
6. To avoid the formation of clumped *A. fumigatus* conidia, vortex and pipette vigorously before use. Before infecting the cells with fungal conidia, control with the microscope to ensure that the conidia no longer form clumps (*see* Fig. 4).
7. Wash coverslips with adherent cells three times with PBS and four times with infection medium.
8. Incubate coverslips with cells with a total conidia number of 1×10^6 at 37 °C and 5% CO_2 for at least 1 h. The conidia number and incubation time are determined by the specific research question. Before proceeding with fixation protocols or live imaging, perform multiple rigorous washing steps with PBS and control under the microscope to ensure conidia-specific binding to micropatterned cell islands.

4 Notes

1. Wash-N-Dry™ coverslip rack (Sigma-Aldrich) can be used as a commercially available alternative to the home-made Teflon holder.
2. UVO-Cleaner from Jetlight (Model 342-220) can be used as a commercially available alternative to the home-made deep UV oven.

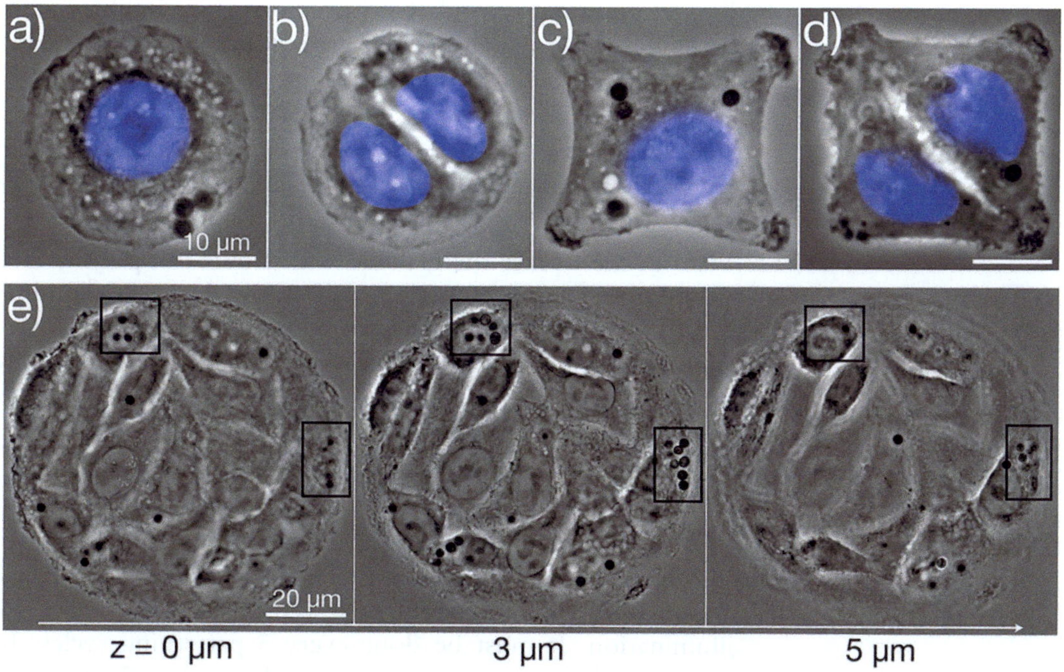

Fig. 4 Alveolar cells cultured on fibronectin-coated micropatterns of different shapes and sizes. Phase contrast and fluorescence microscopy images of a single or two A549 cells adhered for 24 h on (**a**, **b**) a disc-shaped micropattern with a diameter of 28 μm, and (**c**, **d**) a cross-shaped pattern with an edge of 28 μm. Cell nuclei are stained in blue with DAPI. (**e**) Phase contrast of a cellular monolayer island infected with *A. fumigatus* conidia for 3 h. Phase contrast was acquired from the ventral side (substrate plane, $z = 0$ μm) until the dorsal side of a cellular island ($z = 5$ μm). The two framed areas show that conidia can be imaged at different *z* values without the need for more sophisticated methods, like confocal microscopy

3. The design and production of quartz photomasks is critical and has been described by Azioune et al. [14]. It is possible to have different pattern geometries on the same photomask to test several conditions with a single PEG-coated coverslip.
4. Do not culture cells and fungi in the same incubator to avoid contamination.
5. After autoclaving Sabouraud 4% glucose agar, be aware of a possible boiling delay. The finished Sabouraud 4% glucose agar flasks can be stored at 4 °C for up to 6 months.
6. Dispose of Piranha solution and other solvents according to your institutional chemical safety guidelines, including containers for halide-free liquid waste.
7. A bottle of anhydrous toluene can be used several times. Up to three passivation protocols can be made when bottles are flushed with argon and sealed with Parafilm after opening them.
8. Other solvents than water can be used as a spacer between the photomask and the PEG-coated coverslip. The ventilated deep UV oven and the heat generated by the bulb induce undesirable water evaporation. To avoid this, solvents with a higher boiling point, like DMSO, can be used instead of water.

9. It is important to avoid the formation of air bubbles while flattening the water drop on the photomask, as they expand during UV illumination and prevent the patterning of this region.
10. Instead of using a liquid as a spacer, a vacuum mask holder can be used for a more straightforward protocol. The holder design can be obtained from the author of this publication [14].
11. Place in the freezer the glass lid with a frame of polymerized dental glue for at least 15 min before the deep UV illumination. Take it out just before assembling the PEG-coated coverslip and the photomask. Place the sandwich system on the frame of polymerized glue. By doing so, undesired water evaporation is slowed down.
12. Perform deep UV illumination in a fume hood due to the generation of ozone. Equip the device with an ozone killer, if you do not have a fume hood.
13. Photomasks must be regularly cleaned by rubbing isopropanol and ethanol with Kimwipes, followed by 5 min of deep UV illumination. It must be done every 5 patterning cycles. In addition, photomasks must be cleaned with Piranha solution after about 15 patterning cycles.
14. Other protocols can be used to ensure the passivation of glass coverslips. The company SuSoS produces poly(L-lysine)-graft-poly(ethylene glycol) molecules (PLL-g-PEG). For example, PLL(20 kDa)-g[3.3]-PEG(2–5 kDa) has been used in hundreds of peer reviewed publications [3]. Temperature-induced ultra-dense PEG-coated substrates were also shown to be effectively antifouling [15].
15. Deep UV lithography is a versatile method that is compatible with systems other than PEG-coated coverslips and fibronectin. Different types of substrates, like hydrogels, elastomers, and conductive substrates, can be patterned for various applications. Moreover, other ECM proteins, like collagen among others, can be immobilized on micropatterns.
16. Seeding cells with the right conditions (micropattern diameter, cell number, percentage of fetal calf serum, and seeding time of epithelial cells) is important in order to avoid the aggregation of rounded cells attached to the micropatterns.
17. After 2 h in culture on micropatterns, cells must be washed very carefully so that the glass coverslip always remains covered with medium or PBS. If not, liquids slide very fast away from the coverslip, which strongly damages cells and makes the experiment unexploitable. To remove unbound conidia, it is also very important to operate the same way.

References

1. Whitesides GM, Ostuni E, Takayama S et al (2001) Soft lithography in biology and biochemistry. Annu Rev Biomed Eng 3:335–373. https://doi.org/10.1146/annurev.bioeng.3.1.335
2. Blin G (2021) Quantitative developmental biology in vitro using micropatterning. Development 148. https://doi.org/10.1242/dev.186387
3. Azioune A, Storch M, Bornens M et al (2009) Simple and rapid process for single cell micropatterning. Lab Chip 9:1640. https://doi.org/10.1039/b821581m
4. Zhu M, Baffou G, Meyerbröker N et al (2012) Micropatterning thermoplasmonic gold nanoarrays to manipulate cell adhesion. ACS Nano 6:7227–7233. https://doi.org/10.1021/nn302329c
5. Vahey MD, Fletcher DA (2014) The biology of boundary conditions: cellular reconstitution in one, two, and three dimensions. Curr Opin Cell Biol 26:60–68. https://doi.org/10.1016/j.ceb.2013.10.001
6. Li Y, Jiang W, Zhou X et al (2023) Advances in regulating cellular behavior using micropatterns. Yale J Biol Med 96:527–547. https://doi.org/10.59249/UXOH1740
7. Théry M (2010) Micropatterning as a tool to decipher cell morphogenesis and functions. J Cell Sci 123:4201–4213. https://doi.org/10.1242/jcs.075150
8. Hu X, Bao M (2024) Advances in micropatterning technology for mechanotransduction research. Mechanobiol Med 2:100066. https://doi.org/10.1016/j.mbm.2024.100066
9. Feng Y, Wang S, Liu X et al (2023) Geometric constraint-triggered collagen expression mediates bacterial-host adhesion. Nat Commun 14:8165. https://doi.org/10.1038/s41467-023-43827-6
10. Bastounis EE, Radhakrishnan P, Prinz CK et al (2022) Mechanical forces govern interactions of host cells with intracellular bacterial pathogens. Microbiol Mol Biol Rev 86:e0009420. https://doi.org/10.1128/mmbr.00094-20
11. Schiefermeier-Mach N, Polleux J, Heinrich L et al (2025) Biological boundary conditions regulate the internalization of Aspergillus fumigatus conidia by alveolar cells. Front Cell Infect Microbiol 15:1515779. https://doi.org/10.3389/fcimb.2025.1515779
12. Schiefermeier-Mach N, Heinrich L, Lechner L et al (2023) Regulation of surfactant protein gene expression by Aspergillus fumigatus in NCl-H441 cells. Microorganisms 11. https://doi.org/10.3390/microorganisms11041011
13. Thau N, Monod M, Crestani B et al (1994) rodletless mutants of Aspergillus fumigatus. Infect Immun 62:4380–4388. https://doi.org/10.1128/iai.62.10.4380-4388.1994
14. Azioune A, Carpi N, Tseng Q et al (2010) Protein micropatterns: a direct printing protocol using deep UVs. Methods Cell Biol 97:133–146. https://doi.org/10.1016/S0091-679X(10)97008-8
15. Ogaki R, Zoffmann Andersen O, Jensen GV et al (2012) Temperature-induced ultradense PEG polyelectrolyte surface grafting provides effective long-term bioresistance against mammalian cells, serum, and whole blood. Biomacromolecules 13:3668–3677. https://doi.org/10.1021/bm301125g

Chapter 13

Bronchoalveolar Lavage Fluid Collection in Invasive Pulmonary Aspergillosis Murine Model for Downstream Proteomic Analysis

Sarah Sze Wah Wong

Abstract

The immunocompromised populations are susceptible to invasive fungal infections, such as invasive pulmonary aspergillosis (IPA). *Aspergillus* spp. produce airborne spores, called conidia. The small size of conidia allows them to reach the most proximal airways, such as the alveoli. The bronchoalveolar lavage fluid represents an important source of proteins, either from the host or the fungal pathogen, that are associated with the pathobiology of *Aspergillus fumigatus* and, thus, serve as crucial factors to elucidate virulence factors and/or promising novel antifungal drug targets. This chapter describes the method to obtain bronchoalveolar lavage fluid from an immunosuppressed murine model of IPA and the protein concentration estimation required for downstream proteomic analysis of the bronchoalveolar lavage fluid.

Key words *Aspergillus fumigatus*, Immunocompromised, Mouse model, Humoral immunity

1 Introduction

Aspergillus fumigatus is an ubiquitous fungus that is capable of causing invasive infections in humans [1]. Asexual reproduction of *A. fumigatus* generates airborne spores with a diameter of 2–3 μm. The small size of these spores facilitates their entry into the alveoli of the lung. In general, the inhaled spores are efficiently cleared by the innate immune cell population in the alveoli (alveolar macrophages, recruited neutrophils and monocytes). In the case of failure to clear the inhaled spores, such as in immunocompromised hosts, the inhaled spores could germinate, invading the tissues, leading to invasive pulmonary aspergillosis.

Besides cell-mediated immunity, increasing studies have uncovered the role of the humoral arm of innate immunity in host defense against *Aspergillus* infection [2, 3]. Innate humoral factors are a group of diverse pattern-recognition molecules (complement system, pentraxins, lectins, etc.), which possess host-defense

Natalia Mach (ed.), *Aspergillus fumigatus: Methods and Protocols*, Methods in Molecular Biology, vol. 3028,
https://doi.org/10.1007/978-1-0716-5222-0_13,

functions, such as complement activation, opsonization, and agglutination [4].

We have previously identified the various humoral factors that interact with *Aspergillus* spores [5]. Importantly, the interacting humoral factors vary, depending on the sources of the humoral factors (serum or bronchoalveolar lavage fluid), highlighting the importance of studying innate humoral factors in the relevant biological niche. For invasive pulmonary aspergillosis, the lung is the target organ, and thus, the use of bronchoalveolar lavage fluid (BALF) is of relevance.

Since the collection of BALF is an invasive procedure in humans, typically performed to diagnose infections in the lower respiratory tract. Mouse models serve as an ideal and valuable tool for studying the pathobiology of *Aspergillus* infections, especially for the humoral responses in the lungs (Fig. 1).

2 Materials

2.1 Fungal Culture and Conidial Suspension

1. Glycerol stock of *A. fumigatus*. Here, strain CBS 144-89 is used as wild-type [6].
2. Malt extract 2% agar: Mix 20 g of Malt extract and 15 g of agar in 1 L of water. Autoclave at 110 °C for 20 min, then cool to 45 °C, distribute 10 mL into each 17 cm glass culture tubes with cap, and let the agar solidify at an angle to create a slant (*see* **Note 1**).
3. Sterile inoculation loops.
4. Sterile physiological saline: 0.9% NaCl in distilled water.
5. Sterile water with 0.1% Tween-20 (Tween-water): Add 915 μL Tween-20 to 1 L of water and mix thoroughly. Autoclave at 121 °C for 15 min.
6. Sterile cell strainer, pore size 40 μm.
7. Sterile 50 mL Falcon tubes.
8. Hemocytometer or automated cell counter (e.g., LUNA-FL™ Automated Fluorescence Cell Counter) and cell counting slides (e.g., Cat. No. L20001 and L12001, respectively).

2.2 Invasive Pulmonary Aspergillosis in Immunosuppressed Mice

1. Mouse strain: C57BL/6 or BALB/c, male or female, 8–12 weeks old, around 20 g of body weight.
2. Sterile physiological saline: 0.9% NaCl in distilled water.
3. Immunosuppressant: cyclophosphamide monohydrate, use 40 mg/kg of mouse weight. Reconstitute cyclophosphamide monohydrate in sterile physiological saline to 4 mg/mL. Use 200 μL per mouse of 20 g body weight (0.8 mg).

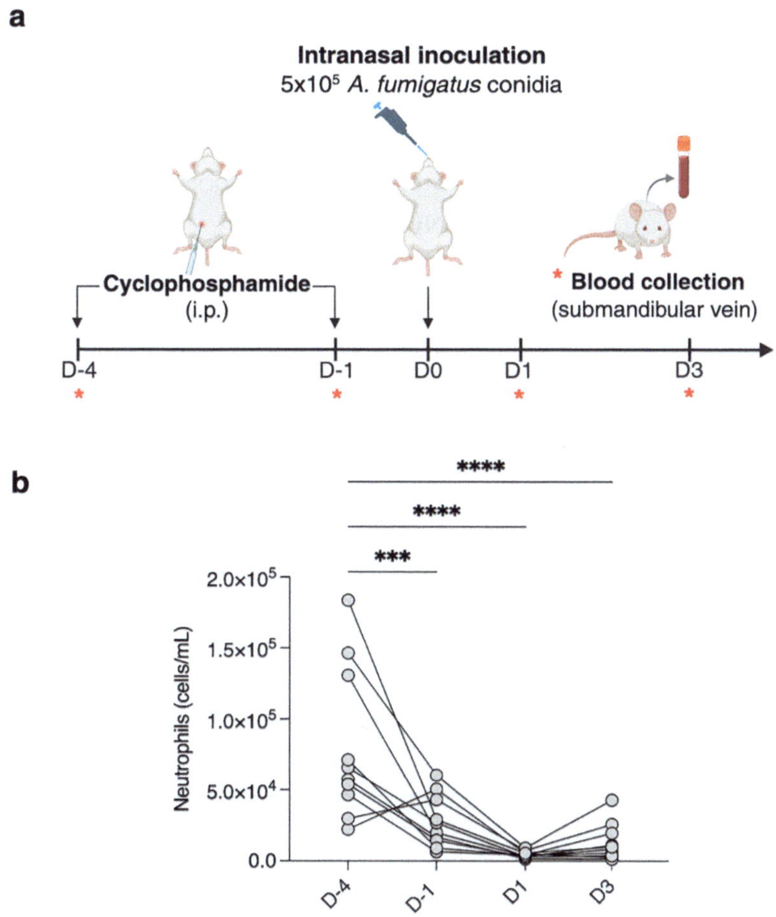

Fig. 1 Experimental timeline and neutrophil counts in immunosuppressed mice. (**a**) Timeline of experimentation. Mice were immunosuppressed by intraperitoneal injections of cyclophosphamide on day -4 (D-4) and day -1 (D-1). On day 0 (D0), intranasal inoculation of either conidia (infected group) or PBS (uninfected group) is given to mice. Blood samples are collected on day -4 (D-4, baseline), day -1, day 1, and day 3 (D-1, D1, D3) for examination of the immunosuppression status. (**b**) Number of neutrophils per mL of blood of mice immunosuppressed with cyclophosphamide. The neutrophils in blood (collected from the submandibular vein) on day -4 (D-4, baseline), day -1, day 1, and day 3 were quantified by flow cytometry. Each dot represents one mouse (10 mice in total). GraphPad Prism is used to interpret data, and p values are calculated using repeated measures one-way ANOVA with Bonferroni's multiple comparison test. *** $p < 0.001$ and **** $p < 0.0001$

4. 1 mL sterile syringe.
5. Anesthetics: isoflurane. Use isoflurane chamber: 3–3.5% of isoflurane in 100% oxygen.

2.3 BALF Collection

1. Ice-cold sterile 1× phosphate-buffered saline (PBS).
2. I.V. catheter, 18 G × 32 mm (e.g., Surflo™, Cat. No. SR +OX1832C1).
3. 1 mL sterile syringe without needle.
4. Sterile dissecting tools, forceps, and scissors.
5. Nylon string.
6. Ethanol, 70%.
7. Ice bucket.
8. Sterile 15 mL Falcon tubes.

2.4 Bicinchoninic Acid (BCA) Protein Concentration Estimation

1. Bovine serum albumin (BSA) stock solution: BSA lyophilized powder, dissolve in PBS to 2 mg/mL.
2. BCA Reagent A: 50 mg/mL of bicinchoninic acid in 2% sodium carbonate solution.
3. BCA Reagent B: 4% cupric sulfate solution in 0.4 N sodium hydroxide.
4. 96-well flat bottom microtiter plate.
5. Spectrophotometer, with wavelength available at 562 nm (optionally, 540–590 nm range).

3 Methods

All experimental steps are described relative to the day of infection (day 0). Thus, fungal culture is initiated on day -12, meaning 12 days prior to the day of infection. Mice are immunosuppressed on days -4 and -1, and infection is performed on day 0, followed by BALF collection and analysis at the indicated time points thereafter.

3.1 Fungal Culture (Day -12)

1. Streak *A. fumigatus* wild-type strain from frozen glycerol stock on malt extract agar slant using sterile inoculation loop. Close the lid of the agar slant, but do not tighten it completely.
2. Incubate the agar slant at 37 °C for 18 h. The *A. fumigatus* culture will appear white due to hyphal growth.
3. Keep the *A. fumigatus* culture at ambient temperature for around 12 days. The agar slant will gradually appear dark green as conidiation takes place (*see* **Note 2**).

3.2 Immuno-suppression of Mice (Day -4 and Day -1)

1. Record weight of mice every day.
2. Immunosuppress mice by intraperitoneal injections of 200 μL cyclophosphamide at 40 mg/kg of body weight on day -4 and repeat the procedure on day -1, which will lead to neutropenia (*see* **Note 3**).

3.3 Preparation of Fungal Inoculum (Day 0)

1. Collect the conidia (*see* Subheading 3.1) by adding 10 mL of sterile Tween-water in the slant and vortex vigorously for around 10 s (*see* **Note 4**).
2. Place a 40 μm cell strainer on a sterile 50 mL Falcon tube and pour the conidial suspension through the cell strainer to remove hyphal fragments and agar residuals.
3. Centrifuge the conidial suspension at 3000 rpm for 5 min and remove the supernatant.
4. Resuspend the conidia in sterile Tween-water and repeat the washing step (**step 3**) twice.
5. Resuspend the conidia in 1 mL sterile Tween-water, dilute the conidial suspension by 1:1000, add 10 μL of the diluted conidial suspension to the cell counting slide, and count conidia with the cell counter (*see* **Note 5**).
6. Dilute the conidial suspension to 2×10^7 conidia/mL in sterile physiological saline of required volume. Use 25 μL, which contains 5×10^5 conidia, per mouse (*see* **Note 6**).

3.4 Infection (Day 0)

1. Anesthetize mice in an isoflurane chamber (3–3.5% of isoflurane in 100% oxygen).
2. Vortex or agitate the conidial suspension and collect 25 μL with the pipette (*see* **Note 7**).
3. Once mice are deeply anesthetized, restrain and hold the mouse vertically and inoculate the conidial suspension gently drop by drop alternatively on each nare. The droplets will be rapidly inhaled by the mouse (*see* **Note 8**).
4. Continue holding the mouse vertically for around 10 s after the inoculation.
5. Place the mouse back in the cage on its back to recover (*see* **Note 9**).

3.5 BALF Collection

Depending on the experimental setup, BALF collection can be performed hours to days postinfection.

1. Euthanize mice by CO_2 asphyxiation at desired time point (*see* **Notes 10** and **11**).
2. Place and immobilize the mouse on its back on the dissection board.
3. Sterilize the neck area with 70% ethanol.
4. Lift the skin at the center of the neck of the mouse and make an incision with sterile scissors.
5. The trachea of the mice is aseptically exposed by pulling the salivary glands to the sides with sterile forceps.

6. Insert an 18-G lavage catheter into the trachea, paying attention not to pierce through the dorsal wall of the trachea.
7. Pass a nylon string under the trachea with the help of a pair of forceps and tie the trachea with the inserted catheter.
8. Remove the needle inside the catheter and attach the catheter to a 1 mL syringe with 1 mL ice-cold PBS (*see* **Note 12**).
9. Gently inflate the lung with the PBS and recover the lavage fluid (*see* **Note 13**).
10. Repeat **step 9** for a total of 3 times (*see* **Note 14**).
11. Keep the collected BALF on ice.
12. Centrifuge the BALF at 13,000 rpm for 10 min at 4 °C, and store the cell-free supernatant at −20 °C until further analysis.

3.6 BCA Protein Concentration Assay

1. Prepare standards with known concentrations of bovine serum albumin (BSA) in diluent (in this case, PBS), ranging from 0 to 2000 μg/mL (*see* Table 1). 100 μL of each standard will be prepared (*see* **Note 15** and Fig. 2).
2. Prepare the BCA working reagent (WR) fresh by mixing 50 parts of BCA Reagent A with 1 part of BCA Reagent B. 200 μL of WR reagent is required for each sample (*see* **Note 16**).

   Use the following formula to determine the total volume of WR required:

$$(9\ \text{standards} + \text{no. of samples}) \times 2\ \text{replicates} \times 200\ \mu\text{L} = \text{total volume of WR required}$$

Table 1
Dilution scheme for bovine serum albumin (BSA) standards used in BCA protein concentration assay

Vial	Volume of diluent (μL)	Volume of BSA	Final BSA concentration (μg/mL)
A	0	100 μL of stock BSA	2000
B	37.5	112.5 μL of stock BSA	1500
C	100	100 μL of stock BSA	1000
D	50	50 μL of vial B	750
E	100	100 μL of vial C	500
F	100	100 μL of vial E	250
G	60	60 μL of vial F	125
H	80	20 μL of vial G	25
I	100	0 μL	0

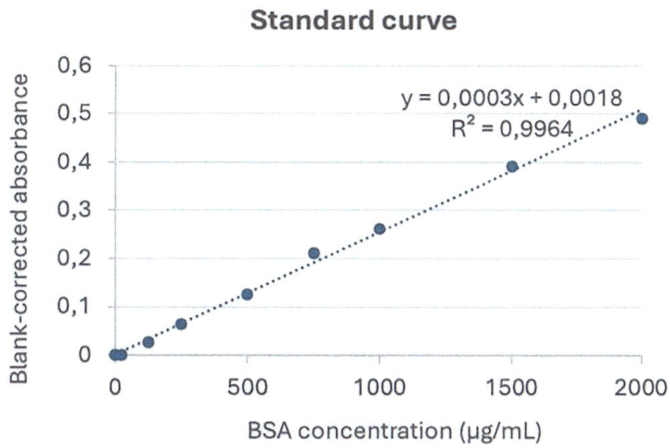

Fig. 2 Standard curve of bovine albumin serum in BCA protein estimation assay. The blank-corrected absorbances are plotted against the known concentrations of the BSA standards. The protein concentrations of the BALF samples can be estimated by the equation of the best-fit line

3. Pipette 5 μL of each standard or BALF sample into separate wells of the microtiter plate. Make two replicates for each standard or sample.
4. Add 200 μL of the WR to each well and mix thoroughly by gently tapping the plate.
5. Incubate the plate at 37 °C for 30 min.
6. Cool plate to room temperature.
7. Measure the absorbance at 562 nm using a spectrophotometer (*see* **Note 17**).
8. Subtract the average absorbance values of the blank standard replicates from those of the other individual standards and samples. With Microsoft Excel, plot a standard curve using the blank-corrected absorbance values of the BSA standards, and calculate the protein concentration of the samples using the best-fit line of the standard curve (*see* **Note 18**).

4 Notes

1. Other culture media, such as Sabouraud agar, potato dextrose agar, or a mix of yeast extract, peptone, and dextrose agar, can also be used.
2. Conidiation is allowed for around 12 days, as some strains, especially mutant strains, have a slower rate of growth and conidiation, as compared to wild-type strains.

3. Two injections of cyclophosphamide in a given concentration lead to neutropenia in mice by significantly reducing the numbers of neutrophils. The induced neutropenia sustains until at least day 3 (*see* Fig. 1). If a longer experimental duration is needed, extra injections of cyclophosphamide will be needed.
4. Resting conidia of *A. fumigatus* are hydrophobic and are difficult to be suspended in water; therefore, surfactants, such as Tween-20, are needed for preparing conidial suspension.
5. Generally, 1×10^9 conidia can be recovered per agar slant from this method of culturing and collecting.
6. Prepare the conidial suspension freshly before the infection. As resting conidia become metabolically active when in contact with water, the viability could decrease [7].
7. Conidia sediment rapidly, therefore, it is important to thoroughly homogenize the conidial suspension before each usage.
8. The whole inoculation process takes around 30 s. Make sure that the mice are properly anesthetized; otherwise, the mice could swallow the conidial suspension instead of inhaling.
9. Placing the mice on its back to avoid contact of the nares with the bedding. Mice should wake up within 2 min.
10. With the immunosuppression protocol and inoculum size proposed here, Balb/c mice would succumb to infection on day 5 ± 1 day.
11. Avoid cervical dislocation as a sacrifice approach, as it may break the trachea.
12. PBS should be ice-cold to minimize the protease activity. A protease inhibitor could also be added to the PBS.
13. If there is a leakage of the instilled fluid, tighten the nylon string around the trachea.
14. The repeat rinsing is to ensure the maximal recovery of the proteins in the lung.
15. The diluent used for preparing the standard curve should be the same buffer as in BALF collection.
16. The work reagent should be prepared freshly before the assay and mixed thoroughly to yield a clear green solution without any precipitates.
17. As the BCA protein estimation assay is not end-point measurement, measure the absorbance within 30 min.
18. The absorbance values of the samples should be within those of the standard curves; if not, repeat the assay with adjusted volume or dilution of samples.

References

1. van de Veerdonk FL, Carvalho A, Wauters J et al (2025) Aspergillus fumigatus biology, immunopathogenicity and drug resistance. Nat Rev Microbiol. https://doi.org/10.1038/s41579-025-01180-z
2. Wong SSW, Dellière S, Schiefermeier-Mach N et al (2022) Surfactant protein D inhibits growth, alters cell surface polysaccharide exposure and immune activation potential of Aspergillus fumigatus. Cell Surf 8:100072. https://doi.org/10.1016/j.tcsw.2022.100072
3. Dellière S, Chauvin C, Wong SSW et al (2024) Interplay between host humoral pattern recognition molecules controls undue immune responses against Aspergillus fumigatus. Nat Commun 15. https://doi.org/10.1038/s41467-024-51047-9
4. Mantovani A, Garlanda C (2023) Humoral innate immunity and acute-phase proteins. N Engl J Med 388:439–452. https://doi.org/10.1056/NEJMra2206346
5. Wong SSW, Daniel I, Gangneux J-P et al (2020) Differential interactions of serum and bronchoalveolar lavage fluid complement proteins with conidia of airborne fungal pathogen Aspergillus fumigatus. Infect Immun 88. https://doi.org/10.1128/IAI.00212-20
6. Thau N, Monod M, Crestani B et al (1994) rodletless mutants of Aspergillus fumigatus. Infect Immun 62:4380–4388. https://doi.org/10.1128/iai.62.10.4380-4388.1994
7. Novodvorska M, Stratford M, Blythe MJ et al (2016) Metabolic activity in dormant conidia of Aspergillus niger and developmental changes during conidial outgrowth. Fungal Genet Biol 94:23–31. https://doi.org/10.1016/j.fgb.2016.07.002

Chapter 14

Galleria mellonella to Study Aspergillus Infections and Antifungal Efficacy by Traditional Readouts and Novel BL-Imaging Techniques

Jakob Scheler and Ulrike Binder

Abstract

Galleria mellonella (greater wax moth) larvae have emerged as a valuable alternative in vivo model for studying fungal infections, including aspergillosis, due to their advantages such as ease of use, low cost, and ability to survive at mammalian body temperatures. *Galleria* larvae are useful for evaluating fungal virulence, host-pathogen interactions, and antifungal toxicity and efficacy. Recent studies have incorporated bioluminescence (BL) imaging, which enhances the utility of the model by enabling noninvasive, real-time monitoring of fungal burden and disease progression in individual animals. Together, these advantages make *G. mellonella* an effective and ethically favorable platform for preclinical aspergillosis research, perfectly aligning with 3R principles.

Key words *Galleria mellonella*, *Aspergillus fumigatus*, Virulence potential, Antifungal efficacy testing, Host response

1 Introduction

Research on infectious diseases and potential treatment options still largely depends on animal models. Because of ethical concerns, high costs, and the need for special facilities and trained personnel, alternative models have emerged in studying fungal infections. For aspergillosis studies, specifically studies involving *Aspergillus fumigatus*, the use of *Galleria mellonella* (greater wax moth larvae) has been proven one valuable alternative model [1–3]. This invertebrate model offers several advantages that make it increasingly attractive in fungal pathogenesis and antifungal research. One of the primary benefits of *G. mellonella* is its practicality and ethical suitability compared to traditional mammalian models. As an invertebrate, it is not subject to the same ethical regulations that govern vertebrate animal use, allowing for more flexible experimental design, such as high-throughput screening assays. The larvae, if

Natalia Mach (ed.), *Aspergillus fumigatus: Methods and Protocols*, Methods in Molecular Biology, vol. 3028,
https://doi.org/10.1007/978-1-0716-5222-0_14,

commercially available, are inexpensive, can also be reared easily, are easy to handle, and are amenable to large-scale studies, making them well-suited for high-throughput screening of antifungal compounds and fungal virulence factors. Larvae can be incubated at 37 °C, the temperature at which *Aspergillus fumigatus* thrives, and they possess an innate immune system with hemocytes, antimicrobial peptides, and melanization responses that are similar to key features of mammalian innate immunity [4–7]. These similarities explain why findings on the virulence potential of certain strains or deletion mutants correlate well with those in murine models, allowing researchers to use *Galleria* as a bridge before moving to vertebrate systems [1]. The model has also proven particularly useful for screening drug efficacy [3, 8–11] for testing licensed as well as experimental antifungals [12–14]. Nevertheless, *Galleria* larvae cannot fully replace mammalian models. It lacks lungs and adaptive immunity, which makes it unsuitable for investigating the inhalational route of infection or T and B cell responses that are critical to human disease. Survival curves and "health indices" can vary, depending on larval source, size, and handling practices, which shows that standardization is essential.

Despite these drawbacks, the model has yielded important findings for *Aspergillus* research. Classic virulence factors such as conidial melanin, siderophore biosynthesis, and fumagillin production [6] have been validated in *Galleria*, with mutants consistently showing reduced virulence in line with murine outcomes, e.g., lack of siderophore production [1].

Traditional readouts still monitor survival rates of infected larvae, supplemented by measuring the rate of melanization or the classification into groups by so-called "health scores." Fungal burden can be determined by PCR, which is tedious work, or by histological analysis of formalin-embedded larvae [3, 9]. Recently, a particularly significant advance has been the adaptation of bioluminescence imaging for the larval mini-hosts [8, 9]. By injecting engineered *A. fumigatus* strains expressing firefly luciferase, one can follow infection in real time in individual animals after luciferin is injected. This approach transforms the model from one that relied mainly on endpoint survival into one capable of delivering quantitative, longitudinal data on fungal growth and treatment response [9]. It improves reproducibility and statistical power, allows earlier detection of therapeutic effects, and aligns the larval model more closely with mammalian imaging approaches.

Here, we explain principles of using *Galleria mellonel*la as an infection model for *A. fumigatus*, testing antifungal treatment options, extracting hemolymph, and handling BL-imaging of infected larvae.

2 Materials

2.1 Growth Media and Buffers

1. Aspergillus complete medium (ACM): 20 mL 50× salt solution, 1 mL trace elements, 20 g D-glucose, 2 g peptone, 1 g yeast extract, 1 g N-Z-Amine A (casamino acids), 15 g agar; add distilled water to 1 L.
2. Salt solution: 26 g KCl, 26 g $MgSO_4{\cdot}7H_2O$, 76 g KH_2PO_4, 75 μL chloroform; add distilled water to 1 L.
3. Trace elements: 800 mg $ZnSO_4{\cdot}7H_2O$, 40 mg $CuSO_4{\cdot}5H_2O$, 71 mg $MnSO_4{\cdot}H_2O$, 2.1 mg $Na_2B_4O_7$, 80 mg $Na_2MoO_4{\cdot}2H_2O$, 140 mg $FeSO_4{\cdot}7H_2O$; add distilled water to 100 mL (*see* **Note 1**).
4. Insect physiological saline (IPS): 8.76 g NaCl, 0.36 g KCl, 15.76 g Tris-HCl, 3.72 g EDTA, 4.72 g sodium citrate; adjust to pH 6.9.
5. Phosphate-buffered saline (PBS): 2.88 g Na_2HPO_4, 0.524 g NaH_2PO_4, 18 g NaCl; adjust to pH 7.0.
6. Spore solution buffer: 0.9% NaCl, 0.01% Tween 20.

2.2 Preparation of Spore Solutions

1. Falcon Cell Strainer: 40 μm nylon mesh (e.g., Corning #352340).
2. CellTrics 10 μm filter, sterile single packed (Sysmex #04-004-2324).
3. 50 mL Falcon tubes.
4. Filter tips (1000 μL, 200 μL).
5. Sterile loops.
6. 2 mL Eppendorf tubes, safe lock.
7. Glycerol: 95%, sterile.
8. Liquid nitrogen.
9. Hemocytometer.

2.3 Galleria Infections and Hemolymph Extraction

1. Filter paper: Whatman 602H round filter, 90 mm.
2. 100 mm Petri dishes.
3. Syringes for infection/inoculation: 0.3 mL Terumo Myjector U-100 sterile insulin syringes.
4. Needles for hemolymph extraction: BD Microlance 3 needles, 23 G syringes.
5. *Galleria* larvae (sixth instar larvae), commercially available (e.g., "Kreuch und Fleuch," Austria).
6. 37 °C incubator.

7. Cotton swabs.
8. Metal tweezers.

2.4 Bioluminescent Imaging

1. In vivo imaging system: IVIS Spectrum (from Revvity).
2. D-luciferin.
3. 12-well microtiter plates.
4. Ice.

3 Methods

3.1 Fungal Growth and Preparation of Spore Solutions

1. Inoculate three ACM agar plates with the *Aspergillus* strain of choice, and incubate at 37 °C for 4 days until full sporulation.
2. Harvest spores by adding 5 mL of spore solution buffer to each agar plate and carefully releasing the conidia with a sterile loop. Transfer the suspension into a 50 mL Falcon tube by filtering through a 40 μm cell strainer to remove large hyphae and conidiophores. Combine suspensions from all plates into one Falcon tube.
3. Centrifuge conidia at 3500 rpm for 5 min, and wash 3 times with spore solution buffer.
4. Filter the suspension into a fresh tube through a 10 μm filter to remove remaining larger material.
5. Count conidia using a hemocytometer and adjust to a concentration of 5×10^8/mL. Dilute spores in IPS containing 10% glycerol, aliquot in 2 mL safe-lock tubes, shock freeze in liquid nitrogen, and store at −20 °C until use (can be stored up to several months).

3.2 Galleria mellonella *Infection and Monitoring*

1. Place ten healthy larvae of the same size into a Petri dish with filter paper at the bottom (*see* **Note 2**).
2. Use 10–30 larvae per strain/sample to be tested.
3. If frozen conidial stocks are used, thaw tubes and pre-warm conidial solution to room temperature before injection. Alternatively, use freshly harvested conidia that have not yet swollen.
4. Infection is carried out by injecting 20 μL of conidial suspension (*see* **Note 3**) into the right proleg of each larva (*see* Fig. 1).
5. Prepare control groups consisting of untouched larvae and sham-infected larvae (*see* **Note 4**).
6. Place larvae in Petri dishes in a 37 °C incubator and monitor survival every 24 h. Assess larval viability by gently touching immobile larvae with a cotton swab.

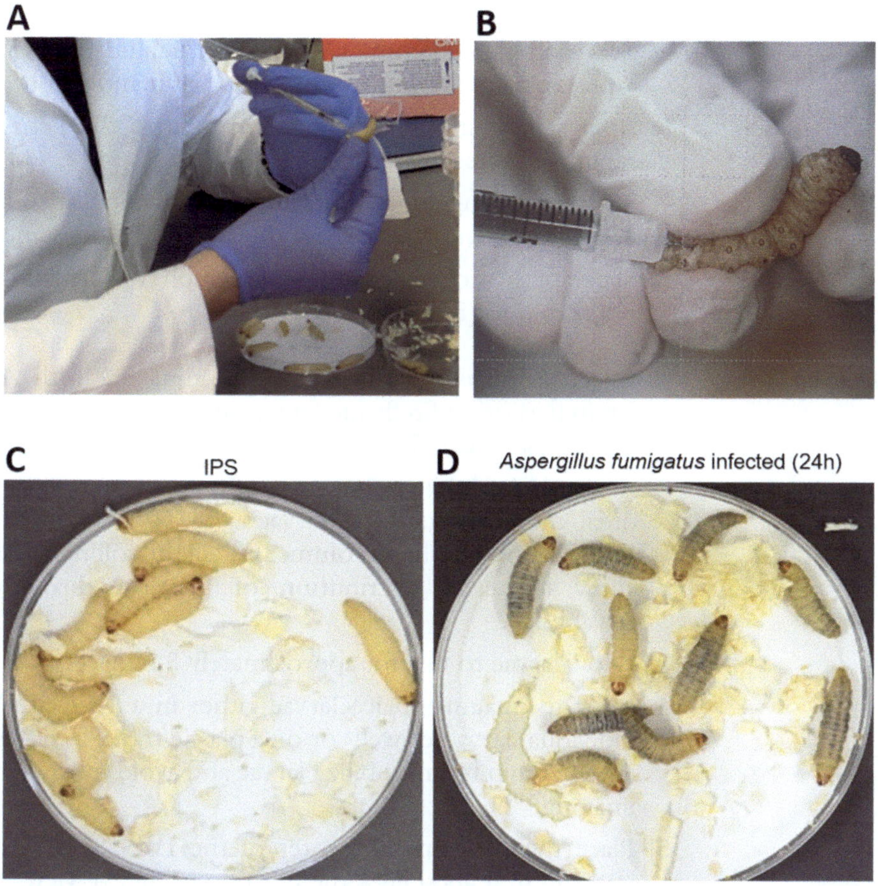

Fig. 1 Infection process of *Galleria mellonella* larvae with a Myjector syringe through one of the hind prolegs (**a**, **b**). Panels **c** and **d** show examples of usual experiments. (**c**) Sham-infected control. (**d**) *Aspergillus*-infected larvae 24 h post infection, which already show signs of melanization

3.3 Antifungal Efficacy Treatment

1. Infect larvae with *A. fumigatus* conidia as described above (*see* Subheading 3.2), using 20 larvae per test sample. Dilute conidia in PBS to avoid cross-effects with antifungal drugs.
2. At 2 h post infection, administer the antifungal agent diluted in PBS in a volume of 20 μL, injected into the opposite hind proleg, where the infectious agent was delivered.
3. Controls include untreated larvae, sham-infected larvae, and larvae injected with antifungal agent only (toxicity control).
4. Place larvae in Petri dishes in a 37 °C incubator and monitor survival every 24 h. Assess larval viability by gently touching immobile larvae with a cotton swab.

3.4 Isolation of Hemolymph and Hemocytes

1. Prepare collection tubes (2 mL Eppendorf) on ice for hemolymph collection.

2. Hold the larvae that need to be sampled tightly between fingers, and pierce gently with a needle behind the head region. Squeeze gently to induce the release of hemolymph.
3. After sampling, larvae can be returned to the Petri dish and resampled once more after 24 h.
4. To analyze drug concentrations in hemolymph or count hemocytes, pool hemolymph from three larvae of the same sample.
5. Separate hemocytes from hemolymph by centrifugation at 5000 rpm for 5 min at 4 °C.

3.5 BL Imaging of Infection

1. Inject larvae as described in Subheading 3.2 with *Aspergillus* strain harboring firefly luciferase, such as *Aspergillus fumigatus* 2/7/1 (*see* **Note 5**).
2. At the desired time point, or at 24 h post infection for longitudinal monitoring, inject larvae with 60 μL of 8.3–10 mM D-luciferin. The high volume and slight rolling of the larvae allow for even distribution of luciferin throughout the larval body.
3. Allow larvae to rest for approximately 5 min.
4. Prior to imaging, place larvae either in a 12-well plate (one individual larva per well) or on a plastic tray. Place the plate on ice for 2 min to immobilize larvae and avoid movement during imaging.
5. Place the plate in the center of the IVIS Spectrum in vivo imaging unit and check the position using live view.
6. In acquisition mode, set the emission filter to "open," filter position to 1, binning factor to 8″ and subject height to "2 cm." For optimal exposure time, acquire images at two time points: 30 s and "Auto."
7. For analysis of multiple images, open them together. In the "Tool Palette," navigate to "Image Adjust." Deselect "Individual" in the color scale limits to create a global scale for all images. Adjust the "Min" value under "Aggregate Color Scale" to remove unwanted noise. Under "Color Table," set the scale to logarithmic to improve readability and enhance differences. In the top left of "Sequence View," set "Units" to "Radiance (photons)." This unit will normalize the photons to acquisition time and field of view (square centimeter and steradian) and allows for comparison of results from different experiments.
8. In the "Tool Palette," select "ROI Tools," choose the grid matching the plate size (e.g., 3 × 4). Adjust the grid to align with the wells. Once all grids are in place, click "Measure ROIs." The "ROI Measurements"—Window will pop up. Inside, navigate to "Grid ROI Measurements." Change

"Measurement" to "Average Radiance [p/s/cm^2/sr]." Click "Export" and save the file in. txt or .csv format (*see* **Note 6**).

4 Notes

1. Dissolve all components separately in 50 mL Falcon tubes in a small volume of distilled water, then mix, and adjust to 100 mL with distilled water. Add HCl dropwise to $FeSO_4{\cdot}7H_2O$ until the solution looks clear. Sterile filter and store at 4 °C.
2. Larvae in their sixth instar must be used within 15 days and always have to be kept at 15–18 °C in the dark. Only healthy larvae can be used: they should be beige in color, moving, without black spots and with healthy legs and prolegs. Larvae should all have the same size/weight: usually 0.3–0.5 g. Don't keep larvae at 4 °C or 37 °C before infection, as temperature changes can cause changes to their immune system [15].
3. A concentration of 5×10^8/mL is frequently used, but the ideal inoculum has to be determined for every strain used and for every larval batch.
4. Untouched larvae serve as controls for the general quality of the batch used. Sham-infected larvae serve as quality controls for contamination-free buffer and as an operator control to verify that the injection procedure itself does not harm the animal.
5. This strain, kindly provided by Mathias Brock, was used in the images shown in Fig. 1. Recently, a strain with engineered red-shifted firefly luciferase was shown to have superior capabilities to be used for BL imaging in *Galleria* and mice [8].
6. If the position of the plates deviates between individual images and requires individual adjustment, then right-click on one ROI-grid and left-click on "Copy ROI." Delete the grids, then right-click on each individual image and select "Paste ROI" to receive an individually adjustable ROI per image.

References

1. Slater JL, Gregson L, Denning DW et al (2011) Pathogenicity of Aspergillus fumigatus mutants assessed in Galleria mellonella matches that in mice. Med Mycol 49:S107–S113. https://doi.org/10.3109/13693786.2010.523852
2. Curtis A, Binder U, Kavanagh K (2022) Galleria mellonella larvae as a model for investigating fungal—host interactions. Front Fungal Biol 3. https://doi.org/10.3389/ffunb.2022.893494
3. Maurer E, Browne N, Surlis C et al (2015) Galleria mellonella as a host model to study Aspergillus terreus virulence and amphotericin B resistance. Virulence 6:591–598. https://doi.org/10.1080/21505594.2015.1045183
4. Fallon JP, Reeves EP, Kavanagh K (2010) Inhibition of neutrophil function following exposure to the Aspergillus fumigatus toxin fumagillin. J Med Microbiol 59:625–633. https://doi.org/10.1099/jmm.0.018192-0

5. Fuchs BB, Mylonakis E (2006) Using non-mammalian hosts to study fungal virulence and host defense. Curr Opin Microbiol 9:346–351. https://doi.org/10.1016/j.mib.2006.06.004
6. Fallon JP, van Reeves EP, Kavanagh K (2011) The Aspergillus fumigatus toxin fumagillin suppresses the immune response of Galleria mellonella larvae by inhibiting the action of haemocytes. Microbiology 157:1481–1488. https://doi.org/10.1099/mic.0.043786-0
7. Bergin D, Reeves EP, Renwick J et al (2005) Superoxide production in Galleria mellonella hemocytes: identification of proteins homologous to the NADPH oxidase complex of human neutrophils. Infect Immun 73:4161–4170. https://doi.org/10.1128/IAI.73.7.4161-4170.2005
8. Vanhoffelen E, Win T, van Braeckel E et al (2025) Combinations of posaconazole and tacrolimus are effective against infections with azole-resistant Aspergillus fumigatus. Front Cell Infect Microbiol 15. https://doi.org/10.3389/fcimb.2025.1550457
9. Vanhoffelen E, Michiels L, Brock M et al (2023) Powerful and real-time quantification of antifungal efficacy against triazole-resistant and -susceptible Aspergillus fumigatus infections in Galleria mellonella by longitudinal bioluminescence imaging. Microbiol Spectr 11. https://doi.org/10.1128/spectrum.00825-23
10. Jemel S, Guillot J, Kallel K et al (2021) In vivo efficacy of voriconazole in a Galleria mellonella model of invasive infection due to azole-susceptible or resistant Aspergillus fumigatus isolates. J Fungi (Basel) 7:1012. https://doi.org/10.3390/jof7121012
11. Blatzer M, Blum G, Jukic E et al (2015) Blocking Hsp70 enhances the efficiency of amphotericin B treatment against resistant Aspergillus terreus strains. Antimicrob Agents Chemother 59:3778–3788. https://doi.org/10.1128/AAC.05164-14
12. Rochford G, Molphy Z, Browne N et al (2018) In-vivo evaluation of the response of Galleria mellonella larvae to novel copper (II) phenanthroline-phenazine complexes. J Inorg Biochem 186:135–146. https://doi.org/10.1016/j.jinorgbio.2018.05.020
13. Štěpánek O, Parigger M, Procházková E et al (2025) Prodrugging fungicidal amphotericin B significantly decreases its toxic effects. Eur J Med Chem 283:117157. https://doi.org/10.1016/j.ejmech.2024.117157
14. Lackner M, Obermair J, Naschberger V et al (2019) Cryptic species of Aspergillus section Terrei display essential physiological features to cause infection and are similar in their virulence potential in Galleria mellonella. Virulence 10:542–554. https://doi.org/10.1080/21505594.2019.1614382
15. Mowlds P, Kavanagh K (2008) Effect of pre-incubation temperature on susceptibility of Galleria mellonella larvae to infection by Candida albicans. Mycopathologia 165:5–12. https://doi.org/10.1007/s11046-007-9069-9

Part V

Methods for Investigating Fungal Cell Wall and Secreted Components

Chapter 15

Molecular Biosensing of the *Aspergillus fumigatus* Cell Wall Integrity Pathway Using the *(p)agsA*::luc (Luciferase) Reporter System

Jonatas Erick Maimoni Campanella, Mark Arentshorst, Arthur F. J. Ram, Cees A. M. J. J. van den Hondel, and Iran Malavazi

Abstract

In this protocol, we describe a robust luciferase-based biosensor assay to monitor the activity of the cell wall integrity (CWI) pathway in *Aspergillus fumigatus* in real time. The method relies on the stable integration of a markerless, synthetic reporter cassette *(p)agsA*::luc at the *aft4* Safe Haven (Sh) genomic locus using CRISPR-Cas9. This cassette comprises a modified *A. niger agsA* promoter containing three tandem RlmA-binding sites, which drives the expression of the luciferase gene. Upon exposure to cell wall stress, the endogenous transcription factor RlmA activates the reporter, generating a luminescent signal proportional to promoter activity. The protocol includes the construction of a*ft4* locus-specific CRISPR-Cas9 plasmids, *A. fumigatus* transformation and candidate selection, as well as the setup of the luminescence bioassay in white 96-well microplates. This system enables highly sensitive, nondestructive, and time-resolved quantification of CWI pathway activation during early fungal growth or biofilm under various genetic or chemical perturbations. Moreover, it supports comparative studies across wild-type and mutant strains, offering a powerful platform for dissecting stress response signaling and identifying antifungal compounds that target the CWI pathway.

Key words *Aspergillus fumigatus*, Cell wall integrity, *rlmA*, Luciferase real-time reporter, Transcriptional activity

1 Introduction

The quantification of gene expression has been widely used as a powerful tool in molecular and cell biology to understand the regulatory networks that govern nearly all aspects of cell physiology, including metabolic processes and, in particular, signaling transduction and associated transcriptional output. In fungal pathogens, such as *Aspergillus fumigatus*, gene expression analysis is crucial for elucidating the mechanisms underlying virulence, stress adaptation, and antifungal resistance. Given its clinical

Natalia Mach (ed.), *Aspergillus fumigatus: Methods and Protocols*, Methods in Molecular Biology, vol. 3028, https://doi.org/10.1007/978-1-0716-5222-0_15,

importance as the predominant airborne mold responsible for invasive pulmonary aspergillosis in immunocompromised individuals, a detailed understanding of *A. fumigatus* gene regulation is of both biological and medical significance [1, 2].

Traditional methods for analyzing gene expression in filamentous fungi, such as quantitative PCR (qPCR) or RNA sequencing, typically offer reliable measurements of transcript abundance and have provided foundational insights into gene regulation. However, despite their analytical power, these techniques often lack the temporal resolution and in vivo contextual fidelity needed to detect rapid or dynamic transcriptional responses to environmental stimuli. This limitation becomes particularly relevant in the context of filamentous fungal pathogens, such as *A. fumigatus*, where growth and dissemination occur through filamentous growth, spore development, and the formation of biofilms in host tissues. Biofilms exhibit a distinct physiological state characterized by unique gene expression, altered metabolism, and modified susceptibility profile to antifungal agents [3, 4]. Consequently, creating tools to track gene expression within biofilms remains technically challenging in molds due to their multicellular morphology and intricate growth dynamics, yet it is essential for an expanded understanding of fungal physiology in clinically relevant conditions.

Reporter gene biosystems based on the luciferase enzyme offer a robust and sensitive alternative for the real-time monitoring of signaling pathway-specific gene expression in various microorganisms, enabling nondestructive and quantitative assessment of promoter activity in engineered biosensors [5–7]. Additionally, the use of optimized luciferase constructs offers enhanced stability and an improved signal-to-noise ratio. Furthermore, they can be used to monitor other parameters, such as cell health or cell death, both in vitro and in vivo within the animal host [1, 2, 8, 9].

A luciferase reporter fused to the promoter of a selected stress response gene provides a powerful approach to quantify the amplitude of stress responses. Such a biosensor engineering strategy is based on the fusion of regulatory sensor elements to the luciferase reporter gene within a synthetic circuit to dissect the function of cis-acting genetic elements, such as promoters or trans acting elements, such as transcription factors (TF). Finally, the biosensor constructs can be stably integrated into the *A. fumigatus* genome, enabling dynamic and temporally resolved measurements of gene expression within the complex architecture of fungal biofilms, with a high correlation between light emission and fungal biomass [9]. The key to the success of these assays lies in the application of well-characterized promoters that can reliably reflect the activity of TF and signaling pathways of interest, as well as the integration of reporter constructs into defined genomic loci. In this regard, the locus *aft4* Safe Haven (Sh) previously characterized is a site of an inactive Tc1/mariner type transposable element (*aft4*). It provides a strategy for inserting transgene expression cassettes into *A. fumigatus* without any significant impact on the growth, cytotoxicity, or pathogenicity of *A. fumigatus* using CRISPR-Cas9 technology [10].

The method presented here describes a sensitive and highly specific luciferase-based gene expression assay designed to monitor the activity of the *A. fumigatus* cell wall integrity (CWI) pathway. Disruption of the CWI pathway in *A. fumigatus* severely impairs growth and virulence, making it a key focus for antifungal target discovery. The CWI pathway is a conserved signaling cascade that senses various stress sources and, importantly, responds to cell wall stress, orchestrating the expression of genes critical for maintaining cell wall structural integrity while also activating genes related to other functions, such as the production of secondary metabolites [11–18]. In brief, upon activation, the signal is relayed by the apical Protein Kinase C (PkcA) to a three-component MAPK cascade comprised of the proteins Bck1 (a MAP kinase kinase kinase), Mkk2 (a MAP kinase kinase), and MpkA (a MAP kinase) (*see* Fig. 1). The most well-characterized downstream CWI-pathway target is the transcription factor RlmA. *A. fumigatus rlmA* genetically and physically interacts with *mpkA*, which is required for RlmA phosphorylation, resulting in the transcriptional activation of RlmA-dependent genes related to CWI. RlmA is also required for full virulence and for *A. fumigatus* to achieve maximal fungal burden in the lung [13, 19, 20].

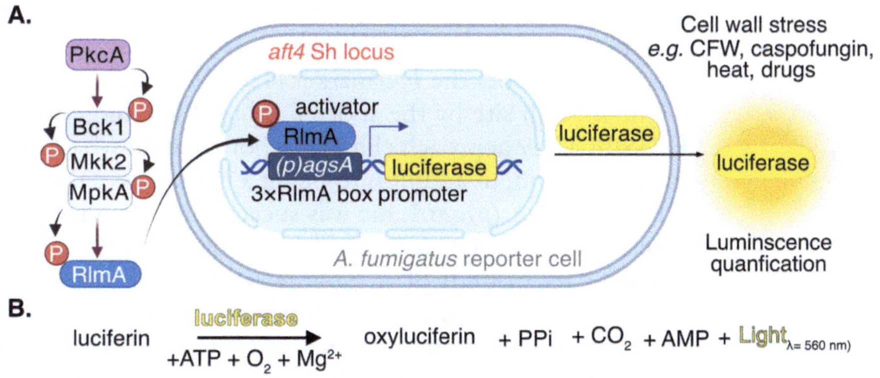

Fig. 1 (**a**) The activity of the CWI (CWI) pathway is initiated by the function of the apical kinase PkcA, followed by the downstream MAPK cascade in the cytosol, ultimately enabling the activity of RlmA in the nucleus. The luciferase gene is placed under the control of the *A. niger agsA* (α-glucan synthase) gene promoter, which contains three RlmA-binding motif recognition sequences (3× RlmA boxes) to maximize RlmA biding. (Created in BioRender. https:/BioRender.com/6k61u03). (**b**) Firefly luciferase from *Photinus pyralis* catalyzes the oxidation of the luciferin substrate to oxyluciferin, producing light in an oxygen and ATP-dependent reaction that also requires Mg^{2+} as a cofactor. The 560 nm chemiluminescence from this reaction develops over time, with the light output being proportional to the luciferase concentration when the substrates are present in excess [28]. Normalized luminescence is the readout of the experiment and is compared across different experimental conditions

Both in *A. niger* and *A. fumigatus*, the α-glucan synthase gene A (a*gsA*) was described as a transcriptional target of RlmA [7, 13, 21]. Additionally, the *agsA* gene serves as an excellent and fungus-specific marker for monitoring cell surface integrity, as its product, the (1,3)-α-D-glucan synthase, is absent in bacteria, plants, mammals, and most yeasts, with the exception of *Schizosaccharomyces pombe* [22]. Therefore, our biosensor strategy leverages the ability of the *A. fumigatus* RlmA, transcribed from the unchanged endogenous locus, to bind and activate the conserved promoter of *A. niger agsA,* which conveniently drives the transcription of the functional luciferase (luc) gene at the *aft4* Sh site (*see* Fig. 1). Notably, to enhance the luminescent signal detection, the *A. niger agsA* promoter was modified to contain a synthetic promoter segment containing multiple RlmA-binding sites (3×RlmA boxes; *see* Fig. 2). The complete circuit referred to as *(p)agsA*::luc also includes the *A. nidulans trpC* terminator. As a result, the luminescence signals are measured to indicate the amplitude of the activation of the endogenous CWI pathway in response to user-defined stimuli.

This system was employed to investigate the function of *A. fumigatus* RlmA as the downstream transcription factor of the CWI pathway in this fungus [13]. The original backbone vector was designed to contain approximately 1 kb of 5′ and 3′ flanking regions of the *A. fumigatus pyrG* gene to allow targeted genomic integration at the endogenous *pyrG* locus of the recipient *pyrG*-deficient strain [13]. Although this integration ensures stable, single-copy insertion and eliminates variation arising from ectopic insertions or copy number effects, it is still integrated in the *pyrG* locus and harbors the *hph* marker ($HygR^+$). Here, the substitution of the *aft4* Sh site by the shorter *(p)agsA*::luc biosensor using the CRISPR-Cas9 approach eliminates potential drawbacks of the *pyrG* locus insertion, in addition to being a markerless DNA cassette. In our hands, the *(p)agsA*::luc was successfully introduced into wild-type strains CEA10, Δ$akuB^{KU80}$, A1160, Δ*rlmA*, and $pkcA^{G579R}$ mutant backgrounds of *A. fumigatus* [13, 14], allowing for functional comparative analysis of RlmA-dependent expression.

This protocol offers several advantages. First, it serves as a functional proxy for RlmA dynamics. Notably, the use of the 3× RlmA box synthetic promoter adds a level of mechanistic precision, allowing for the dissection of the specific contributions of the CWI pathway to gene expression. This is especially relevant in the study of stress-responsive pathways, where rapid and transient transcriptional changes are expected. Second, it enables the screening of genetic or chemical modulators of the CWI pathway under different experimentally relevant conditions. It is especially amenable to high-throughput experimental designs using 96-well microplates, which facilitate the inclusion of numerous controls and technical replicates on a single plate. Potential uses of this system include

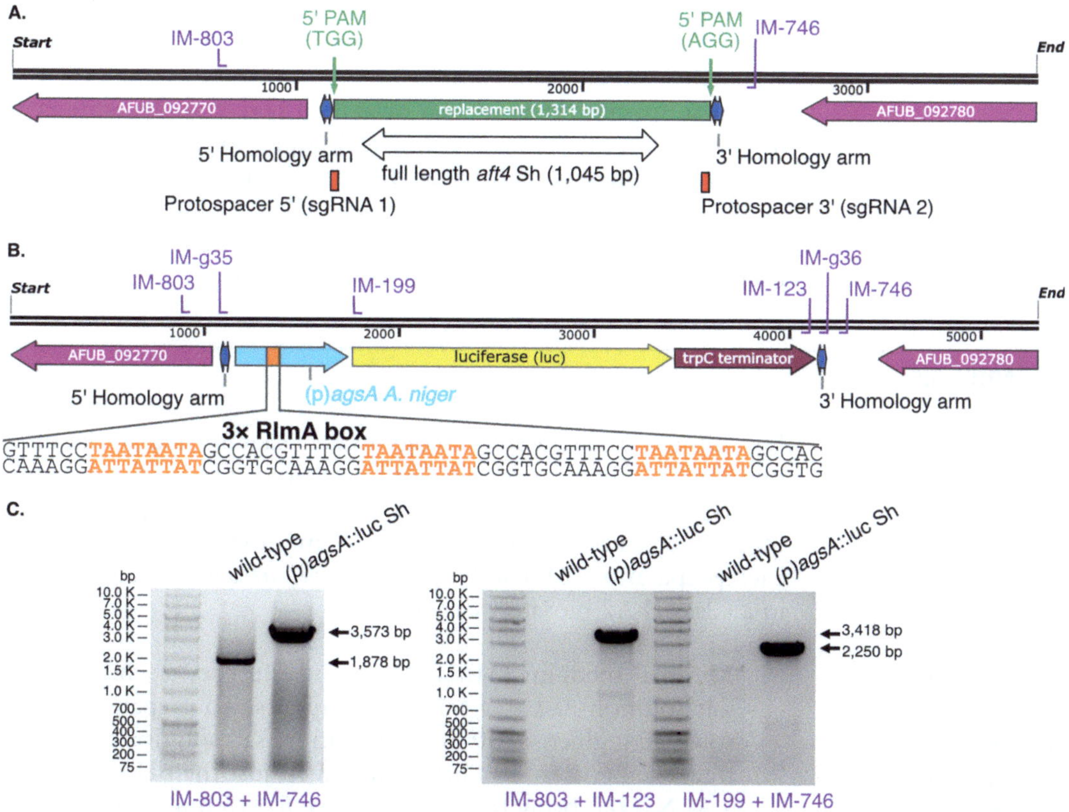

Fig. 2 CRISPR-Cas9-mediated replacement of the *aft4* Safe Haven (Sh) locus by the *(p)agsA*::luc reporter cassette in *Aspergillus fumigatus*. (**a**) Schematic representation of the native *aft4* locus in the *A. fumigatus* genome, highlighting the location of the sgRNA protospacers (red boxes) used for CRISPR-Cas9 cleavage, PAM sequences juxtaposed to the 5′ and 3′ homology arms (blue boxes), and primers used for validation (IM-803 and IM-746 flank the integration site). The full-length *aft4* region (1314 bp) is replaced by a 3012 bp markerless *(p)agsA*::luc cassette. (**b**) Structural organization of the final integrated reporter construct at the *aft4* Sh site, showing the biosensor components: 5′ and 3′ homology arms, the synthetic *agsA* promoter carrying three tandem RlmA-binding motifs (3× RlmA box, highlighted below in orange), the luc gene, and the *trpC* terminator. Primer binding sites for diagnostic PCR validation (IM-803, IM-199, IM-123, and IM-746) are indicated. (**c**) Diagnostic PCR results confirming the successful integration of the *(p)agsA*::luc construct at the *aft4* Sh locus. Gel images show expected band sizes for wild-type versus *aft4* replaced strains using the indicated primer pairs: IM-803 + IM-746 (3573 bp band in the *aft4* replaced transformants and 1878 bp in the recipient wild-type strain); IM-803 + IM-123 (3418 bp, specific for positive *aft4* replaced transformants); IM-199 + IM-746 (2250 bp specific for positive *aft4* replaced transformants)

approaches to identify novel cell wall-targeting antifungal molecules that operate directly or indirectly through the activity of the apical kinase PkcA, the MAP kinase MpkA, and the transcription factor RlmA. Combined with biochemical approaches to detect fluctuations in cell wall carbohydrate content, this method can be used to monitor potential compensatory accumulation of sugars in response to environmental factors, antifungals, host interactions, or immune cell-derived stimuli. Third, the system can be adapted to other genes or regulatory elements by replacing the promoter region upstream of the luciferase gene, thereby extending its utility across diverse regulatory studies.

In summary, the approach described here complements existing transcriptomic and proteomic techniques for functional studies of CWI pathway activity in *A. fumigatus*. This approach can be employed to generate bioluminescent CWI pathway reporters in laboratory strains and different mutant backgrounds derived from them, as well as in clinical isolate backgrounds.

2 Materials

Use ultrapure, molecular biology-grade reagents and sterile ultrapure water (minimum 18 MΩ/cm measured at room temperature) to prepare all solutions, culture media, and DNA preparations. Storage conditions and sterilization methods are indicated for each preparation. All handling of *A. fumigatus* must be performed in a Class II A2 biosafety cabinet.

2.1 Media and Solutions for A. fumigatus Cultivation and Protoplast Regeneration

1. Minimal medium (MM): 1% (w/v) glucose, 1× high-nitrate salt solution, 0.1% (v/v) trace elements. To prepare 1 L: dissolve 10 g glucose, 50 mL high-nitrate salt solution, and 1 mL trace elements in 500 mL of water. Adjust the pH to 6.5 with 4 M NaOH, bring to 1 L. Autoclave for 15 min, store at room temperature. Supplement with 1.2 mg/L uridine and uracil for *pyrG*-deficient strains (*see* **Note 1**).
2. Solid MM: Add 2% (w/v) bacteriological agar directly to the MM before autoclaving. Supplement with 1.2 mg/L uridine and uracil for *pyrG*-deficient strains.
3. 2× Minimal medium (2 × MM): 2% (w/v) glucose, 2× high-nitrate salt solution, 0.2% (v/v) trace elements, 0.006% yeast extract. Dissolve 60 mg yeast extract, 20 g glucose, 100 mL high-nitrate salt solution, and 2 mL trace elements in 500 mL of water. Adjust pH to 6.5 with 4 M NaOH, bring to 1 L, and autoclave.
4. Stabilized minimal medium (SMM): Add 218.6 g/L D-sorbitol (1.2 M final concentration) to solid MM. Autoclave for 15 min, store at room temperature. For hygromycin B selection (*hph*; Hyg$^+$), supplement with 250 μg/mL hygromycin (*see* **Note 2**).
5. To prepare agar plates, pour molten solid MM or SMM (with or without antibiotic) into sterile 90 × 15 mm Petri dishes and allow to solidify at room temperature. Store plates at 4 °C.
6. Stabilized Minimal Medium Top agar (SMM top): Add 218.6 g/L D-sorbitol (1.2 M final concentration) and 0.7%

agar (w/v) to MM to prepare a semisolid medium (*see* **Note 3**). Autoclave for 15 min, store at room temperature. For hygromycin B selection, supplement with 250 μg/mL hygromycin.

7. High-nitrate salt solution (20×): Dissolve 120 g $NaNO_3$, 10.4 g KCl, 10.4 g $MgSO_4 \cdot 7H_2O$, and 30.4 g KH_2PO_4 in 800 mL of water, adjust the volume to 1 L. Store at 4 °C.
8. Trace elements solution (1000×): 22 g $ZnSO_4 \cdot 7H_2O$, 11 g H_3BO_3, 5 g $MnCl_2 \cdot 4H_2O$, 5 g $FeSO_4 \cdot 7H_2O$, 1.6 g $CuSO_4 \cdot 5H_2O$, 1.6 g $CoCl_2 \cdot 6H_2O$, 1.1 g $(NH4)_6$-$Mo_7O_{24} \cdot 4H_2O$, 50 g EDTA. Dissolve each component in the indicated order in 800 mL of water, ensuring each is dissolved before adding the next. Boil the solution, cool until 60 °C. Adjust pH to 6.5–6.8 using KOH pellets. Cool to room temperature and bring to 1 L. Store in a dark bottle at room temperature. Shake before use due to precipitation.
9. Tween-80 solution (0.01%): Carefully pipette 100 μL of Tween-80 into 1 L of water. Autoclave for 15 min, store at room temperature in aliquots (*see* **Note 4**).

2.2 Solutions for Protoplast Preparation and Transformation

A. fumigatus protoplasts generated in this process are sensitive to detergents. All plastics and glassware used to prepare and manipulate transformation solutions must be clean and thoroughly rinsed to remove traces of detergent.

1. $MgSO_4 \cdot 7H_2O$ (1 M): Dissolve 123.2 g of $MgSO_4 \cdot 7H_2O$ in 300 mL water, bring to 500 mL. Autoclave for 15 min and store at 4–8 °C.
2. Tris-HCl (100 mM): Dissolve 6.6 g of Tris base in 300 mL of water. After the complete dissolution of Tris, adjust pH to 7.5 using fuming HCl (*see* **Note 5**) and bring to 500 mL. Autoclave for 15 min, store at room temperature.
3. Protoplast solution 1: 0.8 M $(NH_4)_2SO_4$, 100 mM citric acid, pH 6.0. Dissolve 58.9 g $(NH_4)_2SO_4$ and 9.61 g anhydrous citric acid in 300 mL of water. Adjust the pH to 6.0 with 1 M KOH and bring to 500 mL. Autoclave for 15 min, store at 2–8 °C.
4. Protoplast solution 2: 1% (w/v) yeast extract, 2% (w/v) sucrose. Dissolve 5 g yeast extract and 10 g sucrose in 300 mL of water. Adjust the volume to 500 mL. Autoclave for 15 min, store at 4–8 °C.
5. Lytic solution: In a sterile 50 mL conical tube, combine 20 mL protoplasts solution 1, 20 mL protoplasts solution 2, 6.5 mL of 1 M $MgSO_4$, 400 mg of BSA (e.g., Sigma-Aldrich #B4287), and 600 mg of Lallzyme MMX (Lallemand) (*see* **Note 6**). Mix the components until they dissolve completely. Sterile-filter using 0.22 μm low-protein-binding filter and a sterile syringe (*see* **Note 7**).

6. Protoplast solution 3: 0.4 M $(NH_4)_2SO_4$, 1% sucrose, 50 mM citric acid. Dissolve 26.5 g $(NH_4)_2SO_4$, 5 g sucrose, and 4.8 g anhydrous citric acid in 300 mL of water. Adjust pH to 6.0 with 1 M KOH, bring to 500 mL. Autoclave for 15 min, store at 4–8 °C.
7. Protoplast solution 4: 25% (w/v) PEG 6000 (average molecular weight, e.g., Sigma-Aldrich, #81260), 100 mM $CaCl_2$, 0.6 M KCl, 10 mM Tris-HCl, pH 7.5. Dissolve 25 g PEG 6000, 1.1 g anhydrous $CaCl_2$, and 4.5 g KCl in 50 mL water. Add 10 mL Tris-HCl 100 mM, pH 7.5, bring to 100 mL. Autoclave for 15 min, store at 2–8 °C (*see* **Note 8**).
8. Protoplast solution 5: 0.6 M KCl, 50 mM $CaCl_2$, 10 mM MES, pH 6.0. Dissolve 4.5 g KCl, 0.55 g anhydrous $CaCl_2$, and 2.13 g MES in 50 mL of water. Adjust pH to 6.0 with 1 M KOH, bring to 100 mL. Sterile-filter, store at 4–8 °C.

2.3 Reagents and Stock Solutions

1. EDTA (0.5 M): Dissolve 93.1 g EDTA (disodium dihydrate $Na_2EDTA{\cdot}2H_2O$) in 250 mL water, adjust pH to 8.0 using NaOH pellets, and bring to 500 mL (*see* **Note 9**). Store at room temperature.
2. Conidia lysis buffer: 10 mM Tris-HCl, pH 8.0, 100 mM NaCl, 1 mM EDTA, pH 8.0, 2% Triton X-100, 1% SDS. Dissolve 0.605 g Tris base and adjust pH to 8.0 using fuming HCl. Subsequently, add 2.92 g NaCl, 1 mL of EDTA (0.5 M), 5 g SDS, and 10 mL Triton X-100 (*see* **Note 10**). Bring to 500 mL with DNase/RNAse-free water. Autoclave for 15 min, store at room temperature.
3. Luciferin solution (25 mM): Weigh 12 mg of luciferin (e.g., Promega #E1602 or GoldBio #LUCK-100), and dissolve in 1.5 mL of ultrapure water (without Mg^{2+} and Ca^{2+}). Sterile-filter using a 0.22 μm low-protein-binding filter and a sterile syringe. Separate into small aliquots and store at −80 °C in the dark. The luciferase substrate, D-luciferin, is sensitive to light, oxygen, and moisture (*see* **Notes 11–15**).
4. Stock calcofluor white solution (CFW, 10 mg/mL): Dissolve 10 mg CFW (e.g., Fluorescent Brightener 28, Sigma-Aldrich Cas No #4193-55-9) in 1 mL of water. Prepare a 1 mg/mL working solution by 1:10 dilution. Store both stock and working solutions at −20 °C in aliquots, protected from light.

2.4 Reagents for PCR and Vector Cloning

1. pFC902 template plasmid (*see* **Note 16**).
2. Phusion HF PCR buffer (5×).
3. dNTP mix (20 mM).
4. Primers (20 pmol/μL stock, *see* Tables 1, 2, and 3).

Table 1
Primers used for cloning the sgRNA scaffold into the CRISPR Cas9 vector pFC332 and sequencing of the protospacer (PS)

Primer	Sequence (5′ → 3′)	Purpose	Template/ Size (bp)
IM-g39	atggtcatagctgtttccgctgagggtttaatgggtttaatGATCACATA GATGCTCGGTTGACA	sgRNA Scaffold 5′ PS (FW)	pFC902/ 600 (Fragment 1)
IM-g 5′ PS-Sh REV	**CTGTACAAGTTTTAAATACA** TGCATCATCCGTGAATCGAAC	sgRNA Scaffold 5′ PS (REV)	
IM-g 5′ PS-Sh FW	**TGTATTTAAAACTTGTACAG** GTTTTAGAGCTAGAAATAGCAAGTTAAA	sgRNA Scaffold 5′ and 3′ PS (FW)	pFC902/ 191 (Fragment 2)
IM-g 3′ PS-Sh REV	<u>**AGTTACATTGCGGCTTAGTC**</u> TGCATCATCCGTGAATCGAAC	sgRNA Scaffold 5′ and 3′ PS (REV)	
IM-g 3′ PS-Sh FW	<u>**GACTAAGCCGCAATGTAACT**</u> GTTTTAGAGCTAGAAATAGCAAGTTAAA	sgRNA Scaffold 3′ PS (FW)	pFC902/ 367 (Fragment 3)
IM-g40	ggtgattctgctgtctcggctgaggtcttaatGGTCTTAATACCCT GAGAAGATAGATGTGAATGTG	sgRNA Scaffold 3′ PS (REV)	
IM-617	GATCACATAGATGCTCGGTTGACA	Sequencing of the PS	Assembled pFC332
IM-618	ACCCTGAGAAGATAGATGTGAATGTG	Sequencing of the PS	Assembled pFC332

Lowercase letters indicate homology sequences with the pFC332 vector for NEBuilder High-Fi DNA Assembly. Bold letters indicate the 5′ protospacer sequence inserted as 5′ overhangs in each primer. Underlined bold letters indicate the 3′ protospacer sequence inserted as 5′ overhangs in each primer. *See* Figs. 2 and 3 for primer locations

Table 2
Primers used to amplify the DNA repair fragment *(p)agsA*::luc cassette (3012 bp) from the pNB04 plasmid

Primer	Sequence (5′ → 3′)
IM-g35	gcaggagcaaaaacaggccgggaagatattgcctaggagttctgcgctggCTCTAGAACTAGTGGATCCCC
IM-g36	ctgacagttcatggaatagtagaatggtatgatacattacgtaaccgaccCTAGAAAGAAGGATTACCTCTAAAC

Lowercase letters indicate the 50 bases that comprise the 5′ homology arm (IM-g35) and 3′ homology arm (IM-g36) to the Safe Haven site. *See* Fig. 2 for primer locations

Table 3
Primers used to confirm the replacement of the *aft4* Safe Haven site with the *(p)agsA*::luc cassette

Primer	Sequence (5′ → 3′)
IM-803	GGGTCTCCAACATGTCATTAGC
IM-746	TCCGTTCCATTACGCCTTCC
IM-199	TTAAACACCATGGTCACCG
IM-123	CTAGAAAGAAGGATTACCTCTAAAC

For the expected band sizes, *see* Fig. 2

5. Phusion High Fidelity DNA Polymerase (e.g., Thermo Scientific, #F530).
6. PacI digested CRISPR vector pFC332 (*see* **Note 17**).
7. Agarose-gel purified independent sgRNA PCR inserts.
8. 2× NEBuilder HiFi DNA Assembly Master Mix (e.g., New England Biolabs #E2621).
9. Electrocompetent *E. coli* cells (*see* **Note 18**).
10. LB agar plates supplemented with 100 μg/mL of ampicillin.
11. Phenol:chloroform mix (1:1).
12. Isopropanol.
13. Ethanol, 70%.

2.5 Equipment and Consumables

1. Multimode microplate reader capable of top-read luminescence and optional OD_{600} absorbance detection (e.g., SpectraMax from Molecular Devices, Varioskan from Thermo Scientific, or Spark 10 M from Tecan).
2. White, 96-well microplate with flat, solid bottom and chimney wells for top-read luminescence detection (e.g., Greiner Bio-One #655083). Optional: white, 96-well microplate with flat μClear bottom and chimney wells for simultaneous luminescence and OD600 absorbance detection (e.g., Greiner Bio-One #655095).
3. Sterile filter material for filtration of fungal mass (e.g., Miracloth).
4. Sterile glass wool for filtration of protoplast suspensions.
5. Neubauer chamber (hemocytometer).
6. Microscope.
7. Agarose, molecular biology grade.
8. Gel electrophoresis unit and imaging/documentation equipment.

9. Microvolume spectrophotometers (e.g., NanoDrop UV-Vis from Thermo Fisher Scientific).
10. Orbital shaker incubator.
11. 2 mm acid-washed glass beads.
12. Vortex mixer.
13. Multichannel pipette.

3 Methods

3.1 Conidia Harvesting

1. Inoculate *A. fumigatus* on solid MM agar plates and incubate at 37 °C for 72–96 h.
2. Add 15 mL of 0.01% Tween-80 solution to the plate, and gently dislodge the conidia using a sterile Pasteur pipette or cell scraper.
3. Aspirate the dark green conidial suspension and filter it through sterile filter material into a sterile 50 mL conical tube to remove mycelial debris.
4. Count conidia and adjust the concentration to 3×10^5 conidia/mL for luciferase activity assays.

3.2 CRISPR-Cas9 Plasmid Construction: Cloning the Protospacer for aft4 Safe Haven (Sh) Site Recognition

This method describes the insertion of a cassette containing the *(p)agsA*::luc construct (*see* Fig. 1) into the *A. fumigatus aft4* Safe Haven (Sh) site [10] using the CRISPR-Cas9 methodology. Targeted substitution of the Sh site is achieved by providing a target-specific DNA repair fragment and selecting transformants using hygromycin selection and pFC332 plasmid [23–25]. This CRISPR system has been successfully applied in *A. nidulans*, *A. fumigatus* (Campanella et al., unpublished results), and *A. niger* [26]. To replace the *aft4* Sh site, specific 5′ and 3′ sgRNAs containing the 20 bp protospacer are designed and cloned into pFC332. This plasmid carries the *hph* marker gene for hygromycin resistance, the *AMA1* autonomous replication sequence from *Aspergillus*, which confers episomal localization of the plasmid, and a codon-optimized Cas9 nuclease sequence from *Streptococcus pyogenes* fused to a nuclear localization signal (NLS). The sgRNA components are PCR-amplified from the plasmid pFC902 [25]. The plasmid pFC332 is then assembled with the sgRNA components to generate pFC332-*aft4* Sh, which is used to transform *A. fumigatus*. We constructed this plasmid using the NEBuilder HiFi DNA Assembly Master Mix instead of the original USER cloning approach [23, 24] (*see* **Note 19**).

3.2.1 Preparation of pFC332 for DNA Assembly

1. Digest 10 μg of the purified pFC332 plasmid (empty vector, 250 ng/μL) in a 40 μL reaction volume with 10 U PacI at 37 °C for 18–20 h (*see* **Note 20**).

2. Add 5 U PacI and incubate at 37 °C for 3 h to ensure complete digestion.
3. Heat-inactivate at 80 °C for 20 min.
4. Confirm plasmid digestion by agarose gel electrophoresis (expected band size 15,560 bp). When using NEBuilder HiFi DNA Assembly Master Mix, purification of the restriction endonuclease-digested vector is not necessary unless the same restriction site is present in the insert DNA. In such cases, purify the linearized vector and quantify it using a microvolume spectrophotometer. Purification of the plasmid may also be required if the concentration of the plasmid is necessary.

3.2.2 Generation of Three DNA Insert Fragments Containing the 5′ and 3′ sgRNAs

The sgRNA expression cassette targets the flanking region of the *aft4* Sh site and therefore contains 20 bp target 5′ and 3′ protospacers (*see* red boxes in Figs. 2 and 3). Construction of the pFC332-*aft4* Sh CRISPR vector requires three PCR reactions using a set of primers (*see* Table 1). For these reactions, the pFC902 plasmid serves as a template, as it contains all necessary elements to generate the 5′ and 3′ sgRNAs [25]. These elements include the *A. fumigatus* U3 promoter that drives transcription of *A. nidulans* glycine tRNA fused to the 5′ and 3′ protospacers and the gRNA backbone, followed by the *A. fumigatus* U3 terminator. The three PCR fragments are as follows:

Fragment 1 (PCR 1) contains the U3 promoter, the first glycine tRNA, and the 5′ protospacer (20 nucleotides) added as a 5′ overhang in the IM-g5′-Sh REV primer. The IM-g39 primer includes a 41-bp overlap with the *Pac*I-digested pFC332.

Fragment 2 (PCR 2) contains the gRNA backbone and a second glycine tRNA. Both primers IM-g 5′ PS-Sh FW and IM-g 3′ PS-Sh REV contain overlapping sequences for the 5′ and 3′ protospacers (red boxes in Fig. 3).

Fragment 3 (PCR 3) contains the gRNA backbone followed by the U3 terminator. Primer IM-g 3′ PS-Sh FW includes a 3′ protospacer overhang, and IM-g40 contains a 41-base overlap with *Pac*I-digested pFC332.

1. To amplify each fragment, prepare a 50 μL reaction containing 50–80 ng of pFC902, 10 μL of 5× Phusion HF buffer, 0.5 μL of dNTP mix, 1 μL of each standard desalted primer (*see* Table 1), and 0.5 μL of Phusion High Fidelity DNA polymerase. Adjust the volume to 50 μL with nuclease-free water.
2. Perform PCR under the following cycling conditions: 98 °C for 1 min, 32 cycles of 98 °C for 5 s, 60 °C for 10 s, 72 °C for 15 s. End the reaction by incubating at 72 °C for 3 min and then cool to 8 °C.

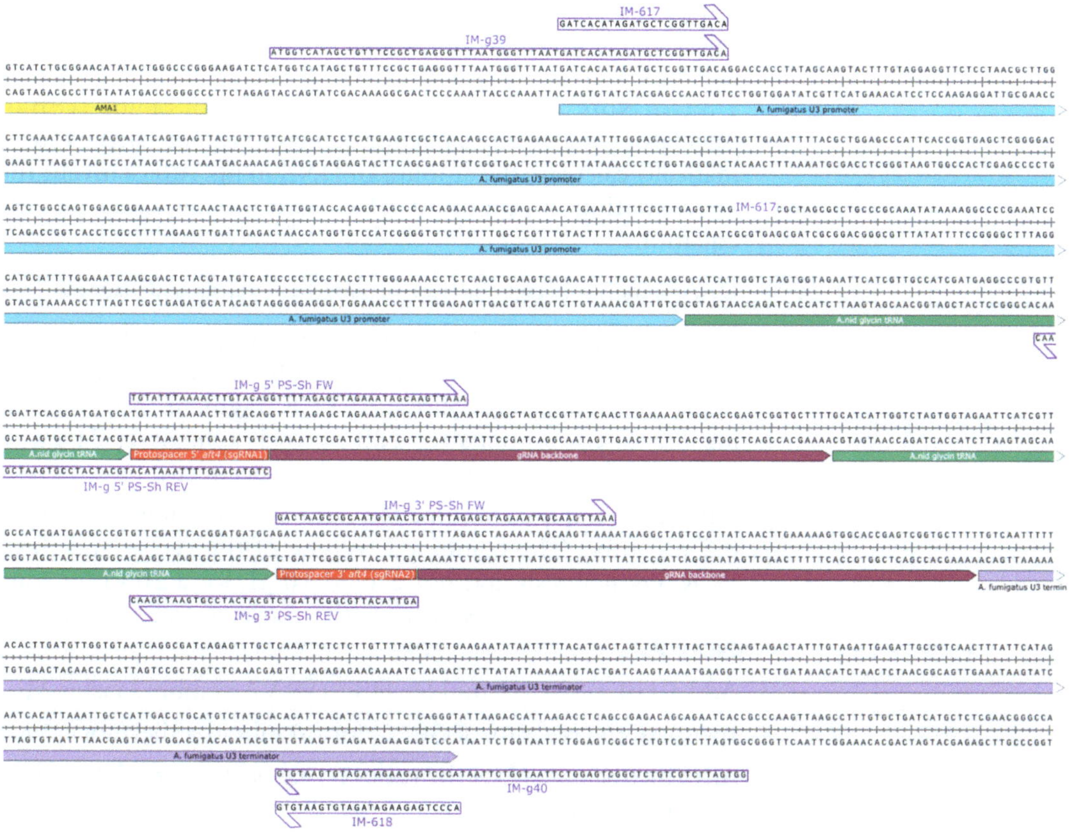

Fig. 3 Organization of the PCR fragments and structural elements that comprise the sgRNA after assembly into the pFC332 plasmid. Scheme of the PCR strategy to generate the three PCR amplicons that will be fused using NEBuilder HiFi DNA Assembly into the CRISPR plasmid pFC332. Fragment 1 is amplified using primers IM-g39 + IM-g5′-Sh REV (600 bp). Fragment 2 is amplified using primers IM-g 5′ PS-Sh FW and IM-g 3′ PS-Sh REV (191 bp). Fragment 3 is amplified using primers IM-g 3′ PS-Sh FW and IM-g40 (367 bp). All three fragments are amplified using the plasmid pFC902 (not shown). 5′ and 3′ Protospacers (red boxes) targeting the aft4 Sh site are introduced as 5′ overlapping sequences within the primers IM-g 5′ PS-Sh FW and REV and IM-g 3′ PS-Sh FW and REV. *See* the main text for further details. Primers IM-617 and IM-618 are used to sequence the sgRNA insert to confirm the presence of the expected sgRNA sequences prior to transformation and CRISPR-Cas9-mediated genome editing

3. Gel-purify PCR fragments using silica-based DNA spin column protocol. Expected product sizes: Fragment 1 (600 bp), Fragment 2 (191 bp), and Fragment 3 (367 bp).
4. Quantify each purified fragment using a microvolume spectrophotometer.

3.2.3 Cloning sgRNA Fragments Into the pFC332 Using NEBuilder HiFi DNA Assembly

We use NEBuilder HiFi DNA Assembly Master Mix to construct the CRISPR vectors. Other seamless cloning kits (e.g., GeneArt, In-Fusion, Gibson Assembly) are also suitable alternatives.

1. Quantify the vector (plasmid) and insert preparations in ng to calculate the required DNA input in pmol to be added to the reaction mixture. Consider a vector:insert DNA molar ratio of 1:3 (*see* **Note 21**) and 0.2 pmol of DNA. An online NEBio-calculator tool (nebiocalculator.neb.com) can be used to automatically calculate the amount of DNA in ng. Alternatively, the concentration in pmol of DNA can be obtained by the equation:

$$\text{DNA (pmol)} = (\text{weight in ng}) \times 1000/(\text{base pairs} \times 650 \text{ daltons}).$$

   In this protocol, 0.2 pmol of *Pac*I-digested pFC332 (*see* Subheading 3.2.1) corresponds to 2000 ng. For a 1:3 vector: insert reaction, 0.6 pmol of each PCR fragment (*see* Subheading 3.2.2) corresponds to: 232 ng of Fragment 1; 74 ng of Fragment 2; 142 ng of Fragment 3, respectively.

2. Pipette the required volume of vector and PCR fragment into a 0.2 mL tube. The combined volume of the vector and three PCR fragments should not exceed 10 μL for a 20 μL of total assembly reaction. Adjust with nuclease-free water (*see* **Note 22**).
3. Add 10 μL of 2× NEBuilder HiFi DNA Assembly Master Mix.
4. Incubate the reaction at 50 °C for 60 min in a thermocycler.
5. After incubation, place the tubes on ice or store at −20 °C until transformation.
6. Transform 1 μL of the assembly mix into electrocompetent *E. coli* cells (*see* **Note 23**).
7. Plate cells on LB agar plates supplemented with 100 μg/mL of ampicillin and incubate overnight at 37 °C.
9. Purify plasmids from selected clones based on comparison to a negative control.
8. Analyze clones by restriction digestion. For example, the *Bsp*EI restriction digestion can discriminate the correct insertion of fragments into pFC332 [23]. Primers IM-617 and IM-618 can also be used to investigate the presence of the insert.
10. Before using the pFC332-*aft4* Sh plasmid in *A. fumigatus* transformation and CRISPR-Cas9 reaction, confirm the sequence of the inserted sgRNAs. Use primers IM-617 and IM-618, which anneal to the U3 promoter and U3 terminator, respectively (*see* Fig. 3).

3.2.4 Amplification of the Target-Specific DNA Repair Fragment

To replace the *aft4* Sh site, the site-specific DNA repair fragment is generated by PCR amplification of the functional *(p)agsA*::luc cassette from plasmid pNB04 (*see* Fig. 2b) using primers IM-g35 and IM-g36 (*see* Table 2), (*see* **Note 24**). This PCR yields a 3012 bp marker-free DNA cassette for insertion at the *aft4* Sh site. Primers

IM-g35 and IM-g36 have 5′ tail sequences of 50 bases corresponding to the 5′ and 3′ homology arms, respectively, shown as blue boxes in Fig. 2a, b. These homology arms are located next to the 5′ and 3′ protospacer sequences at the target site, enabling homologous recombination after the ribonucleoprotein complex (composed of the sgRNA and Cas9 nuclease) induces a DNA double-strand break at the target locus. This cleavage occurs only in the presence of a protospacer adjacent motif (PAM), specifically the NGG sequence required for Cas9 recognition and cutting.

1. For amplification of the DNA repair fragment, set up a 50 μL reaction: Add 50–80 ng of pNB04, 10 μL 5× Phusion HF buffer, 0.5 μL of dNTP mix, 1 μL of each standard desalted primer (IM-g35 and IM-g36, *see* Table 2), and 0.5 μL of Phusion High Fidelity DNA polymerase. Adjust the volume to 50 μL with nuclease-free water. Use the cycling conditions described in Subheading 3.2.2, **step 2**.
2. Gel-purify the amplified repair template and quantify using a microvolume spectrophotometer.

3.3 A. fumigatus Protoplast Transformation of the CRISPR-Cas9 Plasmid

In this protocol, *A. fumigatus* protoplasts are generated by enzymatic digestion using a combination of cell wall-degrading enzymes. After digestion, the protoplasts are harvested, purified, maintained in an osmotically stabilized solution and used in a PEG-mediated transformation with the CRISPR-Cas9 components described above. Transformants are selected based on their resistance to hygromycin.

1. From a fresh conidial suspension of the *A. fumigatus* recipient strain, inoculate 1 × 10^7 conidia in 50 mL MM in a 250 mL Erlenmeyer flask. Incubate with shaking (200 rpm) at 37 °C for 14–16 h.
2. Transfer the mycelia to a 50 mL conical tube and centrifuge at 3200× *g* for 5 min. Discard the supernatant.
3. Resuspend the mycelial pellet in lytic solution in a sterile 250 mL Erlenmeyer flask. Incubate at 30 °C with gentle shaking (100 rpm) for 5 h. Monitor protoplast formation microscopically (*see* **Note 25**). Protoplasts can be identified as round cells containing a large, visible vacuole, which is easily distinguished from germ tubes and undigested hyphae.
4. Filter the suspension through sterile glass wool using a sterile 10 mL syringe in a 50 mL conical tube to remove the undigested mycelium and recover the protoplasts.
5. Centrifuge at 3200× *g* for 10 min at 4 °C. Carefully discard the supernatant without disturbing the protoplast pellet. Keep the protoplast pellet on ice.

6. Resuspend the protoplast pellet in 20 mL of ice-cold protoplast solution 3 and centrifuge at 3200× *g* for 5 min at 4 °C. Discard the supernatant.
7. Repeat **step 6** to remove traces of the lytic solution.
8. Resuspend the protoplast pellet in ice-cold protoplast solution 5. The volume of this solution is adjusted according to the number of planned transformation reactions and the expected yield of protoplasts. Add 100 μL of solution 5 to each reaction. For example, for three reactions including CRISPR-Cas9, negative control, and positive control, use 300 μL (*see* **Note 26**).
9. Incubate the protoplast suspension on ice for 10 min.
10. For each planned reaction, add 50 μL of protoplast solution 4 and mix gently.
11. Transfer 150 μL of protoplast suspension to a sterile 1.5 mL tube. In the example above (**step 8**), prepare three tubes.
12. In the CRISPR-Cas9 reaction tube, add 5 μg of assembled pFC332-*aft4* Sh and 1–1.5 μg of *(p)agsA*::luc DNA repair template in a volume not exceeding 20 μL. For the positive control, add 5 μg of empty pFC332.
13. Incubate the tubes on ice for 20 min.
14. Add 1 mL of protoplast solution 4 to each tube and incubate at room temperature for 20 min.
15. Transfer each transformation reaction to a 15 mL sterile conical tube containing 13 mL of warm SMM Top agar supplemented with 250 μg/mL hygromycin. Mix gently by inversion.
16. Pour the SMM Top agar onto SMM agar plates supplemented with 250 μg/mL hygromycin for selection.
17. Incubate the plates at 37 °C for 3–5 days until discrete colonies appear. Allow the plates to remain upright in the incubator for 1 day before inverting them.
18. Select transformants and analyze CRISPR-Cas9 replacement of the *aft4* Sh site (*see* Subheading 3.4).

3.4 Validation of (p)agsA::luc Reporter Cassette Insertion at the A. fumigatus aft4 Sh Site

Since a CRISPR-Cas9-mediated DNA integration often yields high and specific insertion rates [23], replacement of the *aft4* Sh site can be validated by diagnostic PCR [10]. We routinely use three diagnostic PCRs for this purpose. The primers are listed in Table 3 and illustrated in Fig. 2.

Insertion of the *(p)agsA*::luc reporter cassette at the *aft4* Sh site does not alter colony morphology or pigmentation, so DNA can be reliably extracted from conidia for validation. For this protocol, regular vortexing with acid-washed glass beads is sufficient to disrupt conidia and does not require specialized bead beater equipment or zirconia beads.

1. Harvest conidia from discrete single colonies that grew on SMM agar plates containing hygromycin B (Hyg$^+$) by inserting a sterile toothpick in the center of the colony and transferring the material to a microcentrifuge tube containing 0.01% Tween-80 solution. Select approximately 10 colonies for validation.
2. Prepare tenfold serial dilutions and inoculate 100 μL of the diluted samples onto MM agar plates containing hygromycin B (Hyg$^+$). This step ensures maintenance of the pFC332 vector and facilitates selection of candidates.
3. Select a discrete colony that grew from a single conidium and streak it onto nonselective MM agar plates (without hygromycin) to grow the colony and conidiate (*see* **Note 27**).
4. To prepare genomic DNA, harvest conidia from the purified candidate plate (**step 3**) using 300 μL of Conidia lysis buffer.
5. Transfer 200 μL of the conidial suspension to microcentrifuge tube containing sterile 2 mm acid-treated glass beads. Fill the tube with beads up to the 0.5 mL mark.
6. Vortex the suspension for 30 s and incubate at 70 °C for 30 min in a thermomixer. Vortex for 30 s at 10-min intervals during incubation.
7. Add 200 μL of phenol:chloroform (1:1) to the lysate. Lay tubes horizontally on a vortex using a rack and vortex for 10 min.
8. Centrifuge tubes at maximum speed for phase separation.
9. Transfer approximately 180 μL of the upper aqueous phase to a new tube. Add 200 μL of chloroform to remove residual phenol.
10. Centrifuge tubes at maximum speed for phase separation.
11. Transfer approximately 150 μL of the upper phase to a new tube and add 500 μL of ice-cold isopropanol to precipitate DNA.
12. Incubate at −80 °C for 30 min.
13. Centrifuge for 10 min at maximum speed to pellet DNA.
14. Wash the DNA pellet with 200 μL of cold 70% ethanol. Carefully remove ethanol and air-dry to remove any traces of ethanol.
15. Resuspend the DNA pellet in 50–100 μL of nuclease-free water.
16. Assess DNA concentration and quality using a microvolume spectrophotometer and by electrophoresis on a 0.8% agarose gel (*see* **Note 28**).
17. Run diagnostic PCR with primers flanking the *aft4* Sh site (*see* Table 3 and Fig. 2), following the procedures and cycling

conditions described in Subheading 3.2.2, **steps 1** and **2**. To distinguish between wild-type and correctly edited candidates, use the following primer combinations: IM-803 + IM-746, which yields a 3573 bp amplicon in positive transformants (i.e., *(p)agsA*::luc cassette inserted) and a 1878 bp product in wild-type (negative) candidates. The additional primer pairs IM-803 + IM-123 and IM-199 + IM-746 are also specific for positive transformants, generating 3418 bp and 2550 bp amplicons, respectively. No amplification is expected in wild-type *aft4* Sh site candidates. Always include negative controls using DNA from the wild-type recipient strain. A representative validation profile obtained with these primer combinations is shown in Fig. 2c.

18. After validation, the CRISPR vector pFC332 (carrying the sgRNA, Cas9, and AMA1 sequence) should be removed from the validated strain, as its retention is not desirable. This vector is episomal and not integrated into the genome, so it is lost spontaneously upon growth on nonselective medium. To purge the vector, inoculate conidia from validated strains onto nonselective MM plates and incubate.
19. From the resulting colonies, transfer conidia from the colony edge to fresh MM agar plates. Test these new isolates for loss of hygromycin resistance by plating onto MM agar containing 250 μg/mL hygromycin. Most colonies will have lost the CRISPR vector and will thus be hygromycin-sensitive.

3.5 Plate Preparation for the Luciferase Activity of the (p) agsA*::Luciferase Reporter System*

1. In a sterile, white, 96-well microplate, use a multichannel pipette to dispense 50 μL of 2 × MM supplemented with 0.006% yeast extract into each well.
2. Add 50 μL of a conidial suspension (3 × 10^5 conidia/mL) to each well containing medium, resulting in a final volume of 100 μL per well. Mix gently by pipetting up and down. Perform each condition in at least six technical replicates.
3. Incubate the plate at 37 °C for 5 h in a humidified chamber to allow early conidia germination.
4. After incubation, add following components to each well bringing the final volume to 200 μL: 50 μL of 2 × MM supplemented with 0.006% yeast extract, 4 μL of freshly prepared 25 mM D-luciferin; 8 μL of the CFW working solution, a positive control of the CWI pathway activation. For untreated control wells, replace CFW with 8 μL of sterile water. Adjust volume by adding 38 μL of sterile distilled water.
5. Confirm the final concentrations in each well: 7.5 × 10^4 conidia/mL, 0.5 mM luciferin, and 40 μg/mL CFW in a total volume of 200 μL (*see* **Note 29**).

6. Immediately transfer the plate to a microplate reader pre-equilibrated to 30 °C, and initiate continuous, real-time luminescence measurements (optionally also OD_{600} readings) at 5-min intervals for up to 16 h, or as specified by the experimental design (*see* **Note 30**).

3.6 Data Analysis, Graph Generation, and Interpretation

Luminescence measurements should be performed using at least three independent biological replicates. Graphs are typically based on one representative biological replicate, calculated as the average of all technical replicates measured on the same plate. CFW induces a strong luminescence signal, and a dose-dependent response is observed when using concentrations ranging from 10 to 40 μg/mL (*see* Fig. 4a). The echinocandin caspofungin, a non-competitive inhibitor of β-(1,3)-glucan synthase, also induces luminescence in a dose-dependent manner, within the range of 0.125 to 1 μg/mL (*see* Fig. 4b). Luminescence is expressed in arbitrary units (AU) and plotted over time (in hours). Optional optical density (OD_{600}) readings may be recorded in parallel to normalize luminescence values for fungal growth (i.e., luminescence/OD ratio AU/OD_{600}). If OD_{600} measurements are acquired, calculate the AU/OD_{600} ratio at each time point to generate corrected expression curves.

4 Notes

1. Uracil is insoluble in the medium until heated. Therefore, add the necessary amount directly to the flask prior to autoclaving.
2. This concentration of hygromycin has been used successfully for most of the *A. fumigatus* wild-type recipient strains and some clinical isolates. Hygromycin is purchased as a 50 mg/mL stock in PBS. If using the CRISPR vector pFC330 that carries the functional *A. fumigatus pyrG* gene instead of the *hph* gene, select in MM lacking uridine and uracil.
3. The SMM top is a fluid medium used to spread the transformed protoplasts over a Petri dish containing SMM, thereby avoiding mechanical injury to the protoplasts.
4. Tween-80 is a nonionic, viscous liquid. Use a wide-bore pipette tip to pipette the required amount.
5. The addition of concentrated (fuming) HCl (12.178 M) can cause the solution to heat. Allow the solution to cool at room temperature before final pH adjustment. Avoid contact with the skin and mucosa. Wear gloves.
6. Instead of Lallzyme MMX, Glucanex 200G or VinoTaste Pro can be used as a source of lytic enzymes. The availability of these enzymes may depend on the commercially available

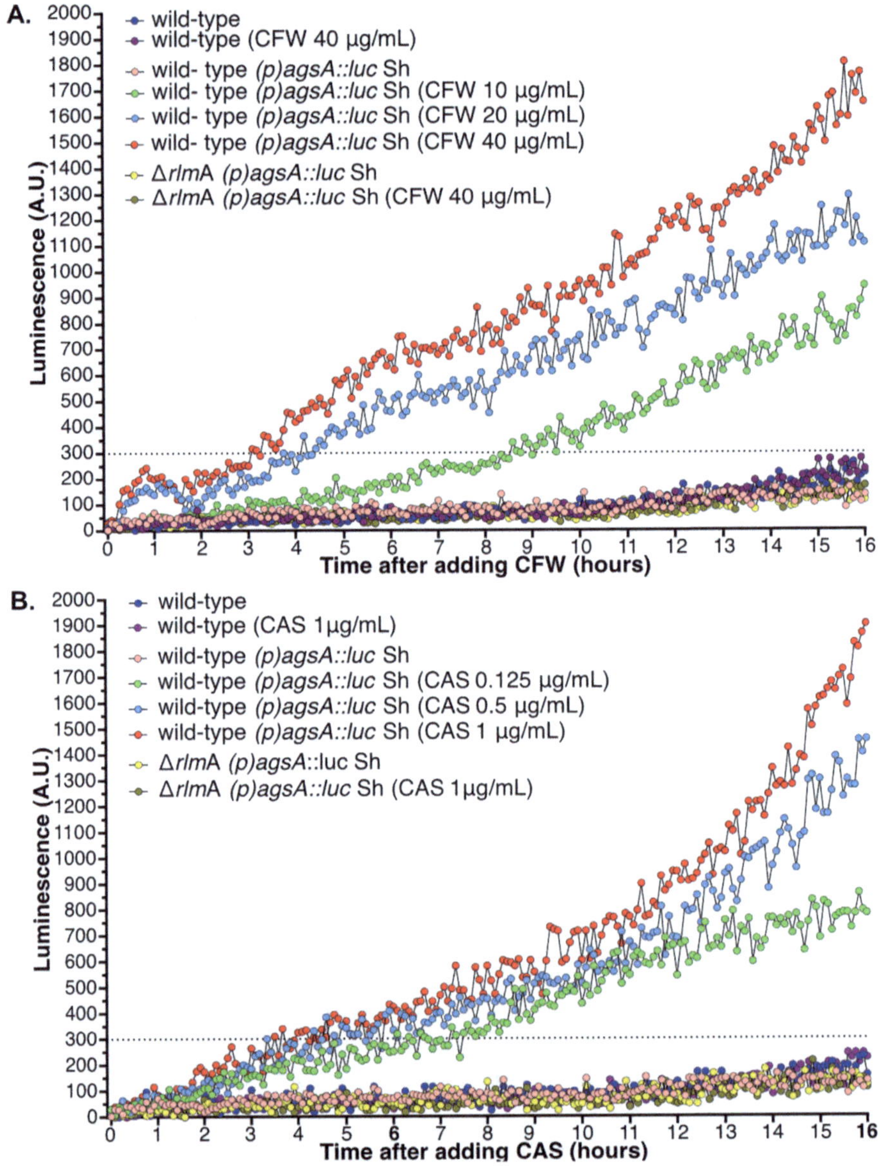

Fig. 4 Typical results for luminescence signals obtained from the wild-type strain transformed with the *(p)agsA*::luc Sh cassette. (**a**, **b**) Luminescence signals obtained from the wild-type strain transformed with the *(p)agsA*::luc Sh cassette, treated with 10–40 μg/mL calcofluor white (CFW) or 0.125–1 μg/mL of caspofungin (CAS). CFW and CAS are potent inducers of the CWI pathway and RlmA activity [21, 13, 29]. The Δ*rlmA* strain transformed with the *(p)agsA*::luc Sh cassette, and the untransformed wild-type strain is used as a negative control for CWI pathway activation

source in different countries. Optimization of the lytic cocktail concentration may be necessary, depending on the source available in the region.

7. Low-protein-binding filters will prevent the removal of the protoplast enzyme and BSA while maintaining aseptic conditions in the lytic solution.
8. This solution is viscous, and the dissolution of PEG 6000 will be slow and requires agitation.
9. The EDTA will not dissolve until the solution is sufficiently alkaline.
10. Wear masks to manipulate SDS. It is irritating to the mucosa and highly dispersible through the air. Triton X-100 is very viscous. Use a wide-bore pipette to pipette, avoiding overcarrying the liquid.
11. The potassium and sodium salts of luciferin are usually used interchangeably and are both commercially available. Sodium salt is more soluble in water (up to 100 mg/mL) than the potassium salt (55–60 mg/mL). Therefore, either substrate will be suitable if your protocol requires a stock solution of approximately 10–15 mg/mL. According to some vendors, D-luciferin potassium salt has slightly better aqueous stability than D-luciferin sodium salt. The free-acid form, although still available from some vendors, is rarely used due to its poor solubility. The free acid form of D-luciferin will not dissolve in water unless a dilute base, such as NaOH or KOH, is added to adjust the pH. However, it is soluble in methanol at a concentration of 10 mg/mL or in DMSO at a concentration of 50 mg/mL.
12. There are conflicting reports concerning the stability of dissolved luciferin, which may be caused by the amount of available oxygen in the water. Dissolving luciferin in degassed water for optimal storage results can also be considered an additional effort to preserve the stability of the solutions.
13. Monitor the activity of luciferase with control reactions to evaluate whether signal loss is occurring due to oxidation of luciferin during storage. One possibility is to use CFW or caspofungin as inducers, as shown in Fig. 4.
14. pH of the luciferin solution should be between 6.1 and 6.5. At a higher pH, luciferin undergoes a base-catalyzed formation of dehydroluciferin, as well as racemization to the L-isomer. Generally, luciferin solutions in reagent-grade water will fall within this pH range. If signal loss is occurring in your conditions, consider investigating the pH of the solution.
15. If luciferin is aliquoted, avoid repeated cycles of freezing and thawing of aliquots. Store luciferin powder desiccated at −20 °C in an airtight container, and consider using an inert gas, such as nitrogen or argon, to purge the container. For the most sensitive experiments, we recommend using freshly prepared luciferin to minimize variables.

16. pFC902 plasmid and complete sequence can be obtained from Addgene (Plasmid #106904) at https://www.addgene.org/106904/.
17. pFC332 plasmid and complete sequence can be obtained from Addgene (Plasmid #87845) at https://www.addgene.org/87845/.
18. In our hands, we achieve better results using electrocompetent bacterial cells, such as *E. coli* DH10B strain, rather than chemically competent cells. A protocol for preparing electrocompetent bacterial cells suitable for this procedure can be found in [27]. We also achieve better results using SOC medium for cell recovery after electroporation.
19. Our cloning protocol avoids the need for uracil-containing primers, which are more expensive than conventional desalted primers, and the use of a DNA polymerase that tolerates uracil in the primer. A detailed handbook protocol for the USER cloning procedures can be found in the ref. 23.
20. pFC330 can also be used and contains the *pyrG* as a selection marker instead of the *hph* gene.
21. The manufacturer recommends using up to a five-fold molar excess of any insert(s) less than 200 bp. 0.2 pmol is the recommended maximum amount of DNA.
22. Although the manufacturer recommends a final volume of 20 μL for the reaction, NEBuilder HiFi DNA assembly can be scaled up for larger volumes, depending on the concentration and yield of the DNA samples, i.e., PacI-digested CRISPR vector and PCR fragments.
23. In our hands, arching is frequent when transforming more than 1 μL of the NEB HiFi DNA assembly reaction. If no colonies are obtained after transformation with 1 μL, dialyze the mixture and transform a larger volume. Colonies should be obtained.
24. pNB04 sequence is available in [13].
25. Do not add too much mycelium to the lytic solution. The amount of mycelium dispersed in the lytic solution should be sufficient to allow the liquid to flow freely. Protoplasting is not effective if too many cells are used, as this can saturate the enzyme. Optimization of the amount of mycelium may be necessary.
26. A positive control reaction can be done using the circular pFC332 plasmid.
27. It is important to keep the number of single colonies in the selective plates low to maximize the effective discrimination of single (homokaryons) colonies. Therefore, if the density of

colonies in these plates is high, perform additional dilutions and streak again to obtain fewer colonies. Occasionally, heterokaryons may arise from multinucleate protoplasts. In this case, additional purification of the colonies may be necessary. If validation fails or spurious bands appear in the diagnostic PCR, consider an additional round of single conidia purification in selective medium before passing it to the nonselective medium.

28. If necessary, warm the sample at 37–45 °C to aid DNA solubilization. Optionally, DNA can be treated with DNAse-free RNase.

29. A master mix containing all components can be prepared in advance to streamline pipetting and enhance experimental reproducibility. The CFW concentration of 40 μg/mL is sufficient to induce the CWI pathway, as previously demonstrated [13]. Alternative concentrations may be tested, as well as other compounds with suspected or confirmed activity on the CWI pathway. If using other agents, we recommend maintaining the total volume per well and adjusting the drug concentration accordingly. For negative (untreated) controls, replace the drug volume with an equivalent volume of sterile water. If DMSO or other solvent is used as vehicle for any tested compound, include vehicle control reactions.

30. OD measurements at 600 nm are optional to correct for luminescence during growth. OD reading can be useful for identifying conditions in which fungal growth is inhibited due to the presence of drugs, as well as for evaluating the growth kinetics of mutants compared to the wild-type strain.

Acknowledgments

We are grateful to Dr. Uffe H. Mortensen (Technical University of Denmark) for developing the CRISPR-Cas9 system for *Aspergillus* and for kindly donating the CRISPR plasmids used in this protocol. Protocol development was funded by the Fundação de Amparo à Pesquisa do Estado de São Paulo (FAPESP) grant numbers 20251/11784-9 and 2021/049777-5 (IM) and the Conselho Nacional de Desenvolvimento Científico e Tecnológico (CNPq) grant 312556/2021-7 (IM). We are also indebted to "The National Institute of Science and Technology" (INCT FUNVIR) (grant number 405934/2022-0 CNPq) from Brazil.

References

1. Brock M, Jouvion G, Droin-Bergère S et al (2008) Bioluminescent Aspergillus fumigatus, a new tool for drug efficiency testing and in vivo monitoring of invasive Aspergillosis. Appl Environ Microbiol 74:7023–7035. https://doi.org/10.1128/AEM.01288-08
2. d'Enfert C, Vecchiarelli A, Brown AJ (2010) Bioluminescent fungi for real-time monitoring of fungal infections. Virulence 1:174–176. https://doi.org/10.4161/viru.1.3.11119
3. Kowalski CH, Morelli KA, Schultz D et al (2020) Fungal biofilm architecture produces hypoxic microenvironments that drive antifungal resistance. Proc Natl Acad Sci U S A 117: 22473–22483. https://doi.org/10.1073/pnas.2003700117
4. Liu S, Le Mauff F, Sheppard DC et al (2022) Filamentous fungal biofilms: conserved and unique aspects of extracellular matrix composition, mechanisms of drug resistance and regulatory networks in Aspergillus fumigatus. NPJ Biofilms Microbiomes 8. https://doi.org/10.1038/s41522-022-00347-3
5. Díaz RD, Larrondo LF (2020) A circadian clock in Neurospora crassa functions during plant cell wall deconstruction. Fungal Biol 124:501–508. https://doi.org/10.1016/j.funbio.2020.03.003
6. Dengler V, McCallum N (2016) Luciferase reporter gene system to detect cell wall stress stimulon induction in Staphylococcus aureus. In: Hong H-J (ed) Bacterial cell wall homeostasis, vol 1440. Springer New York, New York, pp 139–150
7. Richter A, Blei F, Hu G et al (2024) Enhanced surface colonisation and competition during bacterial adaptation to a fungus. Nat Commun 15. https://doi.org/10.1038/s41467-024-48812-1
8. Fabri JHTM, Rocha MC, Fernandes CM et al (2021) The heat shock transcription factor HsfA is essential for thermotolerance and regulates cell wall integrity in Aspergillus fumigatus. Front Microbiol 12. https://doi.org/10.3389/fmicb.2021.656548
9. Opperman TJ, Dhingra S, Gutierrez-Perez C et al (2023) Luciferase-based high-throughput screen with Aspergillus fumigatus to identify antifungal small molecules. In: Krysan DJ, Moye-Rowley WS (eds) Antifungal drug resistance, vol 2658. Springer US, New York, pp 17–34
10. Furukawa T, van Rhijn N, Chown H et al (2022) Exploring a novel genomic safe-haven site in the human pathogenic mould Aspergillus fumigatus. Fungal Genet Biol 161: 103702. https://doi.org/10.1016/j.fgb.2022.103702
11. Rocha MC, Minari K, Fabri JHTM et al (2021) Aspergillus fumigatus Hsp90 interacts with the main components of the cell wall integrity pathway and cooperates in heat shock and cell wall stress adaptation. Cell Microbiol 23. https://doi.org/10.1111/cmi.13273
12. Rocha MC, Fabri JHTM, Silva LP et al (2021) Transcriptional control of the production of Aspergillus fumigatus conidia-borne secondary metabolite fumiquinazoline C important for phagocytosis protection. Genetics 218. https://doi.org/10.1093/genetics/iyab036
13. Rocha MC, Fabri JHTM, Franco de Godoy K et al (2016) Aspergillus fumigatus MADS-box transcription factor rlmA is required for regulation of the cell wall integrity and virulence. G3 Genes Genomes Genetics 6:2983–3002. https://doi.org/10.1534/g3.116.031112
14. Rocha MC, de Godoy KF, de Castro PA et al (2015) The Aspergillus fumigatus pkcAG579R mutant is defective in the activation of the cell wall integrity pathway but is dispensable for virulence in a neutropenic mouse infection model. PLoS One 10:e0135195. https://doi.org/10.1371/journal.pone.0135195
15. Valiante V, Baldin C, Hortschansky P et al (2016) The Aspergillus fumigatus conidial melanin production is regulated by the bifunctional bHLH DevR and MADS-box RlmA transcription factors. Mol Microbiol 102:321–335. https://doi.org/10.1111/mmi.13462
16. Jain R, Valiante V, Remme N et al (2011) The MAP kinase MpkA controls cell wall integrity, oxidative stress response, gliotoxin production and iron adaptation in Aspergillus fumigatus. Mol Microbiol 82:39–53. https://doi.org/10.1111/j.1365-2958.2011.07778.x
17. Valiante V, Jain R, Heinekamp T et al (2009) The MpkA MAP kinase module regulates cell wall integrity signaling and pyomelanin formation in Aspergillus fumigatus. Fungal Genet Biol 46:909–918. https://doi.org/10.1016/j.fgb.2009.08.005
18. Valiante V, Heinekamp T, Jain R et al (2008) The mitogen-activated protein kinase MpkA of Aspergillus fumigatus regulates cell wall signaling and oxidative stress response. Fungal Genet Biol 45:618–627. https://doi.org/10.1016/j.fgb.2007.09.006
19. Rocha MC, Fabri JHTM, Simões IT et al (2020) The cell wall integrity pathway contributes to the early stages of Aspergillus

fumigatus asexual development. Appl Environ Microbiol 86. https://doi.org/10.1128/AEM.02347-19
20. Liu H, Xu W, Bruno VM et al (2021) Determining Aspergillus fumigatus transcription factor expression and function during invasion of the mammalian lung. PLoS Pathog 17: e1009235. https://doi.org/10.1371/journal.ppat.1009235
21. Damveld RA, Arentshorst M, Franken A et al (2005) The Aspergillus niger MADS-box transcription factor RlmA is required for cell wall reinforcement in response to cell wall stress. Mol Microbiol 58:305–319. https://doi.org/10.1111/j.1365-2958.2005.04827.x
22. Ouedraogo J-P, Lagendijk EL, van den Hondel CAMJJ et al (2013) Screening for compounds exerting antifungal activities. In: Gupta VK, Tuohy MG, Ayyachamy M et al (eds) Laboratory protocols in fungal biology. Springer New York, New York, pp 225–230
23. Hoof JB, Nødvig CS, Mortensen UH (2018) Genome editing: CRISPR-Cas9. In: de Vries RP, Tsang A, Grigoriev IV (eds) Fungal genomics, vol 1775. Springer New York, New York, pp 119–132
24. Nødvig CS, Nielsen JB, Kogle ME et al (2015) A CRISPR-Cas9 system for genetic engineering of filamentous fungi. PLoS One 10: e0133085. https://doi.org/10.1371/journal.pone.0133085
25. Nødvig CS, Hoof JB, Kogle ME et al (2018) Efficient oligo nucleotide mediated CRISPR-Cas9 gene editing in Aspergilli. Fungal Genet Biol 115:78–89. https://doi.org/10.1016/j.fgb.2018.01.004
26. Ost KJ, Arentshorst M, Moerschbacher BM et al (2025) Comprehensive phenotypic analysis of multiple gene deletions of α-glucan synthase and Crh-transglycosylase gene families in Aspergillus niger highlighting the versatility of the fungal cell wall. Cell Surf 13: 100141. https://doi.org/10.1016/j.tcsw.2025.100141
27. Green MR, Sambrook J (2020) Transformation of Escherichia coli by electroporation. Cold Spring Harb Protoc 2020:pdb.prot101220. https://doi.org/10.1101/pdb.prot101220
28. Viviani VR (2023) Looking into luciferin. Nat Chem 15:742. https://doi.org/10.1038/s41557-023-01185-2
29. García Carnero LC, Li X, de Castro PA et al (2025) Colistin enhances caspofungin antifungal efficacy against Aspergillus fumigatus by modulating calcium homeostasis and stress responses. Nat Commun 16. https://doi.org/10.1038/s41467-025-60991-z

Chapter 16

Evaluation of *Aspergillus fumigatus* Metabolic Activity In Vitro: A Comparison Between Resazurin and XTT Methods

Thaila Fernanda dos Reis, Iran Malavazi, and Gustavo Henrique Goldman

Abstract

Aspergillus fumigatus is a significant opportunistic pathogen that causes a range of diseases collectively known as aspergillosis. Measurement of filamentous fungal growth in molds, such as *A. fumigatus*, is crucial for the development of antifungal drugs and for clinical and epidemiological monitoring of drug resistance. However, determining their fungal growth is a challenging approach, since these organisms have multinucleate hyphae that make the individual colony-forming unit assessment difficult. Measurement of cellular metabolic activity is, therefore, an alternative method to assess the impact of stressful conditions on fungal growth resulting from drug activity. The current protocol presents a comparative analysis of resazurin and XTT assays as indicators of metabolic viability for *A. fumigatus* under different experimental approaches. Resazurin, a redox-sensitive compound, is widely used due to its high sensitivity in detecting early-stage metabolic activity, particularly in conidia. In contrast, XTT, a tetrazolium-based assay, is particularly effective for evaluating metabolic activity in mature hyphal forms and biofilms, as it allows for shorter incubation times with the fungus, with the formation of a stable and soluble formazan product. The protocol outlines the respective advantages and limitations of each assay. It provides a guide for researchers in selecting the most appropriate assay for evaluating fungal viability, antimicrobial effectiveness, drug interactions through checkerboard assays, and biofilm-related resistance.

Key words *A. fumigatus* metabolic activity, Resazurin, XTT, Minimal inhibitory concentration, Fungal growth inhibition, Antibiofilm activity

1 Introduction

Aspergillus fumigatus is a major opportunistic fungal pathogen responsible for a spectrum of diseases collectively known as aspergillosis, ranging from allergic reactions in atopic patients to life-threatening localized or disseminated infections in immunocompromised individuals [1]. Given its clinical significance and intrinsic resistance mechanisms to various drug classes currently available for clinical use, which are being increasingly identified

Thaila Fernanda dos Reis and Iran Malavazi contributed equally with all other contributors.

Natalia Mach (ed.), *Aspergillus fumigatus: Methods and Protocols*, Methods in Molecular Biology, vol. 3028,
https://doi.org/10.1007/978-1-0716-5222-0_16,

worldwide, accurate and reliable assessment of fungal viability and metabolic activity is crucial for antifungal drug development, therapeutic monitoring, epidemiological surveillance, and understanding fungal pathophysiology [2–4].

Similar to any mold, the dissemination of the fungus occurs through the dispersion of asexual spores (conidia) via air, water flows, and animals. Notably, prolific conidiation is a well-recognized virulence determinant in *A. fumigatus*, making conidia the primary infectious propagules. In contrast, inhaled conidia break dormancy, germinate, and transition into filamentous hyphae within the vulnerable host, thus colonizing the lungs and giving rise to invasive pulmonary aspergillosis (IPA). This form of infection is characterized by the formation of intertwined hyphal structures that organize into fungal biofilms that lack conidiation in vivo. These biofilms not only serve as a physical barrier to host immune defenses, but also allow fungal dissemination to other organs in severe cases [5]. Fungal biofilms are also present in the chronic forms of the disease, Chronic Pulmonary Aspergillosis (CPA), and are named "fungal balls" [6].

Fungal biofilms are therefore metabolically and morphologically distinct from dormant or germinated conidia. Accordingly, experimental approaches to quantify metabolic activity must be tailored to these two fungal growth forms, as traditional methods such as spectrophotometry and dry cell weight are ineffective in filamentous fungi [7]. In addition, it has been demonstrated that biofilms can significantly contribute to enhanced resistance to antifungal agents [8–10]. These findings highlight the importance of accurate assay systems that can reliably detect metabolic shifts during growth, stress response, and antifungal exposure. Therefore, this protocol provides a comparative framework for employing resazurin or XTT as metabolic viability indicators under different experimental conditions relevant to *A. fumigatus* research. By outlining their respective strengths and limitations in the context of conidial or biofilm assays, this protocol aims to help researchers choose the most appropriate assay for evaluating antifungal effectiveness, fungal recovery processes, and biofilm-related resistance. We aim to review the two primary methods used to assess metabolic activity, highlighting their differences and similarities, and provide practical guidance for implementation and data analysis.

Resazurin (7-hydroxy-3H-phenoxazin-3-one-10-oxide sodium salt) is a colorimetric and fluorometric redox-based indicator that requires cellular uptake and is used to assess cell viability, metabolic activity, or cytotoxicity. This molecule has been widely used for over 60 years as a chemical indicator in the dairy industry to monitor microbial contamination in milk and to measure the viability of sperm [11]. It functions by reducing the blue, nonfluorescent resazurin dye to the pink and fluorescent compound resorufin in metabolically active cells. This change can be quantified via absorbance or fluorescence, and the intensity is directly proportional to cellular metabolic activity [12]. The reduction process is thought to primarily occur in the mitochondria, where resazurin acts as an electron acceptor in the electron transport chain, reflecting the reducing environment of viable cells [11]. Resazurin is a versatile, non-toxic compound suitable for kinetic measurements. It can be used in both homemade solutions and commercial reagents (e.g., alamarBlue, Thermo Scientific), which can be used interchangeably, according to some reports [11, 12]. Importantly, resazurin is highly sensitive, particularly effective in detecting early metabolic activity in conidia or during stress responses. Its redox-based mechanism enables the detection of subtle metabolic changes, making it valuable for determining the minimal inhibitory concentration (MIC) of different antifungals against *A. fumigatus* and effective in assessing fungistatic versus fungicidal responses when associated with the determination of colony forming units [12, 13]. It remains stable at room temperature and under light exposure, making it suitable for high-throughput screening and routine viability testing (*see* Table 1).

XTT (2,3-bis(2-methoxy-4-nitro-5-sulfophenyl)-2H-tetrazolium-5-carboxanilide), in contrast, is a colorimetric tetrazolium-based assay also widely used to evaluate cell viability and metabolic activity. This method relies on the reduction of XTT by cellular dehydrogenases, particularly mitochondrial and plasma membrane oxidoreductases, into a highly water-soluble orange formazan product in metabolically active cells [7]. The primary mechanisms of formazan generation from XTT reduction are the mitochondrial succinoxidase and cytochrome P450 systems, as well as flavoprotein oxidases [14]. Unlike other tetrazolium salts, such as MTT, which produce insoluble formazan crystals, thus requiring cell killing to allow for a solubilization step, XTT's soluble product allows for direct and real-time quantification without further manipulation. This makes it especially advantageous for continuous or endpoint measurements in high-throughput formats (*see* Table 1). Unlike resazurin, the use of an electron-coupling reagent, such as menadione, is required to facilitate efficient electron transfer, especially in fungal cells, where outer cell wall structures may limit dye access to the respiratory chain.

In *A. fumigatus* experimentation, XTT is frequently used to evaluate biofilm-associated resistance and mature hyphal forms, as its readout correlates well with biofilm density, hyphal viability, and overall metabolic state [7]. This approach has been employed to characterize drug resistance phenotypes, assess genetic mutants with impaired biofilm formation, and test novel antifungal agents or combinations in both wild-type laboratory strains and clinical isolates [8, 15, 16]. It has also been widely applied to antifungal susceptibility testing and the study of the effects of stressors on fungal biofilms, enabling the generation of reproducible

Table 1
Comprehensive comparison between resazurin and XTT properties for use in *A. fumigatus* experiments

Feature/application	Resazurin	XTT
Chemical nature and detection	Redox indicator; fluorescent and colorimetric	Tetrazolium salt; colorimetric only
Reduction mechanism	Reduced to water-soluble resorufin by cellular redox enzymes	Reduced to water-soluble formazan by mitochondrial/plasma membrane enzymes
Detection mode	Absorbance: Reduced (pink color): 570 nm Oxidized (blue color): 600 nm Fluorescence: (excitation: 530–560 Emission: 590–615 nm nm)	Absorbance at 450–492 nm
Reagent stability	Stable at room temperature and light; degrades post-reduction	Stable after reduction; formazan is chemically stable
Preparation complexity	Low; ready-to-use from stock solution	Moderate; requires freshly prepared solution
Need for electron coupling agent	No	Yes (e.g., menadione)
Time to readout	4–6 h	1–3 h
Sensitivity (early growth)	High, suitable for conidia and early germlings	Lower for dormant conidia, higher in hyphae[a]
Suitability for biofilms	Moderate performance; less accurate in dense hyphae	Excellent performance in mature biofilms
Quantification consistency	Lower; affected by resorufin degradation	High; reproducible colorimetric quantification
Preferred fungal stage	Conidia, early germlings	Hyphae, biofilm
Application in MIC assays	Yes, especially early stages	Limited; less sensitive for conidia
Application in biofilm assays	Possible, but less reliable in thick biofilms	Yes, gold standard for metabolic biofilm assays
Susceptibility to interference	High (e.g., antioxidants, reducing agents)	Low; more robust in complex media
Readout stability over time	Low; resorufin degradation may occur	High; stable chromogen allows delayed reading
High-throughput compatibility	Excellent (96- or 384-well plates)	Excellent (12-, 24-, 48- or 96-well plate formats)
Common interfering factors	Redox-active compounds; strong reducing agents	Extracellular matrix interactions (minor)

[a]The use of the XTT-colorimetric assay to compare biofilm-forming ability among different isolates should be interpreted with caution, as it has been reported that various fungal species and even strains from the same species exhibit marked differences in their ability to metabolize the XTT substrate. In addition, alterations in the metabolic states during the different phases of biofilm formation may lead to fluctuations in the ability of cells within the biofilms to metabolize this dye [18]

dose-response curves and clear distinctions between fungistatic and fungicidal effects [17]. The absorbance of the reduced formazan is typically measured at 450–492 nm, with signal intensity directly reflecting the metabolic activity of the fungal biomass. XTT assays are compatible with standard microplate readers, making them practical for routine use in fungal drug discovery and pathogenesis studies. However, limitations include potential background signal due to non-specific reduction or extracellular matrix interactions, which must be controlled using proper blanks and calibration (*see* Table 1). Additionally, an important caveat to consider when comparing different strains (mutants or clinical isolates) is that not all strains metabolize XTT in an equal manner, thus posing a problem in attempts to compare different fungal strains and species [14]. It is also essential to note that both methods rely on assumptions based on the initial size of the fungal inoculum. As a result, variations in metabolic activity may evolve from discrepancies in fungal load or may be dependent on the ability of a given strain to form biofilm whenever necessary. Therefore, the ability to form multiple, consistent, and reproducible biofilms is essential for the success of these techniques, which utilize XTT as a viability indicator [18].

Both resazurin and XTT assays can be used to assess the metabolic state of different cells, including fungi. While both resazurin and XTT are suitable for evaluating metabolic activity in *A. fumigatus*, each has distinct functional advantages depending on the fungal morphology and assay goal. Resazurin is a low-cost, easy-to-prepare reagent that does not require an electron mediator. Its dual-mode (fluorescent and colorimetric) signal provides high sensitivity, particularly useful for early-stage or low-density cell populations. However, resorufin may degrade or be further reduced over time, potentially affecting assay accuracy in extended incubations. Furthermore, the redox-based nature of resazurin makes it susceptible to interference from reducing agents, antioxidants, or redox-active compounds present in the test environment. For example, in large-scale drug or compound screenings, resazurin may become colorless and chemically modified immediately after it is added to the media at the beginning of the test, where a given interfering compound is present. This limitation underscores the importance of validating compound-resazurin compatibility when screening chemical libraries or unknown compounds.

In contrast, the reduction of XTT yields a chemically stable chromogenic product, enabling reliable quantification in endpoint-based assays. The resulting formazan remains both soluble and stable following incubation, which permits delayed absorbance measurements without compromising signal integrity. XTT is also less prone to interference, particularly in dense mycelial or biofilm-forming cultures, making it especially suitable for assessing viability in mature hyphal networks and structured fungal biofilms. Additionally, because XTT reduction depends on cellular dehydrogenase activity, it provides a sensitive measure of mitochondrial metabolic function, which may be elevated under biofilm growth conditions.

Selection between resazurin and XTT as viability indicators in *A. fumigatus* assays should be based on fungal morphology, the specific goals of the experiment, and the underlying metabolic characteristics of the test system. In summary, resazurin offers high sensitivity in detecting metabolic activity during early fungal development and under redox-sensitive conditions, making it especially useful for MIC testing and time-kill kinetics assays. On the other hand, XTT is better suited for analyzing mature hyphal structures and biofilms that are not conidiated. Conidiation is an interference that needs to be avoided for both methods. Integrating both methods in complementary experimental designs may provide a more comprehensive assessment of antifungal efficacy across the fungal life cycle.

2 Materials

For media and solutions preparations, use ultrapure reagents (molecular biology grade) and double-deionized water (18 MΩ). All manipulation of *A. fumigatus* in solid or liquid cultures must be performed in a Class A2 biosafety cabinet (*see* **Note 1**).

2.1 Media and Solutions

1. Minimal medium (MM): 1% (w/v) glucose, 1× (v/v) salt solution, 0.1% (v/v) trace elements solution. Dissolve components in 900 mL of water, adjust the pH to 6.5 using 4 M NaOH, and bring to 1 L with water. Sterilize the medium by autoclaving at 121 °C for 20 min and store at room temperature (*see* **Note 2**).
2. Solid minimal medium (MM agar): Add 1.5% (w/v) of bacteriological agar to MM. Sterilize the medium by autoclaving at 121 °C for 20 min and store at room temperature.
3. Trace elements (1000× stock, 1 L): Dissolve 22 g $ZnSO_4 \cdot 7H_2O$, 11g H_3BO_3, 5 g $MnCl_2 \cdot 4H_2O$, 5 g $FeSO_4 \cdot 7H_2O$, 1.6 g $CoCl_2 \cdot 6H_2O$, 1.6 g $CuSO_4 \cdot 5H_2O$, 1.1 g $(NH_4)_6MO_7O_{24} \cdot 4H_2O$, and 50 g EDTA in 900 mL of water. The elements should be dissolved exactly in the indicated order; add each compound only after the previous one has fully dissolved. Boil, cool to 60 °C, and adjust the pH to 6.5–6.8 using KOH. Add water to 1 L. Store in an amber bottle at room temperature. Always shake before use.
4. Salt solution (20× stock): Dissolve 120 g $NaNO_3$, 10.4 g KCl, 30.4 g KH_2PO_4, 10.4 g $MgSO_4 \cdot 7H_2O$ in 800 mL water. Adjust the volume to 1 L.

5. Resazurin sodium salt stock solution (0.02%): Carefully weigh 20 mg of resazurin ($C_{12}H_6NNaO_4$; CAS number 62758-13-8), and dissolve in 100 mL of water. Sterilize by filtration using a syringe filter, 0.22 μm. Make aliquots and store in the dark at 4 °C (*see* **Note 3**).
6. XTT solution (0.5 mg/mL): In a 15 mL Falcon tube, dissolve 0.5 mg of XTT ($C_{22}H_{16}N_7NaO_{13}S_2$; CAS number 111072-31-2) in 10 mL of MM or the culture medium that is being used in the experiment. Alternatively, XTT can be dissolved in sterile PBS (*see* **Note 4**). Sterilize the solution by filtration using a syringe filter, 0.22 μm.
7. Menadione stock solution (100 mM)*:* Dissolve 17.218 mg of menadione ($C_{11}H_8O_2$, CAS number 58-27-5) in 1 mL of water. Prepare this solution fresh immediately before use (*see* **Notes 5** and **6**).
8. XTT/menadione working solution: Add 2.5 μL of the 100 mM menadione stock solution to 10 mL of the 0.5 mg/mL XTT solution (final assay concentration 25 μM). Menadione should be added immediately before use, after XTT solubilization. Different volumes of this XTT/menadione working solution will be required per well for various experimental setups (*see* **Note 7**).
9. Tween-80 solution: 0.01% in sterile water.
10. Voriconazole (VOR) stock solution: 0.5 mg/mL in DMSO.
11. Amphotericin B: 2 μg/mL in DMSO.
12. Sterile PBS.

2.2 Consumables

1. 96-well cell culture microplates, black, chimney flat-bottom well with lid (e.g., Greiner Bio-One, 655,086). These plates are used to read the fluorescence of resazurin.
2. 96-well flat-bottomed polystyrene cell culture microplates, transparent with lid. These plates are used to read the absorbance derived from XTT assays and to cultivate conidia and biofilm for resazurin assays.
3. 24-well polystyrene cell culture microplates, transparent, with lid. These plates are used to grow *A. fumigatus* biofilm for XTT assays.
4. Deep well plate 96 square wells, 2.0 mL (e.g., Corning AXYP2MLSQC). These plates are used to prepare drug dilutions for the checkerboard assay.
5. Sterile 15 mL and 50 mL conical tubes (Falcon-type).
6. 10 mL sterile syringes.
7. Sterile glass wool and sterile filter material (e.g., Miracloth).
8. Sterile disposable Pasteur pipettes.

9. Sterile 90 mm × 15 mm Petri dishes.
13. Hemocytometer.
14. Multichannel pipettes and standard pipettes.
10. Sterile swabs.

2.3 Equipment

1. Plate reader capable of reading fluorescence for the resazurin assays or absorbance for the XTT assays. Multimodal plate readers, such as SpectraMax (molecular devices) or Varioskan (Thermo Scientific), and BioTek Synergy HTX Multimode Reader, are well-suited for assays (*see* **Note 8**).
2. Humidified chamber.
3. Centrifuge with swing-bucket/plate rotor.

3 Methods

3.1 Conidia Harvest from Solid MM Plates

1. Inoculate *A. fumigatus* conidia onto MM agar plates in sterile Petri dishes and incubate at 37 °C for 72–96 h.
2. Add 15 mL of a 0.01% Tween-80 solution directly onto the colony surface. Using a sterile swab, gently dislodge the conidia by rubbing the surface.
3. Collect the resulting dark green suspension using a disposable Pasteur pipette, and filter it through sterile Miracloth or glass wool into a sterile 50 mL conical tube to remove hyphal fragments.
4. Quantify the conidia using a hemocytometer and adjust the conidial suspension to 1×10^7 conidia/mL. This conidial suspension will be used in the downstream assays.

3.2 Resazurin-Based Antifungal Assays and Analysis

Perform assays with at least three biological replicates. In the following MIC assay, voriconazole (VOR) is used as an example antifungal agent. The established MIC value for *A. fumigatus* is 0.5 μg/mL. The final assay volume in each well is 150 μL.

3.2.1 Determination of Minimal Inhibitory Concentration (MIC)

1. Mix 9 mL of MM and 1 mL of 0.02% resazurin solution in a 15 mL tube. This yields a final resazurin concentration of 0.002%, equivalent to 10% (v/v) resazurin solution in the culture medium (*see* **Note 9**). Add 25 μL of 1×10^7 conidia/mL suspension to obtain a final concentration of 2.5×10^4 conidia/mL.
2. Dispense 150 μL of the mixture into wells 1 to 7 of a flat-bottomed 96-well transparent polystyrene plate. In well number 8 (well-8), add 300 μL of the mixture (*see* Fig. 1).
3. Add 1.2 μL of the VOR stock solution to well-8 to reach 2 μg/mL concentration.

4. Mix the contents a of well-8 thoroughly and transfer 150 μL to well-7.
5. Repeat the serial dilution by transferring 150 μL to the next well, counting down to well-2. After mixing in well-2, discard 150 μL to maintain a uniform volume. Well-1 serves as the growth control without VOR added. Use a multichannel pipette for replicates or multiple drugs.
6. Prepare two further controls: Well-9 containing MM and resazurin only (no cells added, *see* Fig. 1) and well-10 containing amphotericin B at 2 μg/mL as fungicidal positive control. The positive control well contains a standard fungicidal drug at a concentration ≥ MIC. Amphotericin B at 2 μg/mL is the MIC value for laboratory reference strains of *A. fumigatus* (*see* Fig. 1).
7. Place the plate in a humidified chamber, protect from light and incubate at 37 °C for 30–36 h (*see* **Note 10**). Visually monitor the resazurin color change from violet to pink throughout the incubation (*see* **Note 11**).
8. Centrifuge the plate at 4000 rpm for 5 min. Transfer 100 μL of the supernatant to a flat-bottomed black 96-well plate for fluorescence reading. Avoid bubbles while transferring.
9. Measure resazurin fluorescence at excitation 570 nm (within 540–570 nm) and emission 590 nm (within 580–610 nm) using a plate reader (*see* **Notes 12–15**). Proceed with data analysis (*see* Subheading 3.4).

3.2.2 Antibiofilm Activity Assay

1. Mix 13.5 mL of MM and 1.5 mL of the 1×10^7 conidia/mL suspension in 50 mL tube to obtain 1×10^6 conidia/mL.
2. Dispense 100 μL of this inoculum in each well of a flat-bottomed transparent polystyrene 96-well plate. Place the plate in a humidified chamber and incubate at 37 °C for 12–24 h to obtain young or mature biofilms, respectively (*see* **Note 16**).
3. Add 50 μL of fresh MM containing 3× the final antifungal drug concentration. For example, to achieve 2 μg/mL of final VOR concentration, add 50 μL of MM containing 6 μg/mL VOR. The final volume per well is 150 μL.
4. Include medium-only (no cells added) to untreated biofilm (no drug added) controls to assess background and baseline metabolic activity.
5. Return the plate to a humidified chamber and incubate at 37 °C for an additional 4 h (*see* **Notes 17** and **18**).
6. Add 15 μL of resazurin stock solution to each well (final concentration 0.002% resazurin).

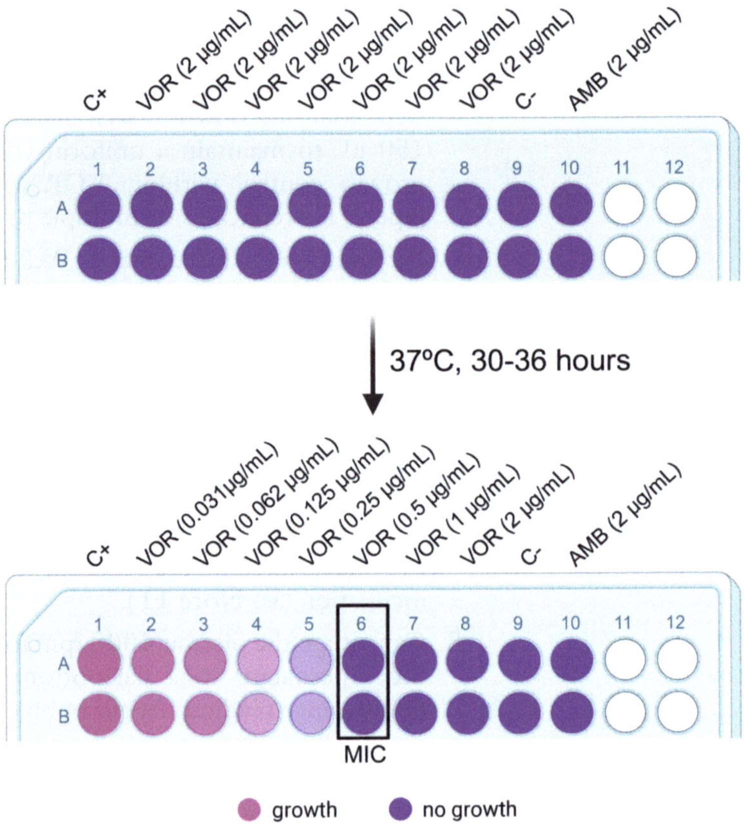

Fig. 1 Schematic representation of the MIC determination for voriconazole (VOR) against *A. fumigatus* using the resazurin-based assay. C+ indicates the positive growth control (no drug added); C– represents the negative control (no conidia added); AMB (positive control of fungal death using the standard fungicidal drug amphotericin B). MIC corresponds to the lowest drug concentration that inhibits fungal growth, as indicated by the absence of resazurin color change. Well number 8 contains 2 µg/mL of VOR and was serially diluted across wells 7 to 2. The MIC was identified in well number 6, corresponding to 0.5 µg/mL of VOR. (Created with BioRender.com (https:/BioRender.com/vzbh2fy))

7. Incubate protected from light in a humidified chamber for 2–4 h at 37 °C. A strong pink color signal in the untreated biofilm control indicates the optimal time for fluorescence measurement. The incubation period may vary, depending on the fungal strain, the maturity of the biofilm, and the type of drug treatment.
8. Centrifuge the plate at 4000 rpm for 5 min. Transfer 100 µL of the supernatant to a flat-bottomed black 96-well plate. Avoid bubbles while transferring.

9. Measure resazurin fluorescence at excitation 570 nm (within 540–570 nm) and emission 590 nm (within 580–610 nm) using a plate reader (*see* **Notes 12–15**). Proceed with data analysis (*see* Subheading 3.4).

3.2.3 Checkerboard Interaction Assay

The checkerboard assay allows a quantitative analysis of the interaction between two antifungal agents. This method involves testing a matrix of drug concentrations in a two-dimensional dilution series to assess potential synergistic, additive, or antagonistic interactions based on fungal growth inhibition [19, 20]. To illustrate this approach (*see* Fig. 2), a representative example using two hypothetical drugs, Drug A and Drug B, is provided.

1. Prepare conidia/resazurin suspension. Mix 13.5 mL of MM, 1.5 mL of resazurin stock solution (final concentration 0.002% equivalent to 10% of resazurin solution in the culture medium), and 37.5 μL of the conidia stock suspension in a 50 mL tube to obtain 2.5×10^4 conidia/mL.
2. Dispense 500 μL of conidia/resazurin suspension in wells 1 to 11 of the deep 96-well plate. In well-12, add 1000 μL.
3. Serially dilute Drug A horizontally (x-axis, well-1 to 12). Add Drug A at 8× MIC to well 12, mix, and transfer 500 μL stepwise to wells 11–2, discarding 500 μL after well 2. Well-1 serves as growth control (no drug added).
4. Transfer 50 μL from each well of the Drug A dilution plate to a new flat-bottomed transparent 96-well plate using a multichannel pipette. This is a checkerboard plate.
5. Dispense 750 μL of the conidia/resazurin suspension in wells B-H of a new sterile deep 96-well plate. In well A, add 1500 μL of suspension.
6. Serially dilute Drug B vertically (y-axis, well A to H): Add Drug B at 4× MIC to well A, mix, and transfer 750 μL stepwise to wells B-G. Discard 750 μL after mixing from well-G. Well-H serves as growth control (no drug added).
7. Transfer 50 μL from each well of the Drug B dilution plate onto the corresponding wells of the checkerboard plate that already contains 50 μL of the mixture containing the serial dilution of Drug A, thus generating the two-dimensional matrix of drug combinations.
8. Protect the checkerboard plate from light, place it in a humidified chamber, and incubate at 37 °C for 30–36 h. Observe the resazurin color change from violet to pink during the incubation.
9. Centrifuge the plate at 4000 rpm for 5 min. Transfer 100 μL of the supernatant to a sterile flat-bottomed black 96-well plate.

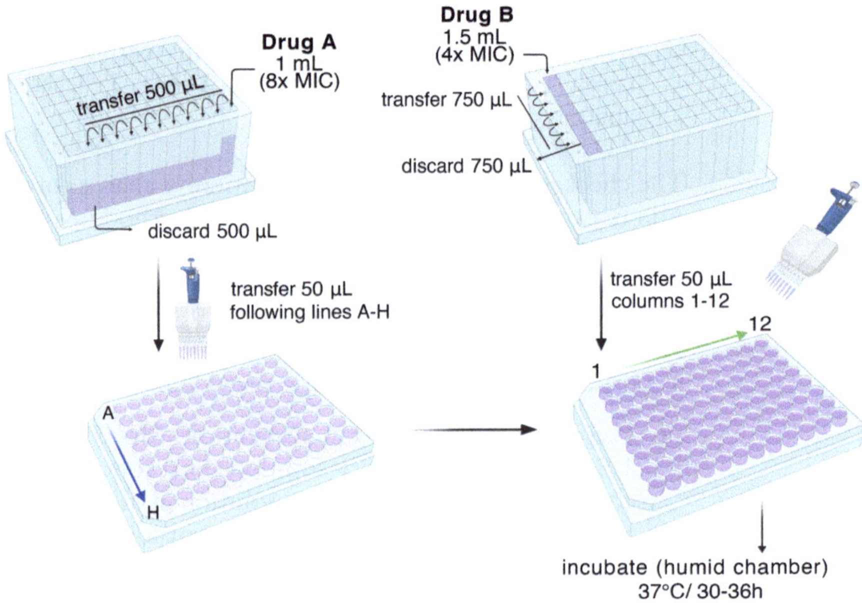

Fig. 2 Schematic representation of a checkerboard assay designed to analyze the interaction between Drug A and Drug B. Use a deep-well 96-well plate to dilute Drug A and Drug B. For Drug A, wells 1–11 are filled with 500 μL of the mixture containing media, conidia, and resazurin, while well 12 has 1 mL of the same mixture. A concentration of 8× MIC is added to well 12 (final volume in the well is 1 mL). Perform a serial dilution (500 μL transfer) until well 2. Subsequently, use a multichannel pipette to transfer 50 μL of each well to a new 96-well polystyrene plate following the horizontal orientation (blue arrow; A-H wells). In parallel, in another deep-well plate, wells B-H were filled with 750 μL of the mixture (media + conidia + resazurin), while well A receives 1.5 mL of the same mixture. A concentration of 4× MIC of Drug B is added in well A (final volume in the well is 1.5 mL). Perform a serial dilution (750 μL transfer) until well B. Then, use a multichannel pipette to transfer 50 μL of each well to the 96-well polystyrene plate following vertical orientation (green arrow; wells 1 to 12). (Created with BioRender.com (https:/BioRender.com/peoueq))

10. Measure resazurin fluorescence at excitation 570 nm (within 540–570 nm) and emission 590 nm (within 580–610 nm) using a plate reader (*see* **Notes 12–15**). Proceed with data analysis (*see* Subheading 3.4).

3.3 Experimental Setup for XTT Metabolic Assays

In this section, we describe two methodological options for conducting XTT-based metabolic viability assays in *A. fumigatus* biofilms: direct addition of XTT/menadione without removing the treatment medium and posttreatment medium removal followed by addition of the XTT/menadione solution. The second approach is most commonly used. However, each approach offers specific advantages depending on the experimental context. Pilot comparisons between the two protocols are recommended to identify the most appropriate workflow for a given application. All assays should be performed with at least three biological replicates.

3.3.1 Measurement of Biofilm Metabolic Activity Assay Without Medium Removal

The direct addition of XTT/menadione without prior biofilm wash retains all the components initially added or produced during the biofilm formation and antifungal challenge. This approach can be advantageous for high-throughput compound screening, as it reduces handling time and preserves biofilm structures. However, the presence of residual drugs, cultivation additives, and/or soluble components of the biofilm matrix may interfere with the assay readout and reduce accuracy.

1. In a 50 mL tube, mix 13.5 mL of MM and 1.5 mL of the 1×10^7 conidia/mL stock suspension to obtain a concentration of 1×10^6 conidia/mL.
2. Dispense 100 μL of the conidia suspension in each well of a flat-bottomed transparent polystyrene 96-well plate. Place the plate in a humidified chamber and incubate at 37 °C for 12 h (young biofilm) or 24 h (mature biofilm).
3. Add 50 μL of fresh MM containing 3× the final concentration of the desired drug. For example, to obtain a final drug concentration of 2 μg/mL, add 50 μL of MM containing 6 μg/mL of the drug. The final volume in each well is 150 μL.
4. Include medium only and untreated (no drug added) controls to assess background signal and baseline metabolic activity in the absence of treatment.
5. Return the plate to a humidified chamber, and incubate at 37 °C for 4 h or as defined by experimental setup (*see* **Note 17**).
6. Prepare fresh XTT/menadione working solution (*see* Subheading 2.1) and add 50 μL of this solution to each well, including controls (*see* **Notes 4** and 7).
7. Protect the plate from light, incubate in a humidified chamber at 37 °C for 1–2 h. Visually monitor color development, where dark orange color indicates higher metabolic activity of the biofilm (*see* **Note 19**).
8. Centrifuge the plate at 4000 rpm for 5 min. Transfer 100 μL of the supernatant to a new sterile flat-bottomed transparent 96-well plate. Avoid bubbles while transferring.
9. Measure the absorbance at 450 nm using a plate reader. Proceed with data analysis (*see* Subheading 3.4).

3.3.2 Measurement of Biofilm Metabolic Activity After Medium Removal (Posttreatment Wash)

This protocol is particularly suited for precision-driven applications, such as quantifying residual metabolic activity after antifungal challenge (e.g., MIC90, the concentration required to reduce viability by 90%) or determining MBEC (minimum biofilm eradication concentration) values. By removing potential interfering substances such as residual drugs, cell debris, or soluble components of the fungal biofilm, this approach increases the accuracy and reproducibility of metabolic readouts. However, it may introduce biofilm disruption, especially in immature or loosely adherent biofilms, and requires careful pipetting and washing steps that can reduce throughput.

1. In 50 mL tube, mix 14.85 mL of MM and 150 μL of the 1×10^7 conidia/mL stock suspension to obtain a concentration of 1×10^5 conidia/mL.
2. Dispense 1000 μL of conidia suspension in each well of a flat-bottomed transparent polystyrene 24-well plate. Place the plate in a humidified chamber, and incubate at 37 °C for 12 h (young biofilm) or 24 h (mature biofilm) (*see* **Note 20**).
3. Include medium only and untreated (no drug added) controls to assess background signal and baseline metabolic activity in the absence of treatment.
4. After biofilm formation, carefully aspirate the medium to avoid disturbing the biofilm structure. Use a multichannel pipette and gently angle the tips toward the borders of the wells, minimizing direct contact with the biofilm surface and maximizing medium removal. Avoid biofilm drying (*see* **Note 21**).
5. Add 500 μL of fresh MM containing the desired concentration of the drug to be tested (*see* **Note 22**). Incubate the plate for 4 h at 37 °C in a humidified chamber. The incubation times of drugs with biofilm vary depending on the drug and the specific experimental objectives (*see* **Note 23**).
6. Carefully aspirate the medium containing the drug, avoiding disturbing the biofilm structure as described in **step 4**.
7. Wash 1 or 2 times with 500 μL of PBS. Centrifuge the plate at 4000 rpm for 5 min. Use a pipette to remove the liquid completely (*see* **Note 24**).
8. Prepare fresh XTT/menadione working solution as described. Add 300 μL in each well containing the fungal biofilm.
9. Protect the plate from light and incubate the plate at 37 °C for 1 h. At this stage, the gradient of orange coloration is visible. A longer incubation time may be required.
10. Centrifuge the plate at 4000 rpm for 5 min. Transfer 150 μL of the supernatant to a new flat-bottomed transparent 96-well plate. Avoid bubbles while transferring.
11. Measure the absorbance at 450 nm using a plate reader. Proceed with data analysis (*see* Subheading 3.4).

3.4 Data Analysis

3.4.1 Analysis of Resazurin Assays

1. After reading the resazurin MIC plate, calculate the percentage of growth (or growth inhibition) relative to the control well without drug added, which represents 100% growth (*see* Table 2). Perform the experiment at least in technical duplicate, and calculate the average fluorescence for each well.

Table 2
Hypothetical values for MIC value calculation in a typical experiment using the resazurin method

Fluorescence reading (MIC assay)	Drug X (μg/mL)								
	0	0.062	0.12	0.25	0.5	1	2	4	C-
replicate 1	37220	39229	37530	21124	2559	2530	2481	2530	2460
replicate 2	37053	37290	37277	22887	2574	2468	3790	2468	2338
average	37136.5	38260	37404	22006	2566.5	2499	3135.5	2499	2399
growth (%)	100.0	103.0	100.7	59.3	6.9	6.7	8.4	6.7	6.5

The color gradient shown in the table mimics the visual change observed for resazurin reduction. Magenta/purple indicates high fluorescence values, corresponding to the reduced (pink) form of resazurin and reflecting high metabolic activity and fungal growth. Darker violet shades indicate lower fluorescence values, corresponding to the oxidized (blue) form of resazurin and reflecting reduced metabolic activity due to drug-mediated growth inhibition

$$\text{Growth (\%) of the analyzed well} = (\text{fluorescence} \times 100)/\text{fluorescence of the control}$$

Example:

Growth (%) of the well containing 4 μg/mL of the "drug X" = (2499 × 100)/37136.5 = 6.72%

2. Repeat this calculation for all wells to obtain the percentage of growth in each well. If the fluorescence of the negative control (no conidia added) reveals 6.45% growth, the MIC value for "Drug X" can be concluded as 0.5 μg/mL, where the growth was 6.91%. At low concentrations (e.g., 0.25 μg/mL), fungal growth increases, confirming that the concentration is below the MIC.
3. For the biofilm assay, perform the data analysis as described above. Consider the untreated biofilm (positive control) as 100% growth and calculate the growth in treated wells accordingly.
4. For checkerboard assays, calculate the percentage of growth (or growth inhibition) for each well as described for the MIC assay. Use the value from the well without any drug (positive control, *see* Table 3) as 100%.

$$\text{Growth (\%)} = (\text{fluorescence of test well} \times 100)/\text{fluorescence of positive control}$$

Example:

Table 3
Hypothetical values for checkerboard analysis in a typical experiment using the resazurin method

Fluorescence reading		Drug A (µg/mL)											
		0	0.078	0.156	0.312	0.625	1.25	2.5	5	10	20	40	80
Drug B (µg/mL)	2	6280	6479	6020	6515	7489	6703	6344	6040	6596	6647	6054	6141
	1	6532	6619	5646	5215	4959	4673	4557	4398	4588	4979	5656	5209
	0.5	31429	10068	11005	4374	4149	4178	3998	3964	3929	4181	4651	5779
	0.25	34368	36005	33340	5733	4032	3988	3909	3812	3790	3869	4266	5068
	0.125	33374	35216	35092	34901	4895	3973	3895	3785	3781	3847	4127	5136
	0.0625	34077	34964	35477	35140	34028	20797	5100	4547	4044	4104	4434	5801
	0.0313	32195	33263	33869	34197	35066	33400	30400	26377	24126	21758	4536	4952
	0	33642	32366	31529	32567	33579	33289	33288	33260	32481	30716	7827	9558
Percentage of growth		**Drug A (µg/mL)**											
		0	**0.078**	**0.156**	**0.312**	**0.625**	**1.25**	**2.5**	**5**	**10**	**20**	**40**	**80**
Drug B (µg/mL)	2	18.7	19.3	17.9	19.4	22.3	19.9	18.9	18.0	19.6	19.8	18.0	18.3
	1	19.4	19.7	16.8	15.5	14.7	13.9	13.5	13.1	13.6	14.8	16.8	15.5
	0.5	93.4	29.9	32.7	13.0	12.3	12.4	11.9	11.8	11.7	12.4	13.8	17.2
	0.25	102.2	107.0	99.1	17.0	12.0	11.9	11.6	11.3	11.3	11.5	12.7	15.1
	0.125	99.2	104.7	104.3	103.7	14.6	11.8	11.6	11.3	11.2	11.4	12.3	15.3
	0.0625	101.3	103.9	105.5	104.5	101.1	61.8	15.2	13.5	12.0	12.2	13.2	17.2
	0.0313	95.7	98.9	100.7	101.6	104.2	99.3	90.4	78.4	71.7	64.7	13.5	14.7
	0	100.0	96.2	93.7	96.8	99.8	99.0	98.9	98.9	96.5	91.3	23.3	28.4

The color gradient shown in the table mimics the visual change observed for resazurin reduction. Magenta/purple indicates high fluorescence values, corresponding to the reduced (pink) form of resazurin and reflecting high metabolic activity and fungal growth. Darker violet shades indicate lower fluorescence values, corresponding to the oxidized (blue) form of resazurin and reflecting reduced metabolic activity due to drug-mediated growth inhibition

Positive control well (no drug added):

Drug A: 0 µg/mL.

Drug B: 0 µg/mL.

Fluorescence value = 33,642.

Drug combination well (test well):

Drug A: 0 µg/mL.

Drug B: 2 µg/mL.

Fluorescence value = 6280.

Growth (%) of the well containing drug A at 0 µg/mL and drug B at 2 µg/mL.

$= (6280 \times 100)/33{,}642 = 18.66\%$

5. Repeat this calculation for all wells to obtain the complete growth profile (*see* Table 3).
6. Analyze the checkerboard assay further by calculating the fractional inhibitory concentration index (FICI), a standard method to evaluate drug interactions [21]. The FICI value can be calculated for all wells of the microtiter plate presenting a MIC value, according to the formula:

$$FICI = FIC_A + FIC_B$$

Where:

$FIC_A = MIC_A$ in combination/ MIC_A alone,

$FIC_B = MIC_B$ in combination/ MIC_B alone.

7. Interpret the average FICI values as follows: synergistic (FICI ≤ 0.5), additive ($0.5 < \text{FICI} < 1.0$), indifferent ($1.0 < \text{FICI} < 2.0$), and antagonistic (FICI >2.0) [22].
8. Alternatively, analyze the checkerboard data using the Synergy Finder [23]. Input data in table or matrix format (examples provided on the website https://synergyfinder.org/). Apply the following parameters: detect outliers *yes*; curve fitting *LL4*; method *Bliss*; correction *ON*.
9. Interpret SynergyFinder results as follows: values ≤ -10 antagonistic interaction, values between - 10 and 10 indicate additivity, and values ≥ 10 suggest a synergistic interaction. Figure 3a–c illustrates examples of synergic, additive, or antagonistic profiles between drugs.

3.4.2 Analysis of XTT Assays

1. After reading the absorbance values at 450 nm, subtract the absorbance of the negative control well (medium + XTT/menadione only) from each treated well.
2. For experiments with replicates, calculate the average absorbance for each condition.
3. Calculate the percentage of growth relative to the untreated biofilm (positive control, defined as 100% growth):

$$\text{Growth (\%) of the analyzed well} = (\text{absorbance} \times 100)/\text{absorbance of the positive control}.$$

4. Apply this calculation to all wells to determine the growth percentage for each condition.

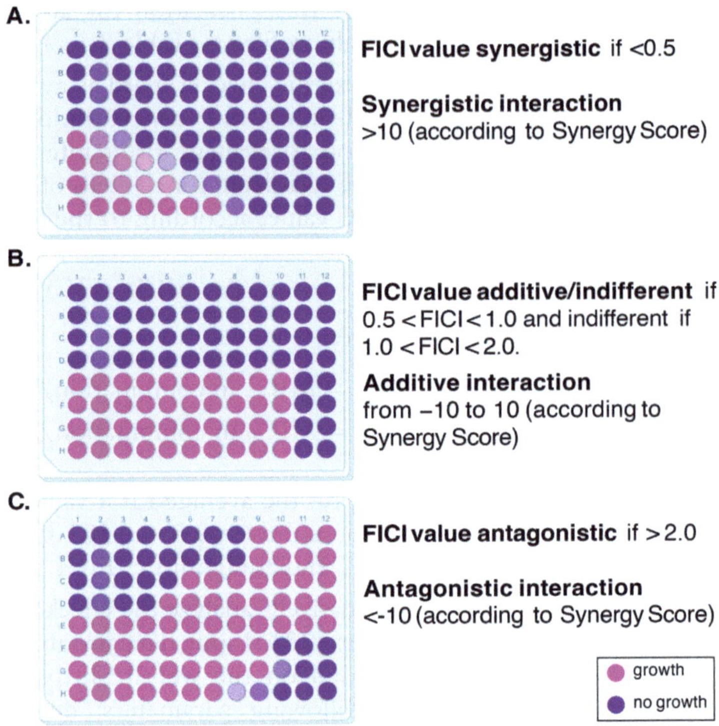

Fig. 3 Hypothetical checkerboard results of Drug A and Drug B interaction profiles using the resazurin method. The images indicate two drug interactions scored as synergistic (**a**), indifferent/additive (**b**), or antagonistic (**c**). (Created with BioRender.com (https:/BioRender.com/a3g1dhl))

4 Notes

1. After harvesting conidia, carefully clean the cabinet and sterilize it with UV for 30 min. Extreme care should be taken when multiple *A. fumigatus* strains are used to avoid cross-contamination of conidia and to minimize aerosol generation during harvest.
2. For large-scale screenings of drugs or compound libraries with unknown chemical properties, MM is recommended, as it is cost-effective and does not interfere with resazurin or XTT-based signal detection.
3. alamarBlue (Thermo-Scientific) can be used interchangeably with resazurin solution.
4. Always prepare the XTT solution fresh before each use. To ensure complete dissolution in aqueous solution, incubate the XTT at 50 °C for 30 min with occasional gentle mixing.
5. Menadione is relatively insoluble in water. Although it can dissolve at the concentrations described here, precipitation

may occur upon freezing. To avoid this, menadione stock solution can be prepared using 100% acetone and stored at −20 °C up to a week without compromising assay performance.

6. Menadione acts as an electron acceptor and can interfere with the electron transport chain in humans. It is potentially toxic at high concentrations. Always wear a lab coat, gloves, and a mask when handling the powder.
7. Preparing the XTT and menadione solution together ensures uniformity and helps prevent signal fluctuations caused by improper XTT preparation, especially when multiple plates are processed simultaneously. Some companies sell XTT reagent already containing an electron-coupling agent, in which case the addition of menadione is not necessary. An example is the "CyQUANT XTT Cell Viability Assay" (Thermo Scientific X12223).
8. Some plate readers can perform fluorescence readings from both the top and bottom of the microplate wells. If bottom-reading is used, a white bottom-clear black plate should be used (e.g., Greiner Bio One 655090, chimney well flat μClear bottom).
9. If another culture medium is required, ensure it is compatible with resazurin and does not interfere with signal detection. Preliminary validation experiments may be necessary. Certain reagents such as fetal bovine serum (FBS) and bovine serum albumin (BSA) can quench fluorescence. In such cases, the same serum concentration should be maintained in all control and test conditions to avoid misleading results.
10. Plates or the humid chamber should be wrapped in aluminum foil to protect resazurin from light exposure during incubation, as prolonged exposure to light may lead to photodegradation of resorufin, reducing the accuracy of fluorescence readings.
11. Resazurin 0.002% (15 μL from the stock) can also be added after incubation. This approach may be preferred when there is concern about potential interactions between the resazurin dye and the tested compounds during incubation, such as redox-active drugs that could nonenzymatically reduce the dye. However, the time required for color development may be extended, and this delay can affect the final results, as overgrowth or conidiation may occur in the well plate during the additional incubation period. See Table 1 for more information.
12. Absorbance measurement at 570 nm with normalization at 600 nm is possible, however, fluorescence detection generally provides more consistent and sensitive results.

13. Although resazurin signals can be detected through fluorescent or colorimetric readings, we highly recommend the use of fluorescence as the primary readout for these experiments. Fluorescent detection of resorufin offers higher sensitivity, better linearity across a wider dynamic range, and reduced background interference compared to absorbance-based methods. This is particularly important when analyzing low-density fungal populations, early growth stages, or subtle metabolic shifts. As a result, the mode of data analysis employing colorimetric measurements will not be described in this protocol.
14. It is important to ascertain whether the plate reader being used performs fluorescence readings from the top, bottom, or both. Top-read mode is recommended, as it minimizes background interference and optimizes signal detection when fluorescence is generated in the solution rather than at the surface of the well, which is exactly the case for resazurin assays. In contrast, bottom-read mode may be more appropriate for adherent cell-based assays where the signal originates near the well surface.
15. The use of clear polystyrene 96-well plates is possible; however, fluorescence signal interference between adjacent wells may occur. This can compromise data accuracy, and therefore, black, flat-bottom plates are strongly recommended for fluorescence-based assays.
16. *A. fumigatus* biofilms incubated for more than 48 h tend to conidiate and form aerial hyphae. Conidiation should be avoided, as it causes post-processing problems and reproducibility issues in both resazurin and XTT assays. If necessary, shorten the incubation period to prevent these effects.
17. Ziplock bags can be used instead of a humidified chamber. If several plates are being analyzed simultaneously, avoid stacking too many, as this can hinder heat exchange for the plates in the center of the stack. We recommend stacking no more than two plates.
18. Drug treatment time may vary according to the target drug. If longer treatment time is required, ensure that conidiation is avoided at the medium-air interface. Adjustment of drug concentration may be an option to reduce treatment time.
19. A 1–2 h incubation period is generally sufficient to observe color development resulting from formazan production in typical *A. fumigatus* wild-type strains in an experiment setup in 96-well plate assay format. See reference [16] for further information. Visual inspection of the plates may help determine the optimal incubation time for each experiment and account for strain-specific differences. If biofilms are visually detectable in the well plate but no orange color develops after incubation,

consider preparing a fresh XTT/menadione working solution, as it may have been incorrectly prepared or inactive.

20. In this 24-well plate assay format, the amount of biofilm produced is larger and can support a wider range of experimental approaches. For instance, biofilms grown in this format using glass-bottom dishes compatible with laser scanning confocal microscopy can be inspected to evaluate biofilm morphology prior to XTT addition and metabolic activity measurement. This setup is also compatible with gas-permeable plates. Additionally, this format can be useful for studying host–pathogen interactions in coculture systems that involve immune cells. The medium removed after the relevant incubations can be forwarded for cytokine determinations. For small-scale biofilm experiments, assays can be performed using standard 96-well plate formats.
21. At this stage, the biofilm can be washed 1–2 times with PBS or fresh medium (300–500 μL per well) to remove any traces of growth media and additives eventually used during biofilm formation. However, not all protocols include these washes. Therefore, evaluate whether this washing step is necessary, as it can be a source of biofilm dislodgment or perturbation. The integrity of the biofilm and the original biomass should be preserved to avoid reproductivity issues between replicated wells or plates.
22. It is also possible to perform serial dilution of the targeted drug over the biofilm-containing wells, as previously reported [18]. In this case, extra care should be taken to avoid touching or otherwise disrupting the biofilms.
23. When testing drugs or compounds with no prior information about their activity against an *A. fumigatus* biofilm, increasing concentrations of the compound and varying exposure times should be combined as experimental variables to obtain conclusive results considering the particularities of the antifungal agents being tested.
24. Proceed with the washes carefully to avoid dislodging or perturbing the biofilm. The integrity of the biofilm and the original biomass should be preserved to prevent reproducibility issues between replicate wells or plates. Additionally, removing the remaining liquid helps prevent dilution of the formazan formed during the in situ reaction. Also, if visibly disrupted biofilm areas are detected to the naked eye after the washing steps, the wells should be excluded from subsequent analyses and calculations.

Acknowledgments

Protocol development was funded by the Fundação de Amparo à Pesquisa do Estado de São Paulo (FAPESP) grant numbers 2021/04977-5 (GHG e IM), 2021/11784-9 (IM) and 2023/07785-5 (TFR), the Conselho Nacional de Desenvolvimento Científico e Tecnológico (CNPq) and Fundação Coordenação de Aperfeiçoamento do Pessoal do Ensino Superior (CAPES) grant number 405934/2022-0 (The National Institute of Science and Technology INCT FUNVIR), and the Conselho Nacional de Desenvolvimento Científico e Tecnológico (CNPq) grant 312556/2021-7 (IM) and 301058/2019-9 (GHG), the National Institutes of Health/National Institute of Allergy and Infectious Diseases from the USA, grant R01AI153356 to GHG. This work was also funded by the Joint Canada-Israel Health Research Program, jointly supported by the Azrieli Foundation, Canada's International Development Research Centre, Canadian Institutes of Health Research, and the Israel Science Foundation (GHG). The views expressed herein do not necessarily represent those of IDRC or its Board of Governors.

References

1. Earle K, Valero C, Conn DP et al (2023) Pathogenicity and virulence of *Aspergillus fumigatus*. Virulence 14. https://doi.org/10.1080/21505594.2023.2172264
2. Fisher MC, Alastruey-Izquierdo A, Berman J et al (2022) Tackling the emerging threat of antifungal resistance to human health. Nat Rev Microbiol 20:557–571. https://doi.org/10.1038/s41579-022-00720-1
3. Lestrade PP, Bentvelsen RG, Schauwvlieghe AFAD et al (2019) Voriconazole resistance and mortality in invasive aspergillosis: a multicenter retrospective cohort study. Clin Infect Dis 68:1463–1471. https://doi.org/10.1093/cid/ciy859
4. Perlin DS, Rautemaa-Richardson R, Alastruey-Izquierdo A (2017) The global problem of antifungal resistance: prevalence, mechanisms, and management. Lancet Infect Dis 17:e383–e392. https://doi.org/10.1016/S1473-3099(17)30316-X
5. Dagenais TRT, Keller NP (2009) Pathogenesis of aspergillus fumigatus in invasive aspergillosis. Clin Microbiol Rev 22:447–465. https://doi.org/10.1128/CMR.00055-08
6. Denning DW, Cadranel J, Beigelman-Aubry C et al (2016) Chronic pulmonary aspergillosis: rationale and clinical guidelines for diagnosis and management. Eur Respir J 47:45–68. https://doi.org/10.1183/13993003.00583-2015
7. Moss BJ, Kim Y, Nandakumar MP et al (2008) Quantifying metabolic activity of filamentous fungi using a colorimetric XTT assay. Biotechnol Prog 24:780–783. https://doi.org/10.1021/bp070334t
8. Kowalski CH, Morelli KA, Schultz D et al (2020) Fungal biofilm architecture produces hypoxic microenvironments that drive antifungal resistance. Proc Natl Acad Sci USA 117:22473–22483. https://doi.org/10.1073/pnas.2003700117
9. Kowalski CH, Kerkaert JD, Liu K-W et al (2019) Fungal biofilm morphology impacts hypoxia fitness and disease progression. Nat Microbiol 4:2430–2441. https://doi.org/10.1038/s41564-019-0558-7
10. Liu S, Le Mauff F, Sheppard DC et al (2022) Filamentous fungal biofilms: conserved and unique aspects of extracellular matrix composition, mechanisms of drug resistance and regulatory networks in *Aspergillus fumigatus*. NPJ Biofilms Microbiomes 8. https://doi.org/10.1038/s41522-022-00347-3
11. O'Brien J, Wilson I, Orton T et al (2000) Investigation of the Alamar blue (resazurin) fluorescent dye for the assessment of mammalian cell cytotoxicity. Eur J Biochem 267:5421–

5426. https://doi.org/10.1046/j.1432-1327.2000.01606.x

12. Monteiro MC, de La Cruz M, Cantizani J et al (2012) A new approach to drug discovery: high-throughput screening of microbial natural extracts against *Aspergillus fumigatus* using Resazurin. SLAS Discov 17:542–549. https://doi.org/10.1177/1087057111433459
13. García Carnero LC, Li X, Castro PA de et al (2025) Colistin enhances caspofungin antifungal efficacy against *Aspergillus fumigatus* by modulating calcium homeostasis and stress responses. Nat Commun 16. https://doi.org/10.1038/s41467-025-60991-z
14. Kuhn DM, Balkis M, Chandra J et al (2003) Uses and limitations of the XTT assay in studies of Candida growth and metabolism. J Clin Microbiol 41:506–508. https://doi.org/10.1128/JCM.41.1.506-508.2003
15. Kerkaert JD, Le Mauff F, Wucher BR et al (2022) An alanine aminotransferase is required for biofilm-specific resistance of *Aspergillus fumigatus* to echinocandin treatment. mBio 13. https://doi.org/10.1128/mbio.02933-21
16. Meletiadis J, Mouton JW, Meis JFGM et al (2001) Colorimetric assay for antifungal susceptibility testing of aspergillus species. J Clin Microbiol 39:3402–3408. https://doi.org/10.1128/JCM.39.9.3402-3408.2001
17. Rocha MC, Minari K, Fabri JHTM et al (2021) Aspergillus fumigatus Hsp90 interacts with the main components of the cell wall integrity pathway and cooperates in heat shock and cell wall stress adaptation. Cell Microbiol 23. https://doi.org/10.1111/cmi.13273
18. Pierce CG, Uppuluri P, Tristan AR et al (2008) A simple and reproducible 96-well plate-based method for the formation of fungal biofilms and its application to antifungal susceptibility testing. Nat Protoc 3:1494–1500. https://doi.org/10.1038/nprot.2008.141
19. Bidaud A-L, Schwarz P, Herbreteau G et al (2021) Techniques for the assessment of in vitro and in vivo antifungal combinations. JoF 7:113. https://doi.org/10.3390/jof7020113
20. Spitzer M, Robbins N, Wright GD (2017) Combinatorial strategies for combating invasive fungal infections. Virulence 8:169–185. https://doi.org/10.1080/21505594.2016.1196300
21. Meletiadis J, Verweij PE, Te Dorsthorst DTA et al (2005) Assessing in vitro combinations of antifungal drugs against yeasts and filamentous fungi: comparison of different drug interaction models. Med Mycol 43:133–152. https://doi.org/10.1080/13693780410001731547
22. Faleiro ML, Miguel MG (2013) Use of essential oils and their components against multidrug-resistant bacteria. In: Fighting multidrug resistance with herbal extracts, essential oils and their components. Elsevier, pp 65–94
23. Ianevski A, Giri AK, Aittokallio T (2022) SynergyFinder 3.0: an interactive analysis and consensus interpretation of multi-drug synergies across multiple samples. Nucleic Acids Res 50:W739–W743. https://doi.org/10.1093/nar/gkac382

Chapter 17

Aspergillus fumigatus Cell Wall: Current Advances in Structural Elucidation Methods

Chaithanya Shetty, Melba Elizabeth-Biju, and Vishukumar Aimanianda

Abstract

Being an extracellular organelle, the fungal cell wall (CW) plays a crucial role in fungal biology by providing mechanical strength and protecting from its hostile environment. In pathogenic fungi, their CW is explored with two main objectives: (a) to develop novel antifungal molecules, as the disruption of CW leads to fungal lysis due to osmotic imbalance, and (b) to decipher fungal pathogenesis, as CW is the first fungal structure to interact with the host immune system. *Aspergillus fumigatus* is an airborne filamentous fungal pathogen, which causes a spectrum of infections ranging from allergic broncho-pulmonary aspergillosis (ABPA), mild aspergilloma (formation of fungal mass in preformed lung cavities, as in post-tuberculosis condition), to life-threatening chronic pulmonary aspergillosis (CPA) and invasive pulmonary aspergillosis (IPA). CPA occurs in those with an underlying lung condition, which facilitates the colonization of *A. fumigatus*. Whereas immunocompromised hosts are highly susceptible to IPA due to their inability to eliminate inhaled conidia reaching lung alveoli, *A. fumigatus* life-cycle spans between airborne spores (conidia) and hyphae. Accordingly, its CW undergoes continuous reorganization, leading to compositional changes, which pose difficulties for antifungal drugs to function efficiently or for the host immune system to mount an effective antifungal defense. Therefore, deciphering the organization of *A. fumigatus* CW during its dynamic state and at the molecular level is essential to study its impact against antifungal drugs and host immune defenses for the better management of aspergillosis. Focusing on these objectives, the current chapter compiles protocols for the extraction of different CW components, their quantification, and the techniques used to study the structural organization of *A. fumigatus* CW.

Key words *Aspergillus fumigatus*, Conidia, Mycelia, Hydrophobins, Melanin, Polysaccharides

1 Introduction

Fungal CW forms an extracellular structure and constitutes ~30% dry mass of the fungal cell. The CW of *A. fumigatus* is composed of four major polysaccharides, namely, β-1,3-branched β-1,3-glucan, α-1,3-glucan, chitin, and galactomannan. These polysaccharides are organized and localized above the plasma membrane. In the conidial (asexual spores) morphotype, these CW polysaccharides are covered, respectively, by a melanin pigment layer and a

Natalia Mach (ed.), *Aspergillus fumigatus: Methods and Protocols*, Methods in Molecular Biology, vol. 3028,
https://doi.org/10.1007/978-1-0716-5222-0_17,

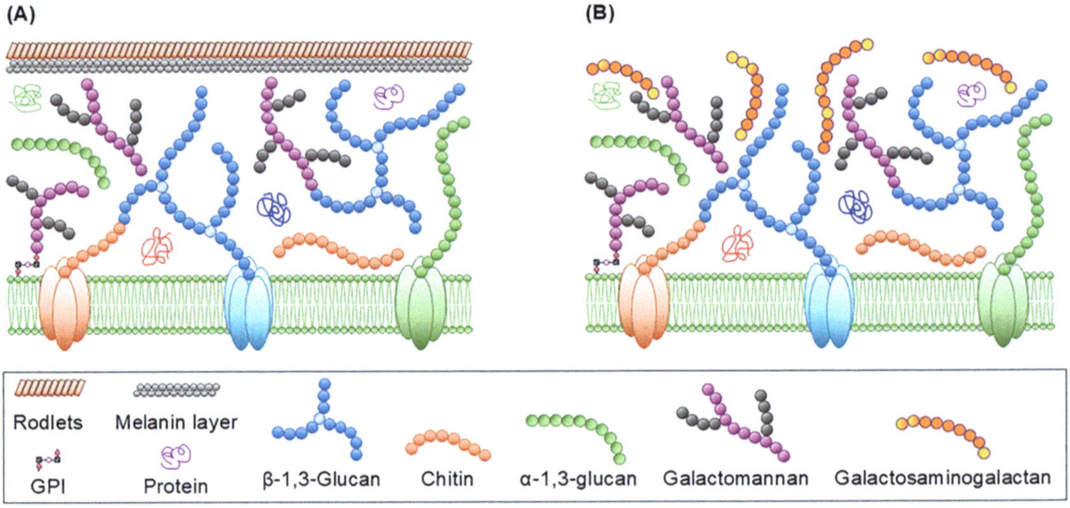

Fig. 1 *A. fumigatus* cell wall organization. (**a**) In addition to the indicated cell wall polysaccharides, in the conidial morphotype, the surface is covered by a superficial rodlet layer and underneath a melanin pigment layer. (**b**) In germinating conidia and mycelial morphotypes, the rodlet and melanin layers disintegrate, and there will be the synthesis of another cell wall polysaccharide, galactosaminogalactan

superficial rodlet layer, composed of dihydroxy-naphthalene (DHN)-melanin and a hydrophobic protein RodA. These two conidial surface layers render hydrophobicity to conidia. When conidia start to germinate upon encountering favorable conditions, the melanin and rodlet layers disintegrate exposing the polysaccharides present underneath. Germination process is also associated with the biosynthesis of another CW polysaccharide, galactosaminogalactan (GAG) (Fig. 1) [1]. β-1,3-Glucan, α-1,3-glucan, and chitin are homopolymers. Glucans are made up of glucose units linked by β-1,3- or α-1,3-glycosidic bonds. Chitin is composed of N-acetylglucosamines as the monosaccharide units linked by β-1,4-bonds. Galactomannan and galactosaminogalactan are heteropolymers. Galactomannan is composed of mannose and galactose as the monosaccharide units, whereas galactosaminogalactan is composed of galactose, galactosamine, and N-acetylgalactosamine. The presence of another glucan in which glucose units are linked by alternative β-1,3–/β-1,4-linkage is also reported in the CW of *A. fumigatus*, but the polysaccharide has been underexplored [2].

2 Materials

2.1 Fungal Materials

1. *A. fumigatus* isolates of choice.
2. Malt extract-agar (MEA) culture slants: prepared by autoclaving malt extract (2%) and agar (1.5%) in distilled water for 20 min at 121 °C, dispensing 10 mL of autoclaved solution

into 20 mL capacity sterile glass tubes, and allowing the solution to solidify at an angle of 30 °C.

3. Isotope-labeled culture slants can be prepared by autoclaving agar (1.5%), ^{13}C-labeled glucose (1%), and ^{15}N-labeled ammonium chloride (5 mM) in distilled water, dispensing and solidifying the autoclaved solution as in **iteam 2**.
4. Sabouraud liquid medium: Mix dextrose (40 g) and peptone (10 g) in distilled water (1 L) and autoclave for 20 min at 121 °C. Bring the solution to room temperature and store at 4 °C until further use.
5. *Aspergillus* minimal medium (AMM): Mix and autoclave 2% glucose in distilled water supplemented with salt-mixture solution at 121 °C for 20 min. Bring the solution to room temperature, store at 4 °C. Before using, the preprepared trace element solution should be added to it (2 mL to 1 L of this autoclaved dextrose-salt solution mixture).
6. Salt solution for AMM: Dissolve sodium nitrate ($NaNO_3$, 6 g), potassium chloride (KCl, 0.52 g), magnesium sulphate ($MgSO_4{\cdot}7H_2O$), and potassium dihydrogen phosphate (KH_2PO_4, 1.52 g) in distilled water (1 L). For convenience, a stock salt solution can be prepared at 50× concentration and added while preparing AMM (*see* **iteam 5**).
7. Trace element solution for AMM: Dissolve ethylenediaminetetraacetic acid (EDTA) disodium salt (50 g in 250 mL), zinc sulphate ($ZnSO_4{\cdot}7H_2O$, 22 g in 100 mL), boric acid (H_3BO_4, 11.4 g in 100 mL), manganese chloride ($MnCl_2{\cdot}4H_2O$, 5.06 g in 50 mL), cobalt chloride ($CoCl_2{\cdot}6H_2O$, 1.61 g in 50 mL), copper sulphate ($CuSO_4{\cdot}5H_2O$, 1.57 g in 50 mL), ammonium molybdate $(NH_4)_6Mo_7O_{24}{\cdot}4H2O$, 1.10 g in 50 mL), and ferrous sulphate ($FeSO_4{\cdot}7H_20$, 4.99 g in 50 mL). All these solutions should be prepared in distilled water. To dissolve EDTA, boil the solution. Ferrous sulfate solution should be prepared last to avoid any oxidation. Except for EDTA, mix all components, boil, then add EDTA solution. Cool to 70 °C, add KOH solution (20%, 85 mL, hot). Make volume up to 1 L with distilled water, pass through a sterile filter, and store at 4 °C until further use.
8. Tween-80 solution for collecting *A. fumigatus* conidia from agar slants: prepare 0.5% stock solution by adding 5 mL Tween-80 to 1 L of distilled water with stirring, autoclave for 20 min at 121 °C. Prepare 0.05% working solution in sterile distilled water.
9. Nylon filter, pore size 0.45 μM.
10. Automated cell counter (e.g., Luna) or hemocytometer.
11. Microcentrifuge tubes (e.g., Eppendorf).

12. Autoclaved culture flasks with sterile cotton plugs for *A. fumigatus* cultures.
13. Orbital shaking incubator maintained at 37 °C (or required temperature).

2.2 Common Buffers and Other Solutions

1. Tris–EDTA–SDS–β-ME buffer, pH 7.4: Add 6.05 g Tris-aminomethane (Tris-base, 50 mM) and 1.17 g ethylenediamine tetraacetate (EDTA, 50 mM) to 800 mL distilled water, mix and adjust the pH to 7.4 [using 1 N hydrochloric acid (HCl) or 1 N sodium hydroxide (NaOH)]. Add 2% sodium dodecyl sulfate (SDS) and 3.49 mL of β-mercaptoethanol (β-ME) to this Tris-EDTA buffer and make up the volume to 1 L with distilled water.
2. TRIS-HCl buffer, 1 M, pH 6.8–7.5: Add 121.14 g of Tris-base to 800 mL of distilled water, adjust the pH to 6.8 or 7.5 with 1 N HCl, and make up the volume to 1 L with distilled water.
3. SDS-polyacrylamide gel electrophoresis (PAGE) sample loading buffer (5×): To 1 mL Tris-HCl buffer (0.5 M, pH 6.8), add 2.5 mL glycerol, 0.08 mL bromophenol blue (100 mg/10 mL), 1.6 mL SDS (1 g/10 mL distilled water), and 2.92 mL distilled water. After homogenizing all the contents, store them as aliquots at −20 °C. Before use, add 0.4 mL β-ME or 3% dithiothreitol as a reducing agent. The sample buffer and the protein samples ratio should be 1: 4.
4. Sodium acetate buffer (0.1 M, pH 4.0 or 6.0): Add 1.36 g of sodium acetate (trihydrate) to 800 mL of distilled water, adjust the pH to 6.0 with glacial acetic acid, and make up the volume to 1 L with distilled water.
5. Phosphate-buffered saline (PBS, 1×, pH 7.4): Add 8 g of NaCl, 0.2 g of KCl, 1.44 g of dibasic sodium phosphate, and 0.245 g of monobasic potassium phosphate to 800 mL of distilled water, adjust the pH to 7.4 (with 1 N HCl or NaOH), and then make up the volume to 1 L with water.
6. Citrate-phosphate buffer (pH 6.2): Prepare 0.1 M citric acid by dissolving 19.21 g per 1 L of distilled water, and 0.2 M sodium phosphate by dissolving 35.6 g of dibasic sodium phosphate per 1 L of distilled water. Then, mix 16.9 mL of 0.1 M citric acid with 33.1 mL of 0.2 M dibasic sodium phosphate solution. This citrate-phosphate buffer solution should have a pH of 6.2. If needed, adjust the pH with 1 N HCl or NaOH.
7. Membrane extraction buffer: This buffer should be composed of HEPES [4-(2-hydroxyethyl)-1-piperazineethanesulfonic acid] (50 mM, pH 7.7), sucrose (1 M), NaCl (50 mM), EDTA (5 mM), dithiothreitol (1 mM), sodium fluoride (NaF, 10 mM), and guanidine triphosphate (GTP, 0.1 mM).

8. Colloidal Coomassie blue solution: To prepare a 1 L solution, add 50 g of ammonium sulphate (5%) to 800 mL of distilled water, followed by the addition of 100 mL ethanol (10%) and 0.2 g of Coomassie brilliant blue G-250 (0.02%). When all the contents are dissolved, add 23.5 mL of orthophosphoric acid (85%, final concentration of 2%). Make up the volume to 1 L with distilled water. At a time, 50 mL of this solution will be sufficient for staining a PAGE gel, and it should be prepared fresh before staining.
9. Lithium methyl sulfinyl carbanion preparation: Slowly dissolve dimethyl sulfoxide (DMSO, 5 mL) in tetrahydrofuran (70 mL) under an inert atmosphere (nitrogen or helium). Cool this reaction mixture (to below 4 °C), and then add 37 mL of N-butyl lithium (1.6 M) solution in tetrahydrofuran, slowly at low temperature.
10. Composition of the solution to assess synthase activity with permeabilized fungal cells: Tris-HCl buffer (0.05 M, pH 7.5), UDP-^{14}C-glucose or UDP-^{14}C-N-acetylglucosamine (0.5 mM, with specific activity of 34 nM/125 nCi; respectively, the substrates for glucan synthase and chitin synthase activities), adenosine triphosphate (ATP, 0.2 mM), GTPs (20 μM), magnesium chloride ($MgCl_2$, 0.2 mM), and N-acetylglucosamine (20 mM).
11. Phenol, 5%: Carefully add 5 mL of non-oxidized phenol to 100 mL of distilled water and stir slowly. Store it in a brown bottle at 4 °C.
12. Urea, 8 M: Dissolve 48.048 g of urea in 80 mL of distilled water by slow stirring, and then make up the volume to 100 mL with distilled water.
13. Trifluoroacetic acid (TFA), 4 N: In a brown bottle, add 30.8 mL TFA (13 N) slowly with stirring to 69.2 mL distilled water.
14. Hydrochloric acid (HCl), 8 N: Slowly with stirring, add 68 mL of concentrated HCl (11.76 N) to 32 mL of distilled water in a brown bottle.
15. Sodium metaperiodate 0.1 M: prepare in 0.1 M sodium acetate buffer, pH 4.0.

2.3 Common Reagents, Chemicals, and Enzymes

1. Acetic anhydride.
2. Acetonitrile (synonym methyl cyanide, MeCN).
3. Ammonium acetate.
4. Ammonium hydroxide (NH_4OH).
5. Bovine serum albumin (BSA).
6. Butanol.

7. Chloroform.
8. Ethanol.
9. Formic acid.
10. Guanidine thiocyanate.
11. Hydrofluoric acid (HF).
12. Hydrazine (anhydrous).
13. Hydrogen peroxide (H_2O_2).
14. Methanol.
15. Meso-inositol.
16. Potassium hydroxide (KOH).
17. Propan-1-ol, pyridine.
18. Sodium borohydride ($NaBH_4$).
19. Sodium nitrite.
20. Sulfuric acid (H_2SO_4).
21. Trichloroacetic acid (TCA).
22. Proteolytic enzyme (proteinase K), CW lysing enzyme (glucanex, novozyme), and endo-β-1,3-glucanase (LamA, from *Thermotoga neapolitana*).

2.4 Common Equipment

1. Probe sonicator with 3 mm microtip.
2. Freeze-dryer (lyophilizer).
3. Bench (3000–5000 × rpm), high-speed (to 15,000 × rpm), and ultra (to 50,000 × g) centrifuges.
4. Vacuum filtration setup and membrane filters.
5. Nitrogen gas drying setup.
6. SDS-PAGE apparatus and setup.
7. Vortex mixer.
8. Orbital shaker or rotator.
9. Heating block or water bath up to 100 °C.
10. Gas chromatography (GC), GC-mass spectrometry (GC-MS), and GC/GC-MS (non-polar, BP-1) column connected to a flame ionization detector.
11. High-performance liquid chromatography (HPLC), HPLC (DIONEX) column (either PA-1 or PA-200) connected to a pulsed electrochemical/radiometric detector.
12. Reverse-phase C18 column for HPLC.
13. NMR spectrophotometer (e.g., 14.1 T NMR spectrometer from Bruker Biospin).
14. MonoQ and Superdex 200 columns for chromatography.

3 Methods

3.1 Extraction of RodA from A. fumigatus *Conidia*

RodA from *A. fumigatus* conidial surface rodlet layer can be extracted upon dislodging the rodlet or using strong acids based on the concepts that (a) RodA is a glycosylphosphatidylinositol (GPI)-anchored CW protein and hydrofluoric acid (HF) cleaves phosphodiester linkage in a GPI-anchored protein and (b) rodlet can be compared to fungal amyloids, and any amyloid can be solubilized by strong acids such as formic acid (*see* **Notes 1–3**).

3.2 Extraction of RodA by Dislodging Conidial Rodlet Layer

1. Culture *A. fumigatus* from −80 °C glycerol stock on MEA slants. Incubate at 37 °C and then at ambient temperature. From this stock, subculture conidia on MEA slants.
2. Harvest conidia with 0.05% aqueous Tween-80 (~5 mL) from a 1-week-old *A. fumigatus* culture on MEA slants. Vortex for 60 s to release conidia from the MEA slants.
3. Pass conidial suspension through a nylon filter to remove any mycelial debris.
4. Suspend conidia in 0.05% aqueous Tween-80.
5. Sonicate conidial suspension (10 mL, collected in a 15 mL Falcon tube) twice for 10 min each time using a sonifier and 50% duty cycle.
6. Centrifuge disrupted conidia at low speed (3000 × rpm, 10 min) to remove conidial debris.
7. Subject the supernatant to ultracentrifugation at 50000 × g for 1 h.
8. Boil the pellet obtained in the SDS-PAGE sample loading buffer', and wash twice with SDS-PAGE loading buffer and thrice with water.
9. Freeze-dry water-washed material (extracted RodA).
10. Treat dried material with 100% TFA for 10 min at an ambient temperature and remove acid by purging a stream of nitrogen; store dried material until further use.
11. Boil dried material in SDS-PAGE sample buffer, and subject to SDS-PAGE on a 15% gel.
12. Reveal protein bands by colloidal Coomassie staining.

3.2.1 Extraction of RodA Using Hydrofluoric Acid (HF)

1. Harvest conidia with 0.05% Tween as described above (Subheading 3.2, **steps 1–3**).
2. Count conidia, freeze-dry conidia, and record conidial dry weight.
3. Weight 10 mg (dry weight) of conidia.

4. Add concentrated HF (48%) to dried conidia taken in an Eppendorf tube and placed in an icebox: to 10 mg of conidia, add 200 μL of HF.
5. Leave the above sample in a cold room for 60 h (*see* **Note 4**).
6. Centrifuge the contents at 5000 × rpm for 10 min and collect the supernatant. If the supernatant still contains a trace of conidia, repeat the centrifugation step.
7. Dry the supernatant under a stream of nitrogen, wash twice with water, and dry each time again under a stream of nitrogen.
8. Reconstitute in water, subject to reverse phase chromatography, dry purified RodA under nitrogen, and store dried material (RodA) at 4 °C until further application (*see* **Notes 5** and **6**).
9. Purity of the extracted RodA could be assessed by suspending dried material in water and subjecting an aliquot to SDS-PAGE (15% gel) as above after boiling this aliquot for 10 min in SDS-PAGE sample buffer. The sample buffer and aliquot ratio should be 1:4.
10. Reveal protein bands by colloidal Coomassie staining (*see* **Note 7**).

3.2.2 Extraction of RodA Using Formic Acid

1. Harvest conidia from malt-extract agar slants using 0.05% aqueous Tween as described above (Subheading 3.1.1), count conidia, freeze-dry, and record conidial dry weight.
2. To 10 mg of dry conidia taken in a 2 mL Eppendorf tube and placed in an ice box, add 200 μL of formic acid (100%) and incubate at 4 °C for 6 h.
3. Collect the supernatant after centrifugation of the content (5000 × rpm, 10 min). If the supernatant still contains traces of conidia, repeat the centrifugation and pool the supernatants of the two centrifugations.
4. Dry the supernatant under a stream of nitrogen.
5. Wash the dried material twice with water, and dry purging a stream of nitrogen.
6. Store the dried material (RodA) at 4 °C until further application (*see* **Note 6**).
7. Purity of extracted RodA can be assessed by SDS-PAGE on 15% gel. To prepare the probe, reconstitute dried RodA in distilled water, take an aliquot, and boil it in SDS-PAGE sample buffer (sample buffer and reconstituted RodA ratio should be 1:4). Reveal protein bands by colloidal Coomassie staining (*see* **Note 7**).

3.3 Extraction of Melanin from *A. fumigatus* Conidia

1. Harvest conidia with 0.05% aqueous Tween-80 from 1-week-old *A. fumigatus* culture on MEA slants. Vortex for 60 s to release conidia from the agar culture.
2. Separate any residing mycelia from the conidial suspension by filtering it through the 0.45 μm filter and count conidia.
3. Freeze-dry conidia and record conidial dry weight.
4. Weigh 10 mg dry weight of conidia.
5. Add proteinase K and glucanex (both at a concentration of 10 mg/mL, prepared in an acetate buffer of pH 6.0) to dry conidia. For 10 mg, add a total volume of 1 mL proteinase-K and glucanex mixture.
6. Incubate at 30 °C overnight. Collect the pellet by centrifugation at 5000 × rpm for 10 min.
7. Wash the pellet with PBS.
8. Incubate the pellet with 4 M guanidine thiocyanate (denaturant) at an ambient temperature overnight, with intermittent vortex.
9. Collect the dark material formed by centrifugation at 5000 × rpm for 10 min and wash the pellet twice with PBS.
10. Incubate the pellet with proteinase K (1 mg/mL) at 70 °C for 6 h.
11. Collect the pellet by centrifugation at 5000 × rpm for 10 min, and wash with PBS.
12. Extract the pellet with chloroform, thrice.
13. Boil the insoluble material obtained after extraction with 6 M HCl for 1 h.
14. Collect the insoluble material by centrifugation at 5000 × rpm for 10 min.
15. Wash the insoluble material extensively with PBS.
16. Dialyze the insoluble materials (melanin pigments as melanin ghosts) against distilled water for a minimum of 10 days, with regular exchange of distilled water (*see* **Notes 8** and **9**).

3.4 Extraction of β-1,3-Glucan from *A. fumigatus*

Branched β-1,3-glucan is a pan-fungal polysaccharide as well as a major component present in the CW of *A. fumigatus*. It is synthesized by a transmembrane protein, *FKS1*, which encodes for β-1,3-glucan synthase that forms a β-1,3-glucan synthase complex along with Rho-proteins. The protocol to isolate CW β-1,3-glucan from the core CW structure is presented here.

1. As above, harvest conidia using 0.05% aqueous Tween-80 from 1-week-old *A. fumigatus* culture on MEA slants. Vortex for 60 s to release conidia from malt-agar culture.

2. Pass conidial suspension through a nylon filter to remove any mycelia.
3. Either freeze-dry conidia and record conidial dry weight or inoculate conidia (1×10^8) in 200 mL culture medium and maintain at 37 °C for 20–24 h in a shaking incubator (with 200 × rpm). Collect mycelia by vacuum filtration, and wash collected mycelia and suspend in distilled water.
4. Subject fungal suspension to probe sonication using a sonifier at 50% duty cycle and probe oscillation rate of 6/s. Perform pulsed sonication with the suspension of conidia (1 min, 4 times) or mycelia (1 min, 3 times).
5. Centrifuge the contents at 5000 × rpm for 10 min, collect the pellet, and wash with aqueous Tween-80 (for conidial pellet) or water (for mycelia). The resultant pellet will have CW polysaccharides, entangled CW proteins, and membrane fraction attached to the CW.
6. Boil the pellet obtained twice with TRIS-EDTA-SDS-β-ME buffer for 20 min. β-ME denatures proteins, while SDS solubilizes the lipids and denatured proteins from the CW and cell membrane, allowing the isolation of only CW polysaccharides. The pellet and buffer ratio should be 1:4 (volume by volume).
7. Collect CW polysaccharides upon centrifugation at 5000 × rpm for 10 min and wash twice with distilled water.
8. Subject the pellet obtained after washing and centrifugation of the CW polysaccharide to alkali-fraction (boiling at 65–70 °C twice). For this fractionation, to approximately 40 mg wet weight of the CW polysaccharide, add 10 mL of alkali (1 M NaOH supplemented with 0.05 M $NaBH_4$). The addition of $NaBH_4$ prevents the leaching of monosaccharides from reducing ends of the CW polysaccharides, which are highly active due to reactive aldehyde or keto-groups.
9. Each time after boiling the pellet with alkali, collect and pool the supernatant obtained after centrifugation of the contents (5000 × rpm, 10 min) as the alkali-soluble fraction.
10. Subject alkali-insoluble fraction to extensive washing with distilled water and freeze-dry.
11. To 10 mg of freeze-dried alkali-insoluble CW fraction, add 1 mL sodium metaperiodate and incubate in dark for 4 days (*see* **Note 10**).
12. Destroy excess of sodium metaperiodate by the addition of ethylene glycol (1 mL per 10 mL sodium meta-periodate, volume:volume).
13. Precipitate polysaccharide from the reaction mixture with 4 volumes of ethanol (at 4 °C overnight), collect the precipitate

by centrifugation at 5000 × rpm for 10 min, wash the pellet obtained twice with ethanol, and freeze-dry.

14. Reduce the freeze-dried pellet with 100 mM ammonium hydroxide containing 0.05 M $NaBH_4$, 10 mL per 1 mg of dried pellet.
15. Remove excess $NaBH_4$ by slowly adding 2% acetic acid dropwise, until frothing stops.
16. Collect the pellet by centrifugation at 5000 × rpm for 10 min.
17. To the pellet, add 10% acetic acid (1 mL to 10 mg wet weight) and incubate at 100 °C for 1 h. Collect the insoluble material by centrifugation at 5000 × rpm for 10 min, wash extensively with distilled water, and freeze-dry (*see* **Notes 11** and **12**).
18. To the freeze-dried material, add 40% NaOH: to 1 mg of dry sample, add 1 mL of NaOH solution supplemented with 0.05 M $NaBH_4$. Incubate at an ambient temperature for 4 days.
19. Dilute NaOH by adding distilled water, add 2 volumes of ethanol, and leave overnight at 4 °C. Separate the precipitate formed by centrifugation at 5000 × rpm for 10 min (*see* **Note 13**).
20. Wash the precipitate twice with ethanol.
21. To the precipitate, add 0.1 M sodium acetate (1 mL/ mg of the pellet, pH 4.0). Aliquots of freshly prepared 1 M sodium nitrite (50 μL per mg of the pellet) should be added three times at hourly intervals at 50 °C.
22. Concentrate the reaction mixture to dryness and suspend in water. Wash the insoluble fraction thoroughly with water and freeze-dry to obtain β-1,3-glucan.

3.5 Extraction of α-1,3-Glucan from A. fumigatus

α-1,3-Glucan is the second major polysaccharide in the CW of *A. fumigatus*, synthesized by the membrane-bound α-1,3-glucan synthases, and there are three α-1,3-glucan synthases identified in the genome of *A. fumigatus*, encoded by *AGS1*, *AGS2*, and *AGS3*. α-1,3-glucan forms a cementing network in the CW, and it can be extracted from the alkali-insoluble fraction.

1. Destroy excess $NaBH_4$ in the collected alkali-soluble fraction (Subheading 3.3, **step 8**) by slowly adding 2% acetic acid.
2. Dialyze against distilled water for 2–3 days and freeze-dry the sample.
3. To freeze-dry alkali-insoluble CW fraction, add 1 mL of sodium metaperiodate per mg of alkali-insoluble fraction and incubate in the dark for 4 days.

4. Destroy excess of sodium metaperiodate by adding ethylene glycol: 1 mL per 10 mL sodium meta-periodate, volume: volume.
5. Precipitate polysaccharide by adding 4 volumes of ethanol (4 °C, overnight), collect the precipitate by centrifugation at 5000 × rpm for 10 min, wash the pellet twice with ethanol, and freeze-dry the pellet.
6. Reduce the freeze-dried pellet with 100 mM ammonium hydroxide containing 0.05 M $NaBH_4$, using 10 mL per 1 mg of dried pellet.
7. Remove excess $NaBH_4$ by slowly adding 2% acetic acid.
8. Collect the pellet by centrifugation at 5000 × rpm for 10 min.
9. To the pellet, add 10% acetic acid; use 1 mL for 10 mg wet weight of the pellet. Incubate at 100 °C for 1 h.
10. Collect the insoluble material by centrifugation at 5000 × rpm for 10 min, wash extensively with distilled water, and freeze-dry to obtain α-1,3-glucan (*see* **Notes 14** and **15**).

3.6 Extraction of Chitin from A. fumigatus

There are eight synthases for chitin biosynthesis in *A. fumigatus*, classified as two families. The family 1 chitin synthases are characterized by their catalytic sites bordered by transmembrane regions on both sides, and family 2 chitin synthases have a central protein core that is bound to the plasma membrane through multiple helices of its C-terminus. Chitin can be extracted directly from conidia or mycelia of *A. fumigatus*.

1. Collect conidia or mycelia as in Subheading 3.3, **steps 1–3**.
2. Boil conidia/mycelia in 5% KOH for 30 min, and wash the insoluble material thrice in distilled water. Add 10 mL of 5% KOH to 10 mg of conidial/mycelial sample.
3. Suspend insoluble material in H_2O_2:glacial acetic acid mix (1: 1) and autoclave at 121 °C for 20 min (*see* **Note 16**).
4. Separate insoluble material by centrifugation at 5000 × rpm for 10 min. Wash thrice with distilled water.
5. Boil in 10 mL of 5% KOH for 30 min, twice (*see* **Note 17**).
6. Separate insoluble material by centrifugation at 5000 × rpm for 10 min, wash thrice with distilled water, and freeze-dry to obtain chitin fibers.

3.7 Extraction of Galactomannan from A. fumigatus

Galactomannan occurs in three different forms in the CW of *A. fumigatus*: (a) anchored to GPI (glycosylphosphatidylinositol), (b) covalently linked to the non-reducing end of branched β-1,3-glucan, and (c) in the free form that is majorly released into the culture medium. Protocols have been developed to purify this polysaccharide in GPI-anchored form (lipo-galactomannan), and that is released into the culture supernatant (galactomannan). However, a strategy has not been developed to purify the CW fraction of galactomannan that is covalently linked to the non-reducing terminus of branched β-1,3-glucan. Recently, *A. fumigatus* conidium-specific mannan has been isolated from the cell wall and characterized [3, 4]. Liquid state NMR analysis indicated that this conidium-specific mannan present in the cell wall contains mannose residues linked by α-1,6-glycosidic bonds and short side chains that is composed of β-1,4-linked glucose (16%), galactose (24%), galactose linked to N-acetylglucosamine (52%), whereas 8% of the mannose residues have been found without any side chains [4].

3.7.1 Purification of Galactomannan from Culture Supernatant

1. Culture *A. fumigatus* in a fermenter maintained at 25 °C and 500 × rpm for 15 min in Sabouraud medium, which is ideal for the purification of galactomannan.
2. Remove mycelia from the culture by filtration, add 4 volumes of ethanol to the culture filtrate, and incubate at 4 °C overnight.
3. Separate the ethanol precipitate by centrifugation at 5000 × rpm for 10 min. Wash the precipitate thrice with ethanol, suspend in distilled water, and freeze-dry.
4. Add 1 mL of anhydrous hydrazine to 15 mg of ethanol precipitate and incubate overnight at 105 °C.
5. From the slurry formed, precipitate insoluble material by adding 4 volumes of ethanol.
6. Collect the precipitate by centrifugation at 5000 × rpm for 10 min, wash twice with ethanol and dialyze against distilled water.
7. Centrifugate at 5000 × rpm for 10 min and subject the pellet to nitrous acid hydrolysis by adding sodium nitrite prepared in acetate buffer (pH 4.0) to a final concentration of 1.5 N nitrous acid.
8. After overnight incubation of the above mixture at ambient temperature, dialyze it against distilled water and precipitate by adding 4 volumes of ethanol.
9. Separate the pellet by centrifugation at 5000 × rpm for 10 min. Wash with ethanol and freeze-dry.
10. Suspend freeze-dried sample in distilled water, pass it through a MonoQ column, and collect the unbound fraction eluted with 25 mM Tris-HCl buffer, pool, and freeze-dry to obtain galactomannan (*see* **Note 18**).

3.7.2 Extraction of Lipo-Galactomannan

1. Culture *A. fumigatus* in a fermenter maintained at 25 °C and 5000 × rpm for 15 h in Sabouraud medium and collect the mycelia by filtration.
2. Extract mycelial membrane fraction by freeze-thaw method: Transfer mycelia into a mold made up of stainless steel and sequentially freeze at −7, −15, −40, and −80 °C for 48 h. Thaw frozen mycelia at room temperature and once again subjected to sequential freezing at above temperatures for 24 h. Repeat freeze-thaw cycle 3 times. Subject frozen material to thawing upon layering the material between absorbent filter paper and applying pressure (1 mPa, pressure thaw). Dry obtained membrane should be stored for 48 h at 45 °C and stored until use.
3. Add chloroform-methanol (1:1, volume/volume) to the dried membrane fraction so that the final chloroform-methanol-water ratio will be (10:10:3, volume-wise).
4. Stir the content at an ambient temperature for 2 h, centrifuge at 15000 × rpm for 10 min.
5. Resuspend the pellet in chloroform-methanol-water mix (10: 10:3 ratio) and repeat the extraction once.
6. Dry the pellet, grind, and suspend the pellet in 50 mL of 50 mM Tris-HCl buffer supplemented with 5 mM sodium azide.
7. Add 50 mg of protease from *Streptomyces griseus* (~5 units/mg, pre-incubated for 30 min) at 37 °C for 48 h.
8. Stop the enzymatic reaction by heating at 100 °C for 5 min.
9. Add propan-1-ol to the enzyme digest to a final concentration of 5%, and separate the insoluble material by centrifugation at 10000 × rpm for 10 min.
10. Fractionate soluble material by chromatography based on the principle of hydrophobic interaction on a column. Equilibrate the column with 5% propan-1-ol in 100 mM ammonium acetate at a flow rate of 10 mL per hour.
11. Elute the bound fractions with a gradient of 5–60% propan-1-ol in 100 mM ammonium acetate.
12. Detect the presence of carbohydrates in the collected fraction by the phenol-sulfuric acid method for total sugar. Pool the fractions positive for carbohydrates, dialyze against distilled water, and freeze-dry.
13. Solubilize freeze-dried sample in butanol-saturated water (50 mL) and partition five times with water saturated butanol (20 mL each time) to remove any contaminating glycolipids.
14. Dialyze the aqueous phase containing lipo-galactomannan against water and freeze-dry.

3.7.3 Extraction of Conidial Specific Mannan

1. Subject alkali soluble fraction obtained from the conidial morphotype, after dialysis against water, to centrifugation at 5000 × rpm for 10 min to separate pellet from the supernatant.
2. Treat the pellet with a recombinant endo-β-1,3-glucanase (LamA, from *T. neapolitana*) in sodium acetate buffer, pH 6.0, overnight.
3. Dialyze the sample, freeze-dry, reconstitute in water, and then submit it to anion exchange chromatography on a Q-Sepharose column; equilibrate the column with TRIS-HCl buffer at a flow rate of 1 mL/min.
4. Elute bound fraction with a gradient of NaCl: 30 min—0–0.25 M, 30–40 min—0.25–0.5 M and 40–50 min—isocratic 0.5 M NaCl in 10 mM Tris-HCl buffer, pH 7.5.
5. Pool fractions positive for mannan by colorimetric method (*see* **Note 19**).
6. Subject pooled mannan-positive fractions to gel filtration chromatography on a Superdex 200 column to obtain conidia-specific mannan.

3.8 Purification of A. fumigatus Galactosaminogalactan

Galactomannan is a germination-specific polysaccharide, synthesized and mainly released into the culture medium, while only a fraction can be found in the alkali-soluble fraction. Therefore, culture supernatant is ideal for the purification of galactosaminogalactan (*see* **Notes 20** and **21**).

1. Collect the culture supernatant of *A. fumigatus* grown in Brian medium after 72 h of culture.
2. Add 2.5 volumes of ethanol to the culture supernatant and allow leave overnight at 4 °C.
3. Collect the pellet by centrifugation at 5000 × rpm for 10 min and wash the pellet thoroughly with 150 mM NaCl.
4. Extract the pellet with 8 M urea, twice, at ambient temperature with regular agitation.
5. Pool the supernatant of the urea extract, dialyze against distilled water, and freeze-dry to obtain soluble galactosaminogalactan (*see* **Note 22**).
6. Wash urea insoluble material with distilled water and freeze-dry to obtain urea insoluble galactosaminogalactan (*see* **Note 22**).

3.9 Quantification of A. fumigatus CW Polysaccharides

Alkali-insoluble fraction (*see* Subheading 3.3, **steps 1–8**) is composed of β-1,6-branched β-1,3-glucan linked to chitin and galactomannan, whereas α-1,3-glucan and galactomannan constitute alkali-soluble fraction that form cementing network. In the mycelial morphotype of *A. fumigatus*, the germination specific CW polysaccharide, GAG, is also a constituent of the alkali-soluble fraction.

The overall composition of alkali-insoluble and alkali-soluble fractions can be determined by two different methods: (i) enzymatic digestion of the CW fractions followed by the quantification of the constituent monosaccharides released, exploiting colorimetric methods or HPLC (specifically DIONEX), or (ii) gas chromatography upon acid depolymerization and derivatization of the constituents present in the CW. The enzymatic method fails to ensure complete depolymerization of polysaccharides, and high-performance chromatography may have batch-to-batch variation due to manual preparation of the solvents required. Gas chromatography, by far, is thus the best method for quantifying the fungal CW polysaccharides. Therefore, the steps involved in the quantification of CW polysaccharides by the gas chromatography technique are presented here.

1. Suspend alkali-insoluble or soluble fractions in distilled water by pulsed probe sonication. Prepared a stock solution of alkali-insoluble or soluble suspension at 10 mg/mL.
2. Transfer 20 μL (200 μg) of the suspended fractions into 4 mL glass vials meant for acid hydrolysis at elevated temperature.
3. Prepare a standard mixture of monosaccharides at 100 μg/mL concentration, take 100 μL in another glass vial.
4. Add 20 μL of meso-inositol, corresponding to 4 μg, to all the samples.
5. Dry all the samples in a desiccator using NaOH pellets.
6. Add 400 μL of 4 N TFA or 8 N HCl to the vials, cap the bottles tightly, and incubate for 4 h at 100 °C (*see* **Note 23**).
7. Unscrew the cap and dry the sample in the vials in a desiccator using NaOH pellets.
8. Add 1 mL of 0.1 M ammonium hydroxide containing 10 mg of $NaBH_4$ to each vial, incubate at ambient temperature in the dark for 7–8 h.
9. Dry the samples in a desiccator using NaOH pellets.
10. Add methanol (300 μL per vial) and dry in a desiccator (under NaOH) and repeat it once.
11. To the dried samples in the vials, add 200 μL of acetic anhydride and 50 μL of pyridine, screw the vials with caps, and incubate overnight at ambient temperature and in the dark.
12. Add 500 μL of chloroform to each vial, fill the rest of the volume in the vials with distilled water, vortex thoroughly, and remove the water later. Repeat this extraction 4–5 times until the chloroform extract is clear.
13. Dry the chloroform extract under a stream of nitrogen. Reconstitute the dried sample in 200 μL of methanol and inject 1 μL into the gas chromatography column.

14. Identify the peaks corresponding to the standard sugar mixture in the alkali-insoluble or alkali-soluble fractions (*see* **Note 24**).
15. Calculate the concentrations of each monosaccharide peak based on the area occupied by meso-inositol (the internal standard) and the standard monosaccharide mixture (*see* **Note 25**).

3.10 Quantification of Synthase Activity Using *A. fumigatus* Membrane Preparation

The plasma membrane (transmembrane) bound synthases are involved in the biosynthesis of β-1,3-glucan, α-1,3-glucan and chitin. A protocol has been standardized to study β-1,3-glucan synthase activity using *A. fumigatus* mycelial membrane fraction.

1. Disrupt *A. fumigatus* mycelia using 0.5 mm glass beads at 4 °C, and subject the disrupted contents to low-speed centrifugation at 5000 × rpm for 10 min.
2. Collect the supernatant and subject it to high-speed centrifugation at 100000 × rpm for 30 min to separate the pellet (membrane fraction).
3. Homogenize the pellet with the membrane extraction buffer, and determine protein concentration in the homogenate by Bradford assay.
4. Adjusting the protein concentration suitably, add UDP-^{14}C-glucose as the substrate for β-1,3-glucan synthesis [5]. The reaction mixture is composed of 2 mg/mL α-amylase, 3 mg/mL bovine serum albumin, 0.5 mM UDP-glucose, and 3.14 μM UDP-^{14}C-glucose (corresponding to 1 μCi/mL) in membrane extraction buffer. Then, add 100 μL of membrane preparation (containing 100 μg protein) with 33 μL of the reaction mixture. Incubate the reaction mixture-membrane preparation for 15 min, and then, stop the reaction by adding 1 mL of cold 5% trichloroacetic acid. This radiolabeled substrate incorporated into the acid-insoluble fraction is the measure of β-1,3-glucan biosynthesis.
5. Recover the precipitate by vacuum filtration using a membrane filter, wash the filter twice with 5% trichloroacetic acid, and measure the radioactivity on the filter using a scintillation counter.

3.11 Quantification of Synthase Activity by Cell Permeabilization Protocol

Although the quantification of glucan or chitin synthase activity using membrane preparation of the *A. fumigatus* mycelia is effective, this protocol could lead to the synthesis of only linear β-1,3-glucan or chitin. However, these polysaccharides undergo remodeling in the CW space. The major remodeling processes are the branching of β-1,3-glucan and its cross-linking with chitin as well as galactomannan. The incapability of membranes to synthesize remodeled CW components could be because of the distortion of the components essential for the synthesis of a remodeled CW

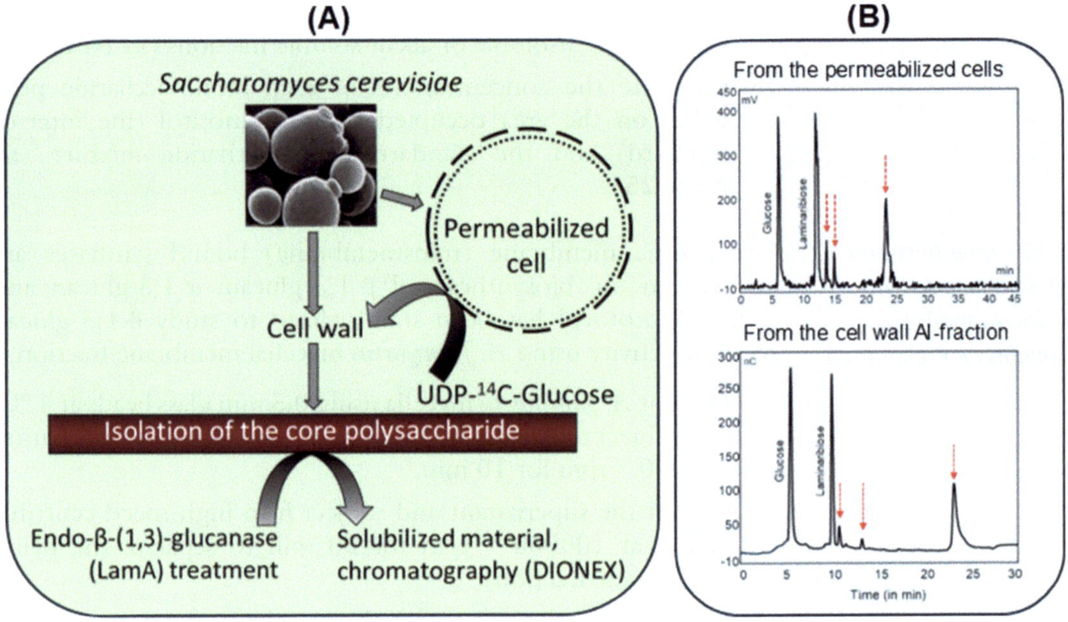

Fig. 2 Strategy to study fungal CW remodeling activity. (**a**) Protocol for fungal cell permeabilization, extraction of neo-synthesized polymers, and their enzymatic solubilization. (**b**) High-performance liquid chromatography profiles of the enzyme-solubilized neo-synthesized polymers (radiolabeled-above and unlabeled-below); peak eluted at ~23 min is a neosynthesized B-1,6-glucan, and those peaks between laminaribiose/β-1,6-glucan are β-1,6-branched β-1,3-oligosaccharides (trimer and tetramer, respectively)

due to mechanical shearing of the mycelia. To overcome the limitation of membrane fraction, we developed an in situ protocol, where we permeabilize fungal cells by osmotic shock and then incubate permeabilized cells with the substrate essential for CW polysaccharide biosynthesis (Fig. 2). Permeabilization of *A. fumigatus* hypha is complicated. Therefore, *Saccharomyces cerevisiae* is used as a model fungus to standardize this strategy. To deduce a specific gene function potentially involved in the CW remodeling, *S. cerevisiae* gene deletion mutant library could be screened, and the homologous gene deletion mutants of *A. fumigatus* can be compared.

1. Incubate *S. cerevisiae* (10^{10} yeasts) in EDTA (0.14 M, 10.5 mL) and β-ME (14.3 M, 72 μL) in a shaking incubator for 1 h at 30 °C.
2. Recover the yeasts by centrifugation at 5000 × rpm for 10 min.
3. Wash the yeasts with 15 mL of 0.8 M sorbitol.
4. Incubate the yeasts at 30 °C for 1 h in 20.2 mL of citrate-phosphate buffer containing 1 mM EDTA and 0.8 M sorbitol.
5. Collect the yeasts by centrifugation at 3000 × rpm for 10 min.
6. Suspend the yeasts in 15 mL of 0.05 M Tris-HCl buffer, pH 7.5, and place on ice for 15 min.

7. Collect the yeasts by centrifugation at 3000 × rpm for 10 min. Suspend them in 1 mL of 0.05 M Tris-HCl buffer. This will be a stock of permeabilized yeasts.
8. To 5×10^8 permeabilized yeasts, add 67 μL of the reaction mixture composed of 0.05 M Tris-HCl buffer, the glucan synthase substrate UDP-^{14}C-glucose (0.5 mM, with specific activity of 34 nM or 125 nCi), ATP (0.2 mM), GTPs (20 μM), $MgCl_2$ (0.2 mM) as well as N-acetylglucosamine (*see* **Notes 26–28**).
9. After incubation at 30 °C for 1 h, add two volumes of ethanol, and incubate the mixture for at least 1 h, at −20 °C.
10. Pellet the contents by centrifugation at 5000 × rpm for 10 min. Wash the pellet with 500 μL distilled water for three times.
11. Incubate the pellet with 500 μL 1 N NaOH containing 0.5 M $NaBH_4$ at 65 °C for 1 h and repeat this incubation for one more time upon pelleting the content between incubations by centrifugation at 15000 × rpm for 10 min.
12. Recover the precipitate by vacuum filtration using a membrane filter, wash the filter twice with 5% trichloroacetic acid, and measure the radioactivity on the filter using a scintillation counter.
13. To characterize neo-synthesized polysaccharides, collect the alkali-insoluble fraction after alkali-fractionation (*see* **step 11**) by centrifugation at 15000 × rpm for 10 min. Wash it with water, thrice. Subject collected alkali-insoluble fraction to specific enzymatic digestion and subject the digest to high-performance liquid chromatography (DIONEX) to characterize enzymatic digestion products formed.
14. In the DIONEX profile, the peaks appearing for the enzymatic digest could be compared to the standard oligosaccharides. Then, unidentified peak fractions could be collected and subjected to liquid state NMR analysis to identify their structures. For NMR analysis, permeabilization and neo-synthesis protocols will be repeated using unlabeled substrates.

3.12 Deducing the Linkage Pattern (Structural Elucidation) of CW Polysaccharides

3.12.1 Enzymatic Method

1. Suspend alkali-insoluble or soluble fractions in distilled water by pulsed probe sonication. Prepare a stock solution of alkali-insoluble or soluble suspension at a concentration of 10 mg/mL.
2. Transfer 20 μL (200 μg) of suspended fractions into an Eppendorf tube, and add 10 μL of CW hydrolytic enzymes of known activity and 170 μL of 0.1 M acetate buffer, pH 6.0.
3. Incubate the reaction mixture at 30 °C for 2 h, and collect the supernatant by centrifugation at 5000 × rpm for 10 min.
4. Subject the supernatant (enzyme digest) to DIONEX. The peaks could be compared to the standard oligosaccharides,

whereas unidentified peak fractions could be collected and subjected to liquid-state NMR analysis to identify their structures.

5. Based on enzyme specificity, the linkage patterns of different CW polysaccharides can be determined (*see* **Note 29**).

3.12.2 Chemical Method

1. Subject 2 mg of purified CW polysaccharides into methylation using 1 mL of lithium methyl sulfinyl carbanion in inert atmosphere. Quench excess of lithium methyl sulfinyl carbanion by slowly adding water, which leads to the formation of methane (evaporates) and sulfone that can be removed by distillation/flash evaporation.
2. Subject distilled material to acid hydrolysis as presented under Subheading 3.8, **steps 2–14**, but subject the sample to gas chromatography (GC) as well as gas chromatography (GC)-mass spectrometry (MS) (*see* **Note 30**).
3. From GC, the purity of the derivatization could be deduced, whereas by GC-MS analysis, based on the ionization pattern and mass/charge ratio of the peaks, the specific glycosidic linkages in a CW polysaccharide could be deduced.

3.12.3 Solid-State Nuclear Magnetic Resonance Spectrometry

Chemical methods for the structural elucidation of CW polysaccharides have disadvantages as they require CW destruction and isolation of the constituent components. During this process, fine structural details could be lost, and the harsh chemicals used to isolate CW components could destroy non-covalent/weak interactions between CW components and derivatize or degrade them. In an arduous search for a technique to characterize an intact fungal CW, we and others exploited the solid-state nuclear magnetic resonance (ssNMR) technique to study CW polysaccharide structure, functional assembly, and their polymorphism at atomic levels.

Exploiting ssNMR, we demonstrated that while breaking dormancy, (i) molar ratios of the major CW polysaccharides of *A. fumigatus* undergoes significant changes and rearrangements, and (ii) exploiting water-polysaccharide polarization transfer analysis, we observed that the surface exposure of CW glucan increases and that of chitin occupies the location deeper in the CW. Also, for the first time, we identified that the conidial CW matrix consists of triglycerides in high abundance. The chemical method failed to identify triglycerides in the CW matrix, as SDS used in the CW fractionation solubilizes this component from the CW.

1. Subculture *A. fumigatus* conidia collected from a week-old 2% malt-extract agar tube on 2% agar slant supplemented with 1% ^{13}C-glucose, 5 mM $^{15}NH_4Cl$, and supplemented with salt solution and trace elements (*see* **Note 31**).

2. Incubate the slants at 37 °C overnight and then at an ambient temperature for 15–18 days.
3. Harvest conidia from the slants as described earlier (Subheading 3.1.1, **steps 1–4**).
4. Inoculate 1 × 10^8 conidia into liquid *Aspergillus* minimal medium composed of 1% ^{13}C-glucose and 5 mM $^{15}NH_4Cl$.
5. Incubate the inoculum at 37 °C in a shaking incubator (250 × rpm) for 5 h and 8 h. Collect the contents by centrifugation at 5000 × rpm for 10 min. Wash twice with distilled water. Conidia collected from agar slants will be dormant conidia, whereas those cultures in liquid medium for 5 h and 8 h, respectively, are the swollen and germinating morphotypes.
6. Acquire the NMR spectrum of the different morphotypes of *A. fumigatus* using 14.1 T NMR spectrometer using 3.2-mm or 4-mm HCN MAS probes. Analyze spectral profile for the identification of CW polysaccharides and their linkage patterns.

4 Notes

1. The initial protocol developed to extract RodA from the conidia was to dislodge rodlet layer by sonication (using a probe sonicator) and solubilizing RodA from rodlets [6]. However, this purification protocol is not suitable for the quantification of RodA concentration, as the sample buffer used for extraction interferes with protein quantification methods. Moreover, probe sonication may not dislodge rodlets from conidia completely, and extracted RodA is often contaminated with other CW components due to the high energy/frequency of sonication.
2. Based on the prediction that RodA is a glycosylphosphatidylinositol (GPI)-anchored CW protein, HF was used to extract RodA from conidia. HF cleaves the phosphodiester linkage in a GPI-anchored protein [7]. Indeed, RodA could be extracted using HF [7]. In addition, this extraction protocol using HF has been standardized with conidial count, the concentration of hydrofluoric acid, temperature, and duration of conidial incubation with hydrofluoric acid, which allows the quantification of RodA per conidium [7].
3. Later, proteomic analysis indicated that RodA is not a GPI-anchored protein. On the other hand, *A. fumigatus* RodA-rodlet layer can be compared to amyloids. Therefore, formic acid, which is known to solubilize an amyloid protein, was exploited for the extraction of RodA [8]. Formic acid concentration and incubation duration with conidia have

been optimized to extract RodA from conidia with optimum purity [8].

4. Duration of conidial incubation with HF or formic acid has been optimized. At least 60 h are required for the extraction of RodA from conidia using HF, while 6 h incubation is optimum for its extraction using formic acid. Prolonged conidial incubation with HF or formic acid results in a gradual degradation of RodA.
5. HF treatment results in the extraction of a small quantity of CW polysaccharide (specifically galactomannan) along with RodA. Therefore, an additional purification step is necessary to remove the other CW contaminant. This could be achieved by a high-performance liquid chromatograph equipped with a reverse-phase C18 column. RodA bound to the C18 column can be eluted using a gradient of acetonitrile (solvent A, 10% acetonitrile with 0.1% TFA; solvent B, 100% acetonitrile with 0.1% TFA; gradient run: 0–10 min solvent A; 10–20 min, to 40% solvent B; and 20–60 min, to 100% solvent B).
6. RodA after HF or formic acid extraction should not be reconstituted in water and stored, as this will lead to a gradual degradation of RodA.
7. Dislodging from conidial surface or extraction of from conidia upon incubation with HF or formic acid has indicated that there are two isoforms of RodA in the conidial rodlet layer of *A. fumigatus*, with apparent masses of 14 kDa and 12 kDa (truncated form) as determined by their migration on SDS-PAGE (Fig. 3).
8. Melanin ghost is an electron-dense layer without a surface rodlet layer and underneath cell components, including CW and cytoplasm (Fig. 4).
9. Conidial counting, conidial weight used for the extraction of melanin, and the dry weight of the melanin ghost will allow quantification of the melanin pigment per conidium.
10. The composition of alkali-insoluble fractions will be β-1,6-branched β-1,3-glucan, chitin, and galactomannan. Sodium meta-periodate acts on vicinal diols [any molecules with hydroxyl (-OH) groups on adjacent carbons], forming cyclic periodate, which further undergoes rearrangement, cleaving the carbon-carbon linkage, leading to the formation of aldehyde or ketone end-groups [9].
11. Smith-degradation (acid hydrolysis) will then cleave the modified polysaccharide (having polyalcohol groups) into smaller fractions [10].

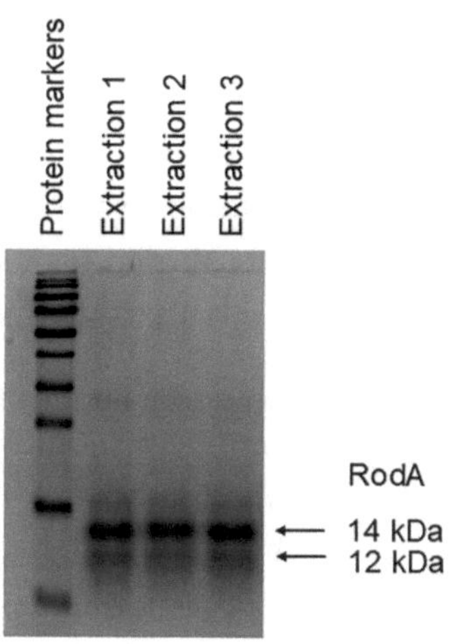

Fig. 3 RodA extracted from the *A. fumigatus* conidial surface. RodA extracted from conidia when subjected to SDS-PAGE, it resolves into two distinct bands with apparent molecular masses of approximately 14 and 12 kDa

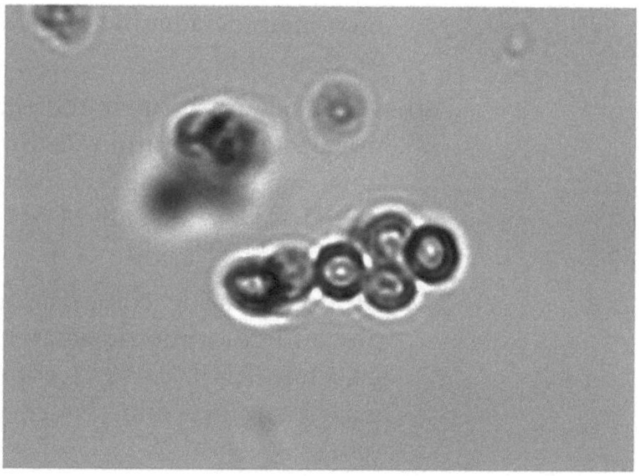

Fig. 4 Melanin ghosts. Processing of conidia for the extraction of melanin pigments results in the formation of *A. fumigatus* melanin ghosts

12. Together, periodate oxidation-Smith degradation leads to the destruction of galactomannan present in the alkali-insoluble fraction, leaving β-1,3-glucan-chitin.
13. Incubation with 40% NaOH results in the deacetylation of N-acetylglucosamine residues of chitin, whereas treatment with sodium nitrite leads to nitrous deamination and thereby the removal of chitin bound to β-1,3-glucan.
14. We observed that the α-1,3-glucan obtained following the above steps contains 3–5% of β-1,3-glucan that is not destroyed during alkali fractionation.

15. A brief treatment with any recombinant endo-β-1,3-glucanase should remove β-1,3-glucan contaminant from α-1,3-glucan preparation. Addition of 2 volumes of ethanol to this mixture will precipitate α-1,3-glucan, which can be separated by centrifugation (5000 × rpm, 10 min).
16. Care should be taken while autoclaving the sample suspended in H_2O_2/glacial acetic acid (chitin preparation). For about 10 mg of mycelial sample, 10 mL of H_2O_2/glacial acetic acid should be added. Bottle containing the sample- H_2O_2/glacial acetic acid should be kept loose.
17. An additional KOH treatment will ensure the production of chitin with the highest purity (>98%).
18. Repeating the steps of hydrazinolysis and nitrous acid hydrolysis one more time will ensure that the galactomannan purified is devoid of any other CW contaminants. The purity of this preparation can be assessed by gas chromatography.
19. Upon acid hydrolysis, mannan can be depolymerized into constituent monosaccharides. Mannose can be quantified by a nicotinamide adenine dinucleotide (NAD)-coupled enzymatic method [11].
20. It has been demonstrated that the production of galactosaminogalactan is optimum in Brian medium (composed of asparagine, glucose, NH_4NO_3, KH_2PO_4, $MgSO_4{\cdot}7H_2O$, $ZnSO_4{\cdot}7H_2O$, $CuSO4{\cdot}5H_2O$, $Co(NO_3)_2{\cdot}6H_2O$, $CaCl_2$; pH 5.4) [12].
21. Monosaccharide compositional analysis by gas chromatography and colorimetric assay had indicated that galactosaminogalactan is a heteropolysaccharide with galactose, galactosamine, and N-acetylgalactosamine as the building blocks.
22. Insoluble and soluble galactosaminogalactan differ in their galactosamine and galactose ratio, which are ~60/40 and ~15/85 in insoluble and soluble GAG, respectively.
23. TFA hydrolysis is suitable for determining the concentrations of hexoses/pentoses, as it destroys amino sugars, whereas HCl hydrolysis could destroy hexoses/pentoses, leaving the amino sugars untouched and thus suitable for quantifying amino sugars.
24. Glucose content in the alkali-insoluble fraction represents β-1,3-glucan, whereas that in the alkali-soluble fraction represents α-1,3-glucan. Glucosamine content indicates chitin content, whereas mannose and galactose are representatives of galactomannan. Galactosamine in combination with galactose quantifies galactosaminogalactan in the CW of *A. fumigatus*.

25. Enzymatic hydrolysis using specific exoenzymes followed by subjecting them to high-performance chromatography is an alternate for the quantification of CW polysaccharides. A standard mixture of monosaccharides for identifying specific monomers in the CW polysaccharide and comparing the area occupied by standard monosaccharide to quantify polysaccharide composition is an alternate to the gas chromatography method. However, an internal standard (meso-inositol) used in the gas chromatography ensures accurate quantification of the monosaccharides in the acid hydrolysate, whereas in high-performance liquid chromatography quantification, there could be handling errors (such as batch-to-batch variation due to manual solvent preparation, or while preparing standard monosaccharide mixtures), leading to batch variations and thereby in the quantification.
26. We adopted this protocol to measure chitin synthase activity by replacing UDP-^{14}C-glucose with UDP-^{14}C-N-acetylglucosamine [13].
27. We modified the fractionation of the CW after co-incubation of the membrane fraction with UDP-^{14}C-glucose/UDP-^{14}C-N-acetylglucosamine using NaOH instead of trifluoracetic acid.
28. Although N-acetylglucosamine is the building block for the biosynthesis of CW chitin, its presence in the reaction mixture for β-1,3-glucan neo-synthesis was found to be essential, as, otherwise, neo-synthesized β-1,3-glucan might get released into the medium instead of being a part of the alkali-insoluble fibrillar fraction. ATP and GTP are also required in the reaction mixture, though their specific functions have not been studied.
29. BgIB and LamA, the endo-β-1,3-glucanases from *Thermotoga neapolitana* (anaerobic bacterium) display laminaribiase and laminarinase activities, respectively [14]. An endo-chitinase from *Rasamsonia emersonii* (a fungus) is associated with chitobiosidase activity [15]. An endo-α-1,3-glucanase from *Flavobacterium* sp. EK14 (a gram-negative bacterium) can hydrolyze α-1,3-glucan into di- and tri-saccharides [16]. Endo-β-1,4-mannanase, β-mannosidase, and α-galactosidase from *A. niger* have been purified and characterized [17]. These enzymes could be instrumental in the structural elucidation of CW polysaccharides by the enzymatic method.
30. In the chemical method for the structural analysis of the *A. fumigatus* CW, an ionization energy of 70 eV can be used for recording the electron ionization spectrum in GC-MS analysis. The TRI mass spectrometry capillary column (30 m x 0.25 mm) would be ideal for the GC-MS. Helium as the mobile phase at a flow rate of 1.2 mL/min and a column

temperature of 100–240 °C, with a rise of 5 °C, will be ideal for recording the ionization spectrum on GC-MS [18].

31. The NMR spectrum can be acquired by growing *A. fumigatus* in unlabeled medium and based on the abundance of naturally occurring ^{13}C label. This requires large number of fungal samples unlike the samples that are labeled with $^{13}C/^{15}N$. However, currently, NMR cups that require smaller quantities of the samples are available.

Acknowledgments

We thank all our colleagues/collaborators who helped us in performing different techniques including atomic force microscopy, transmission electron microscopy and solid-state NMR.

References

1. Beauvais A, Fontaine T, Aimanianda V et al (2014) Aspergillus cell wall and biofilm. Mycopathologia 178:371–377. https://doi.org/10.1007/s11046-014-9766-0
2. Fontaine T, Simenel C, Dubreucq G et al (2000) Molecular organization of the alkali-insoluble fraction of aspergillus fumigatus cell wall. J Biol Chem 275:41528
3. Latgé JP, Kobayashi H, Debeaupuis JP et al (1994) Chemical and immunological characterization of the extracellular galactomannan of *Aspergillus fumigatus*. Infect Immun 62: 5424–5433. https://doi.org/10.1128/iai.62.12.5424-5433.1994
4. Liu Z, Valsecchi I, Le Meur RA et al (2023) Conidium specific polysaccharides in *Aspergillus fumigatus*. J Fungi (Basel) 9. https://doi.org/10.3390/jof9020155
5. Beaulieu D, Tang J, Zeckner DJ et al (1993) Correlation of cilofungin in vivo efficacy with its activity against *Aspergillus fumigatus* (1,3)-beta-D-glucan synthase. FEMS Microbiol Lett 108:133–137. https://doi.org/10.1111/j.1574-6968.1993.tb06088.x
6. Paris S, Debeaupuis J-P, Crameri R et al (2003) Conidial hydrophobins of *Aspergillus fumigatus*. Appl Environ Microbiol 69:1581–1588. https://doi.org/10.1128/AEM.69.3.1581-1588.2003
7. Aimanianda V, Bayry J, Bozza S et al (2009) Surface hydrophobin prevents immune recognition of airborne fungal spores. Nature 460: 1117–1121. https://doi.org/10.1038/nature08264
8. Valsecchi I, Dupres V, Michel J-P et al (2019) The puzzling construction of the conidial outer layer of *Aspergillus fumigatus*. Cell Microbiol 21:e12994. https://doi.org/10.1111/cmi.12994
9. Simon J, Fliri L, Drexler F et al (2023) Debugging periodate oxidation of cellulose: why following the common protocol of quenching excess periodate with glycol is a bad idea. Carbohydr Polym 310:120691. https://doi.org/10.1016/j.carbpol.2023.120691
10. Baird JK, Holroyde MJ, Ellwood DC (1973) Analysis of the products of Smith degradation of polysaccharides by g.l.c. of the acetylated, derived aldononitriles and alditols. Carbohydr Res 27:464–467. https://doi.org/10.1016/S0008-6215(00)81329-4
11. Soyama K (1984) Enzymatic determination of D-mannose in serum. Clin Chem 30:293–294
12. Fontaine T, Delangle A, Simenel C et al (2011) Galactosaminogalactan, a new immunosuppressive polysaccharide of *Aspergillus fumigatus*. PLoS Pathog 7:e1002372. https://doi.org/10.1371/journal.ppat.1002372
13. Muszkieta L, Aimanianda V, Mellado E et al (2014) Deciphering the role of the chitin synthase families 1 and 2 in the in vivo and in vitro growth of *Aspergillus fumigatus* by multiple gene targeting deletion. Cell Microbiol 16:1784–1805. https://doi.org/10.1111/cmi.12326
14. Zverlov VV, Volkov IY, Velikodvorskaya TV et al (1997) Thermotoga neapolitana bglB gene, upstream of lamA, encodes a highly thermostable beta-glucosidase that is a

laminaribiase. Microbiology 143(Pt 11): 3537–3542. https://doi.org/10.1099/00221287-143-11-3537

15. Dwyer K, Bentley IS, Fitzpatrick DA et al (2023) Recombinant production, characterization and industrial application testing of a novel acidic exo/endo-chitinase from Rasamsonia emersonii. Extremophiles 27:10. https://doi.org/10.1007/s00792-023-01293-4
16. Takahashi M, Yano S, Horaguchi Y et al (2023) α-1,3-Glucanase from the gram-negative bacterium Flavobacterium sp. EK-14 hydrolyzes fungal cell wall α-1,3-glucan. Sci Rep 13: 21420. https://doi.org/10.1038/s41598-023-48627-y
17. Ademark P (2000) Galactoglucomannan-degrading enzymes from *Aspergillus niger*. Dissertation, Lund University
18. Aimanianda V, Clavaud C, Simenel C et al (2009) Cell wall beta-(1,6)-glucan of Saccharomyces cerevisiae: structural characterization and in situ synthesis. J Biol Chem 284:13401–13412. https://doi.org/10.1074/jbc.M807667200

Chapter 18

Isolation and Purification of Extracellular Vesicles from *Aspergillus fumigatus*

Lucas F. B. Nogueira and Fausto Almeida

Abstract

Extracellular vesicles (EVs) are increasingly recognized as key players in fungal biology, mediating intercellular communication, modulating virulence, and shaping host-pathogen interactions. Despite their growing importance, the lack of standardized protocols for EV isolation in fungi has hindered reproducibility and the accurate interpretation of functional data. This chapter presents a comprehensive, optimized protocol for the isolation and characterization of EVs produced by *Aspergillus fumigatus* under both solid and liquid culture conditions. Culture parameters, including medium composition and growth format, are critically evaluated for their impact on EV yield, molecular composition, and biological significance. The protocol encompasses detailed procedures for fungal cultivation, EV isolation via differential ultracentrifugation, and multiparametric characterization through nanoparticle tracking analysis (NTA), transmission electron microscopy (TEM), and molecular marker profiling, all aligned with MISEV2023 guidelines. Designed to support both exploratory and high-resolution investigations, this standardized approach robustly analyzes EV-associated biomarkers. The isolated EVs are suitable for a range of downstream applications, including studies of immune modulation, molecular composition, fungal pathogenesis, and intercellular signaling dynamics.

Key words Extracellular vesicles, *Aspergillus fumigatus*, Ultracentrifugation, Fungal-host interactions, Purification protocols

1 Introduction

Aspergillus fumigatus (*A. fumigatus*) is a ubiquitous filamentous fungus and a major opportunistic pathogen responsible for invasive aspergillosis in immunocompromised individuals [1]. Beyond classical virulence factors, extracellular vesicles (EVs) have emerged as key mediators of fungal-host interactions, facilitating molecular communication through the transfer of proteins, lipids, and nucleic acids [2–4]. According to the latest Minimal Information for Studies of Extracellular Vesicles (MISEV2023), EVs are defined as

Natalia Mach (ed.), *Aspergillus fumigatus: Methods and Protocols*, Methods in Molecular Biology, vol. 3028,
https://doi.org/10.1007/978-1-0716-5222-0_18,

particles released from cells, delimited by a lipid bilayer, and incapable of autonomous replication [5]. In fungi, EVs transport a wide variety of cargo, including virulence factors, immunogens, and molecules with unknown functions, contributing to processes such as biofilm formation, antifungal resistance, and the modulation of host immune responses [3, 4, 6].

Variations in fungal culture conditions can significantly impact EV yield and cargo composition, necessitating the development of standardized isolation protocols [2]. The choice of solid or liquid culture systems influences fungal growth patterns, biofilm formation, and EV secretion [1, 2]. Solid media better mimic biofilm-like structures observed in pulmonary infections, whereas liquid media facilitate higher EV yield and greater purity [1, 2]. Selecting an appropriate medium depends on the research objective—whether prioritizing physiological relevance (solid culture) or maximizing vesicle production (liquid culture) [1, 2]. The MISEV2023 guidelines emphasize the importance of clearly defining experimental conditions, including media composition and culture duration, to enhance reproducibility in EV studies [5].

Among the various isolation strategies available, differential ultracentrifugation remains a widely adopted and robust method for EV enrichment across biological systems, including fungi [4, 5, 7, 8]. MISEV2023 supports its continued use, provided that key parameters such as centrifugation speeds, rotor types, and sample volumes are reported and that downstream characterization steps are employed [5]. Moreover, recent protocols adapted for fungal cultures grown on both solid and liquid media—such as those described by Piffer et al. [7] and Meng et al. [8] demonstrate that ultracentrifugation enables efficient and reproducible separation of EVs with minimal contaminants, preserving vesicle integrity while allowing downstream molecular and functional analyses. These approaches emphasize the importance of combining physiological relevance with methodological rigor, especially when studying EVs secreted under complex growth conditions that mimic host-pathogen interactions [1, 5, 7, 8].

To deepen our understanding of the biogenesis, molecular composition, and functional roles of fungal extracellular vesicles (EVs), this chapter presents a rigorously optimized methodology for the isolation of EVs from *A. fumigatus*. The protocol has been carefully developed with attention to critical variables, including culture medium composition, growth conditions, and purification strategies, all of which significantly influence EV yield and quality. By integrating the latest methodological advancements alongside the MISEV2023 guidelines, this approach ensures high reproducibility and functional reliability, thereby enabling consistent comparative analyses across experimental systems.

2 Materials

Prepare all solutions using ultrapure water (18 MΩ·cm at 25 °C) obtained by purifying deionized water. Use only analytical-grade reagents. Unless otherwise specified, prepare and store all reagents at room temperature. Do not add preservatives such as sodium azide to any of the solutions. Use aseptic techniques throughout all procedures. Sterilize all reagents, glassware, and consumables by autoclaving or filtration through 0.22 μm membranes. Handle *A. fumigatus* in accordance with institutional biosafety guidelines. Discard contaminated material according to local regulations.

2.1 Cultivation Media

Selection of cultivation medium is a critical determinant of EV yield and composition in *A. fumigatus*. The nutritional environment strongly influences fungal physiology, biomass production, and EV output. Choose culture format and media composition according to experimental aims (*see* **Note 1**).

2.1.1 Solid Media

1. YAG agar: 2% (w/v) glucose, 0.2% (w/v) Bacto yeast extract, 2% (w/v) malt extract, 2% (w/v) agar, 0.1% (v/v) trace elements (*see* **Note 2**). Dissolve all components in water, adjust pH to 6.5–7.0, autoclave at 121 °C for 15–20 min, cool to 50–55 °C before pouring into 90 mm sterile Petri dishes (25 mL per dish). Allow to solidify completely at room temperature. Once solidified, seal the plates with parafilm or place them inside sealed plastic bags to prevent desiccation and contamination. Use plates within 1 week for optimal performance.
2. Sabouraud Dextrose Agar (SDA): 1% (w/v) Bacto peptone, 4% (w/v) dextrose, 1.5% (w/v) agar. Dissolve in water, adjust the pH to 6.5–7.0, autoclave at 121 °C for 15–20 min, cool to 50–55 °C before pouring into 90 mm sterile Petri dishes (25 mL per dish). Allow the medium to solidify completely at room temperature. Once solidified, seal the plates with parafilm or place them inside sealed plastic bags to prevent desiccation and contamination. Use plates within one week for optimal performance.
3. Potato Dextrose Agar (PDA): 4% (w/v) potato infusion, 2% (w/v) dextrose, 1.5% (w/v) agar. Dissolve in water, adjust the pH to 6.5–7.0, autoclave at 121 °C for 15–20 min, cool to 50–55 °C before pouring into sterile 90 mm Petri dishes (25 mL per dish). Allow the medium to solidify completely at room temperature. Once solidified, seal the plates with parafilm or place them inside sealed plastic bags to prevent desiccation and contamination. Use plates within one week for optimal performance.

2.1.2 Liquid Media for Submerged Cultivation

1. YAG Broth: 2% (w/v) glucose, 0.2% (w/v) Bacto yeast extract, and 2% (w/v) malt extract; dissolve in water, adjust the pH to 6.5–7.0, and autoclave. Cool to 50–55 °C, add aseptically 0.1% (v/v) of sterile trace element solution, and store at 4 °C.
2. RPMI 1640 (modified): RPMI 1640 base medium supplemented with 2% (w/v) glucose or galactose. Sterile filter if glucose/galactose is added post-autoclaving. Store protected from light.
3. Czapek-Dox Broth (CD): 3% (w/v) sucrose, 0.2% (w/v) $NaNO_3$, 0.1% (w/v) KH_2PO_4, 0.05% (w/v) $MgSO_4{\cdot}7H_2O$, 0.002% (w/v) $FeSO_4{\cdot}7H_2O$; dissolve in water, adjust the pH to 6.5–7.0, and autoclave.
4. Aspergillus Minimal Medium (AMM): 1% (w/v) glucose, 0.6% (w/v) $NaNO_3$ (nitrogen source), 0.052% (w/v) KCl (ionic balance), 0.052% (w/v) $MgSO_4{\cdot}7H_2O$ (divalent cations), 0.152% (w/v) KH_2PO_4 (phosphate); dissolve in water, adjust the pH to 6.5–7.0, and autoclave. Cool to 50–55 °C, add aseptically 0.1% (v/v) of sterile trace element solution, and store at 4 °C.

2.2 Fungal Strain

A. fumigatus strain of choice (e.g., strain CEA17).

2.3 Solutions and Buffers

1. Sterile Phosphate Buffered Saline (PBS), pH 7.4: Dissolve 8.0 g of NaCl, 0.2 g of KCl, 1.44 g of Na_2HPO_4, and 0.24 g of KH_2PO_4 in 1 L of ultrapure water. Adjust the pH to 7.4 using NaOH or HCl. Sterile filter through a 0.22 μm filter. Prepare fresh or store at 4 °C for 1 week.
2. Sterile trace elements solution (10×): 0.5 g $FeSO_4{\cdot}7H_2O$, 0.5 g $ZnSO_4{\cdot}7H_2O$, 0.25 g $CuSO_4{\cdot}5H_2O$, 0.05 g $MnSO_4{\cdot}H_2O$, 0.05 g $CoCl_2{\cdot}6H_2O$, and 1 mL of 0.1 M H_3BO_3; dissolve in 1 L of ultrapure water. Sterile filter through a 0.22 μm filter (*see* **Note 2**).

2.4 Equipment and Consumables

1. Sterile Petri dishes 90 mm diameter.
2. Sterile filter material (e.g., miracloth from Merck Millipore).
3. Erlenmeyer flasks or spinner flasks: sterile, 250–500 mL capacity.
4. Sterile T-shaped or L-shaped cell spreaders.
5. Neubauer counting chamber.
6. Benchtop and refrigerated centrifuges capable of 5000 × g and 15,000 × g.
7. Ultracentrifuge (e.g., Optima XPN-80 from Beckman Coulter).
8. Ultra-clear centrifuge tubes (polycarbonate or polypropylene, sterile, rotor-compatible, e.g., from Beckman Coulter).

9. Pressure-based sample concentration setup equipped with 100 kDa ultrafiltration MWCO membrane (e.g., Amicon Stirred Cell from Merck Millipore).
10. Syringe filters, 0.45 μm pore size, PES or PVDF.
11. Nanoparticle tracking analysis (NTA) system (e.g., NanoSight from Malvern Panalytical with NTA 3.0 software for EV size and concentration).

3 Methods

This protocol presents a streamlined and scalable approach for the high-throughput isolation, purification, and characterization of EVs produced by *A. fumigatus* under diverse cultivation conditions. The workflow is designed for flexibility and can be readily adapted to a broad range of fungal strains, including gene deletion mutants from strain libraries. The protocol is also well-suited for assessing the impact of candidate pharmacological or genetic perturbations, such as candidate inhibitors, on EV biogenesis and release. A schematic overview of the experimental pipeline is provided in Fig. 1 to guide implementation of the methodology.

3.1 Preparation of Solid Cultures

1. Inoculate *A. fumigatus* onto fresh agar plates of choice (*see* **Note 1**). Incubate at 37 °C for 5–7 days to promote sporulation.
2. Harvest conidia by adding sterile PBS to the plate surface and gently dislodging spores with a sterile loop or spreader.
3. Filter the conidial suspension through sterile filter material (miracloth) to remove hyphal fragments and aggregates.
4. Count the conidia using a Neubauer counting chamber and dilute the suspension with sterile PBS to obtain a final concentration of 3.5×10^7 conidia/mL.
5. Allow any condensation droplets formed during the preparation of agar plates to dry completely. Evenly spread 300 μL of the conidial suspension onto sterile agar plates using a sterile T- or L-shaped cell spreader. Incubate the plates statically at 37 °C in the dark for 72 h (*see* **Note 3**).
6. Add 15 mL of sterile PBS to the surface of each agar plate. Gently scrape the fungal biomass using a sterile T-shaped spreader, ensuring thorough recovery of the material without damaging the agar surface.
7. Collect a maximum of 10 plates in 15 mL of sterile PBS. For larger sample sizes, divide into multiple tubes. Pool the PBS washes from multiple plates as necessary to concentrate the harvested material (*see* **Note 4**).

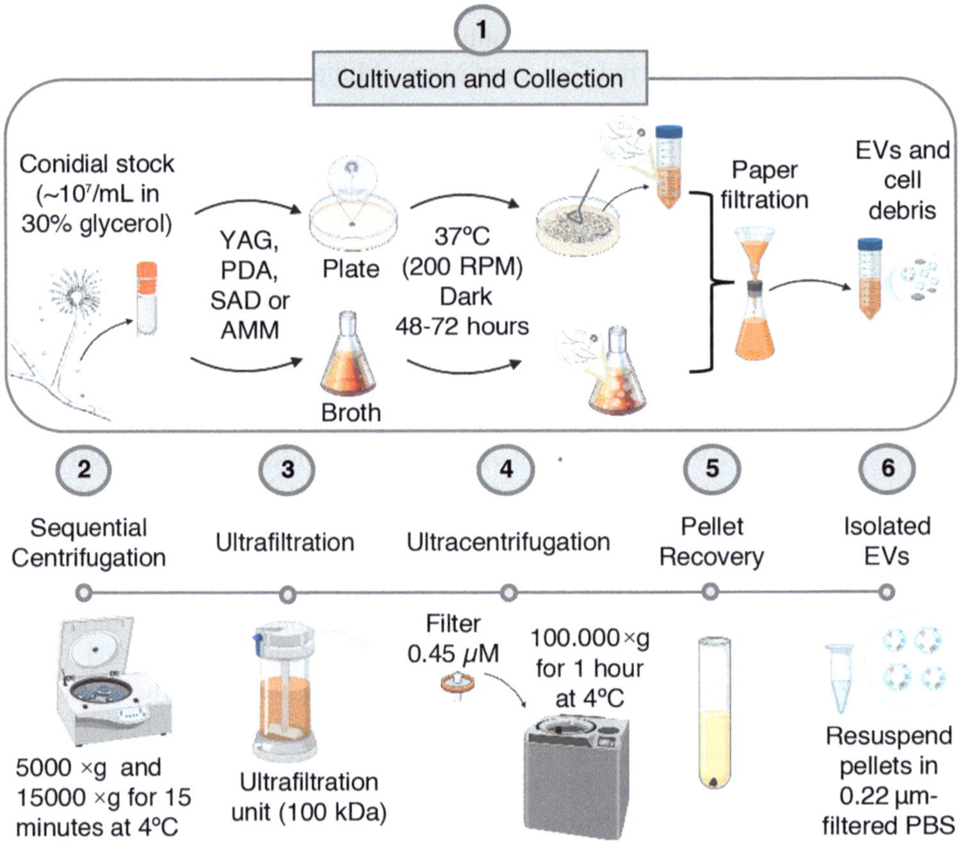

Fig. 1 Workflow for the isolation of EVs from A. fumigatus under different cultivation conditions. The protocol begins with fungal growth either on solid media (1A) or in liquid media (1B), using various culture compositions (e.g., YAG, PDA, SAD, or AMM) for 48–72 h at 37 °C. After incubation, extracellular material is recovered either by plate wash collection or broth collection (2), followed by Miracloth® filtration to remove hyphal cells. From this point onward, the protocol converges into a unified workflow for EV purification: sequential centrifugation steps (3) remove cells and large debris; tangential ultrafiltration with Amicon® stirred cells (4) concentrates EVs and removes small contaminants; final ultracentrifugation (5) isolates EVs based on their size and density. The resulting EV pellet is resuspended in sterile PBS and can be used for downstream analyses. This streamlined approach supports high-throughput and comparative studies, including knockout strain screening and inhibitor testing

3.2 Preparation of Liquid Cultures

1. Prepare a conidial suspension (*see* **steps 1–4**, Subheading 3.1). Adjust the inoculum concentration to 2×10^5 conidia/mL.
2. Inoculate sterile Erlenmeyer or spinner flasks containing liquid medium of choice (*see* **Note 1**) with the conidial suspension. Ensure that the culture volume occupies 10–20% of the total flask volume (e.g., 25–50 mL in 250 mL flasks or 50–100 mL in 500 mL flasks).
3. Incubate the flasks at 37 °C with 150–200 rpm shaking in the dark for 48–72 h (*see* **Note 3**).

4. Filter the fungal suspension through sterile filter material (miracloth) to remove the fungal biomass before proceeding with EVs isolation.

3.3 Harvesting Biomass and Preparing for EV Isolation

1. Centrifuge the collected suspensions (Subheading 3.1) or filtered supernatants (Subheading 3.2) at 5000 × g at 4 °C for 15 min to pellet residual cells and debris.
2. Carefully pipette the supernatant into a new sterile tube for centrifugation without touching the pellet of cells. Despite this centrifugation step, residual cells may remain in the supernatant, especially in strains producing large polysaccharide or when the sample is highly viscous. In such cases, turbidity in the supernatant may indicate the presence of remaining cells. If turbidity is observed, perform an additional centrifugation step using the same speed and time conditions as the previous step. Carefully transfer the supernatant to a new tube without disturbing the pellet (*see* **Note 5**).
8. Resuspend the resulting pellet in sterile PBS by gentle pipetting, using a volume sufficient to enable accurate cell counting and downstream normalization. Typically, use 5–15 mL depending on pellet size, avoiding vortexing. Keep the suspensions on ice during processing.
9. Collect an aliquot of each sample to determine the total number of cells recovered, considering the final volume obtained for each condition (*see* **Note 6**). Count cells using a Neubauer chamber and calculate the number of cells (cells/mL) (*see* **Note 7**).
10. Proceed with EV isolation from clarified supernatants as described in Subheading 3.4.

3.4 Isolation and Purification of EV

1. Carefully transfer the clarified supernatant to a new sterile tube and centrifuge at 15000 × g at 4 °C for 15 min using a refrigerated benchtop centrifuge to remove residual large vesicles, protein aggregates, and smaller debris.
2. Concentrate the filtered supernatant using a stirred cell concentrator equipped with a 100 kDa molecular weight cutoff membrane. Operate the system under inert gas pressure to reduce the sample volume while retaining EV.
3. Filter the concentrate through a sterile 0.45 μm PES membrane to remove residual large particles and debris.
4. Transfer the filtrate into sterile ultracentrifugation tubes appropriate for ultracentrifuge, rotor, and high-speed centrifugation (*see* **Note 8**).

5. Ultracentrifuge the sample at 100,000 × g for 1 h at 4 °C using maximum rotor acceleration and deceleration settings to pellet EVs (*see* **Note 9**).
6. Carefully discard the supernatant by inverting the tube. For swing-bucket rotors, the pellet is located at the bottom of the tube; for fixed-angle rotors, it forms along the outer edge. Place the inverted tube on absorbent paper for a few seconds to remove any residual supernatant without disturbing the EV pellet. Work quickly to prevent the EV pellet from dissociating and drying.
7. Resuspend the EV pellet in up to 300 mL of sterile PBS. Another isotonic buffer can be tested, depending on the intended downstream applications. Maintain vesicle integrity by pipetting gently; do not vortex (*see* **Note 10**).
8. Resuspend the final EV pellet in sterile PBS, ensuring the vesicles remain stable and free from aggregation. Store at 4 °C for immediate use or at −80 °C for long-term storage. Aliquot the EVs to avoid repeated freezing and thawing of the samples, which compromises vesicle integrity (*see* **Note 11**).

3.5 Characterization of EVs Using Nanoparticle Tracking Analysis

1. Dilute the EV suspension in sterile PBS to obtain a particle concentration between 1×10^8 and 1×10^9 particles/mL, which is suitable for optimal detection and analysis.
2. Prepare the nanoparticle tracking analysis (NTA) system by performing a thorough cleaning routine before and between sample runs. Flush the system's flow cell and cannulas three times with 1 mL of distilled water using a clean syringe attached to the syringe pump.
3. Inject the dilutes EV sample into the system using a 1 mL syringe connected to the syringe pump and cannulas. Set the syringe pump to a consistent flow rate, typically around 50, to maintain stable particle movement during analysis.
4. Configure the acquisition of settings in the NTA 3.0 software. Record three 60-second videos per sample. Maintain a constant camera level between 10 and 15 and set the fluid viscosity to that of water. Adjust the detection threshold, typically between 7 and 15, to optimize nanoparticle visualization and sizing.
5. Analyze the size distribution and concentration of EVs in each sample. For example, for this protocol, majority of EVs concentrates between ~80 and 250 nm. YAG broth samples yield higher particle concentrations with broader size distribution, while SDA plate samples show lower concentration and narrower profiles (*see* **Notes 12–15**) (*see* Fig. 2).
6. To meet MISEV2023 minimal characterization criteria, combine physical (size and concentration), morphological, and

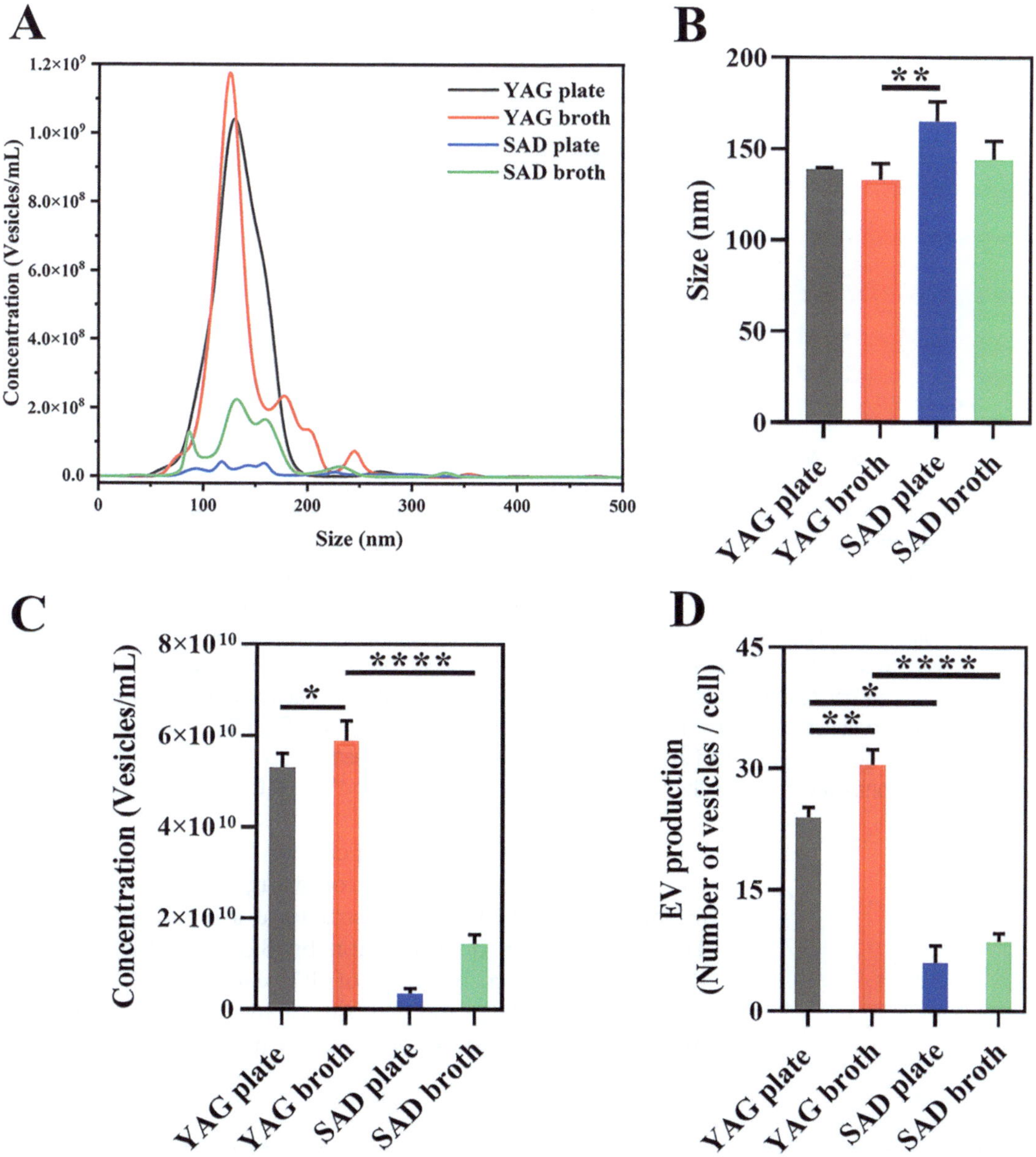

Fig. 2 Characterization and quantification of EVs produced by A. fumigatus under different culture conditions. (**a**) Size distribution profiles of EVs isolated from fungal cultures grown on solid YAG medium (YAG plate, black), in liquid YAG medium (YAG broth, red), on solid Sabouraud medium (SAD plate, blue), and in liquid Sabouraud medium (SAD broth, green), as determined by nanoparticle tracking analysis (NTA). (**b**) Mean particle size (in nm) of EVs from each condition. Statistical significance was determined using one-way ANOVA with multiple comparisons ($p < 0.01$). (**c**) EV concentration (vesicles/mL) measured by NTA, revealing significantly higher yields in broth conditions compared to plates and in YAG versus SAD medium. Data are presented as mean ± SD from three biological replicates. (*$p < 0.05$, ***$p < 0.0001$). (**d**) Number of EVs produced per fungal cell calculated by normalizing EV concentrations to estimated cell counts for each condition. Submerged culture in YAG broth resulted in the highest EV production per cell, while solid growth in SAD medium showed the lowest production. Data are presented as mean ± SD. (*$p < 0.05$, **$p < 0.01$, ***$p < 0.0001$)

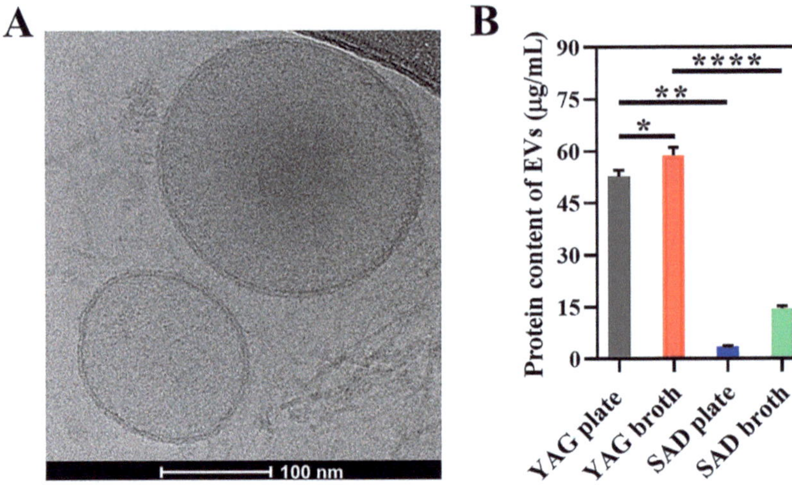

Fig. 3 Morphological and biochemical characterization of Aspergillus fumigatus-derived extracellular vesicles. (**a**) Representative cryo-electron microscopy (Cryo-EM) image of EVs isolated from A. fumigatus grown in liquid YAG medium for 72 h at 37 °C. Vesicles show characteristic spherical morphology with lipid bilayer membranes, consistent with fungal EVs described in the literature. Scale bar: 100 nm. (**b**) Quantification of EV proteins from A. fumigatus. Protein content was measured in EV samples isolated from both solid and liquid cultures using BCA assay. EVs were isolated from cultures grown on YAG and SAD media, both in solid and liquid forms. The bar graph represents protein concentration (μg/mL) of each sample in triplicate. Higher protein levels are observed in EVs from submerged cultures, especially in YAG broth, in agreement with the vesicle yield per cell data presented in Fig. 2d. Data are expressed as mean ± SD. Statistical significance was determined by one-way ANOVA with multiple comparisons (*$p < 0.05$; **$p < 0.01$; ****$p < 0.0001$)

biochemical readouts [5]. Depending on experimental goals, you may optionally include following EV characterization methods: Cryo-EM for ultrastructural assessment (*see* **Note 16**); total protein quantification by BCA (*see* **Note 17**) (*see* Fig. 3b), Western blotting of EV-associated proteins (*see* **Notes 18** and **19**), and Ergosterol measurement using the Amplex Red Cholesterol Assay as an indirect EV marker (*see* **Note 20**) (*see* Fig. 3).

4 Notes

1. Select the right agar and medium formulations carefully. The biogenesis and release of EVs by *A. fumigatus* are predominantly governed by environmental factors, particularly the nutritional composition of the medium. This parameter exerts a greater influence than the cultivation format itself, as it directly modulates fungal physiology, biomass yield, and metabolic activity, which in turn impact EV quantity, size profile, and molecular cargo. Depending on the experimental objective, choose the solid version to harvest conidia or to analyze

surface-associated fungal structures. Use liquid formulation to generate biomass and isolate secreted EVs under submerged conditions. *YAG agar* is optimized for secondary metabolite production, which may influence EV cargo. *SDA agar* is a standard medium for fungal propagation, it yields EVs with high protein content. *PDA agar* supports robust hyphal growth and EV secretion during surface colonization. Using *YAG Broth* results in high biomass yield, it promotes robust mycelial growth and intense conidiation, and EVs can be harvested from supernatant after 48–72 h at 37 °C with 150–200 rpm agitation. *Modified RPMI 1640 medium* mimics host nutrient conditions but may alter EV immunomodulatory properties. *CD Broth* is useful for nitrogen-limitation studies, induces nutrient stress, and potentially enriches EVs with stress-response proteins. The use of *AMM medium* enables controlled physiological studies by minimizing extraneous variables, and it is ideal for evaluating nutrient-specific responses and their impact on EV production.

2. The composition of trace elements solution is critical for experimental reproducibility, particularly considering that the nutritional composition of the medium directly affects fungal physiology, conidial production, and EV biogenesis.
3. The number of agar plates or the volume of liquid culture used may vary depending on the fungal species or strain, cultivation conditions, and the requirements of downstream analyses, particularly when EV production is normalized to the number of cells. For *A. fumigatus*, cultures should be incubated in the dark to avoid potential alterations in fungal morphology and metabolism induced by light. Although *A. fumigatus* is not strongly photo-regulated, light exposure can influence sporulation and secondary metabolism, potentially affecting the composition or quantity of EVs. Incubation in the dark helps minimize environmental variability and improves the reproducibility of EV isolation.
4. If necessary, combine cells collected from multiple plates to achieve a higher EV yield. Use a minimum ratio of 1 agar plate per 1 mL of 1 × PBS to ensure optimal vesicle concentration. When pooling samples for downstream applications, follow MISEV 2023 guidelines: record the number of individual samples combined, the total number of contributing cells, the volume of each individual sample, and the final pooled volume. Provide additional details on the characteristics of each original sample whenever possible.
5. To confirm the absence of viable cells in EV preparation, plate an aliquot onto the agar plate (e.g., YAG, PDA, or SDA) and

incubate for 48–72 h at 37 °C. No colony growth should be observed in sterile preparations.

6. Cell aliquots can be fixed with 4% paraformaldehyde for later counting.
7. Quantification of the recovered cells is important, it allows normalization of extracellular vesicle production relative to the number of cells, enabling accurate calculation of vesicle yield per cell during the cultivation period.
8. Clean and sterilize tubes before reusing. Wash manually with 20% Extran detergent, followed by thorough rinsing with Milli-Q water to remove any detergent residues. Allow the tubes to dry air. Before use, rinse with 70% ethanol and let them air dry in a biosafety cabinet. Perform three additional washes with 0.22 μm-filtered PBS right before adding the sample. Inspect the tubes for any deformation or cracks before reusing. Ensure the tubes are accurately balanced, as an unbalanced rotor may cause equipment damage and personal injury.
9. Optionally, purify the EVs further using density gradient ultracentrifugation by layering the resuspended pellet onto a preformed iodixanol gradient (10–40% w/v) and ultracentrifuging at 100,000 × g at 4 °C for 16–18 h. This step is optional and should be considered when a higher purity of the EVs is required, especially for functional assays or interaction studies, as it separates EVs from contaminating proteins, lipids, and other non-vesicular particles. Collect the fractions corresponding to the density range expected for EVs, dilute in sterile PBS to maintain isotonicity and vesicle integrity, and wash by ultracentrifugation at 100,000 × g at 4 °C for 2 h to further refine the vesicle preparation.
10. To preserve EV integrity, handle all suspensions gently. Avoid vortexing. Perform centrifugation, pipetting, and resuspension steps on ice. Use low-retention pipette tips and minimize repeated pipetting. Aliquot EV samples into sterile tubes to prevent multiple freeze–thaw cycles. Store aliquots at 4 °C for up to 48 h, or at −80 °C for long-term use. The MISEV2023 guidelines emphasize that storage conditions prior to EV isolation can impact vesicle recovery, purity, and downstream interpretation. Specifically, it is recommended to minimize storage time before separation and to avoid repeated freeze–thaw cycles, unless their effects have been properly evaluated. Based on our experience, immediate processing of fungal suspensions collected from solid cultures is strongly advised. When storage is necessary, dilute EVs in sterile, 0.22 μm-filtered PBS, aliquot them, and store at −80 °C. Storage

approaches should always be tailored and validated according to the experimental context.

11. The protocol workflow adheres to MISEV2023 recommendations and yields robust physical/biochemical profiles of *A. fumigatus* EVs. However, some limitations regarding sample purity and heterogeneity should be acknowledged. As the isolation protocol relies primarily on differential centrifugation and ultracentrifugation, it may enrich multiple EV subpopulations along with non-vesicular particles of similar size and density. Particle heterogeneity therefore remains a concern, particularly when higher purity is critical for "omics" applications (e.g., proteomics, transcriptomics). In such cases, consider complementary strategies, such as size-exclusion chromatography and/or iodixanol-based density gradients. These methods alone or combined with ultracentrifugation further increase EV purity and reduce contaminants. These additions typically increase processing time and reduce overall yield, so selection should be guided by downstream goals, sample availability, acceptable losses, and resource constraints. When used, report rotor types, speeds/RCF, volumes, and fractionation details to remain compliant with MISEV2023.

12. Compare EV yields across cultivation conditions. Compare EV yields across cultivation conditions. The number of vesicles produced per cell varied depending on the growth condition and the culture medium. For example, using the NTA analysis, submerged cultivation in liquid YAG yielded approximately 28–32 EVs per cell, whereas growth on solid YAG plates yielded approximately 22–25 EVs per cell. In Sabouraud medium, EV yield per cell was lower overall, with liquid culture yielding around 7–9 EVs per cell, and solid culture yielding approximately 4–8 EVs per cell (Fig. 2d). These results indicate that submerged cultivation in YAG medium supports higher EV production and that medium composition significantly affects vesicle yield. The estimated number of A. fumigatus cells recovered after 48–72 h of growth at 37 °C is approximately 1.5 to 2.5 × 10^9 cells/plate for solid cultures and 1 × 10^9 to 3 × 10^9 cells/flask for liquid cultures.

13. The NTA system operates based on tracking the Brownian motion of particles within the size range of 10–1000 nm and calculates their diameter and concentration using the Stokes-Einstein equation.

14. Importantly, EV concentration was more impacted than size by cultivation conditions such as medium composition (e.g., YAG vs Sabouraud) and physical format (liquid vs solid) (Fig. 2a–c).

15. Although NTA analysis can, in theory, detect particles as small as 10 nm, in practice, it predominantly reports particles larger

than 100 nm, which may impact both the size distribution and the concentration estimates of EVs. Therefore, transmission or scanning electron microscopy is essential, particularly when characterizing EVs released by a specific fungal strain or species for the first time.

16. Cryo-electron microscopy (Cryo-EM) enables ultrastructural analysis of EVs. Apply 3 μL of the EV suspension onto glow-discharged holey carbon grids, followed by blotting and vitrification in liquid ethane using a plunge-freezing system. Use a cryogenic transmission electron microscope to visualize vesicular structures and confirm membrane integrity. For EVs from A. fumigatus grown in liquid YAG, Cryo-EM typically reveals rounded, bilayered vesicles. Some may display fibrillar surface decorations, as previously reported in fungal EVs.
17. Protein quantification can be performed with the bicinchoninic acid (BCA) assay using EV suspensions in sterile PBS. This provides an estimate of total protein content, which reflects EV abundance and cargo diversity. Normalize measurements with a BSA calibration curve, process samples in technical duplicates or triplicates, and interpret protein yield together with NTA and/or ergosterol results (*see* **Note 20**) for a comprehensive EV profile.
18. Western blotting can be used to detect EV-associated proteins, depending on the experimental goals. Prepare EV lysates in an appropriate lysis buffer and load equal protein amounts (based on BCA quantification). Use SDS-PAGE and transfer to PVDF or nitrocellulose membranes. Choose antibodies based on the fungal species or previously reported EV markers. As there are no universal fungal EV markers, select candidate proteins carefully (e.g., tetraspanins, membrane proteins, or species-specific components) and validate with orthogonal methods.
19. In some fungi, proteins such as mannoprotein Mp88 and membrane-associated tetraspanins have been identified in EV preparations and may serve as suitable targets. However, evidence suggests that EV protein composition varies significantly between fungal species, and no universal consensus has been established. Interpret Western blot results in light of these limitations and, where possible, validate findings with complementary characterization methods.
20. Ergosterol quantification can serve as an indirect method to estimate fungal EV abundance. Despite lacking cholesterol, fungal EV membranes are rich in ergosterol, which is detectable using the Amplex Red Cholesterol Assay. Cholesterol esterase and oxidase produce hydrogen peroxide, which reacts with Amplex Red to generate fluorescence. This approach

reflects sterol content and can complement size and protein-based quantifications. While the assay is sensitive and convenient, it quantifies total sterols regardless of their origin, including potential contaminants from culture medium or residual cellular debris. Therefore, it should not be used as a standalone method for EV quantification. A more robust strategy is to combine it with complementary techniques such as NTA or protein content quantification to confirm the presence and relative abundance of EVs.

References

1. Morelli KA, Kerkaert JD, Cramer RA (2021) Aspergillus fumigatus biofilms: toward understanding how growth as a multicellular network increases antifungal resistance and disease progression. PLoS Pathog 17:e1009794. https://doi.org/10.1371/journal.ppat.1009794
2. Cleare LG, Zamith D, Heyman HM et al (2020) Media matters! Alterations in the loading and release of Histoplasma capsulatum extracellular vesicles in response to different nutritional milieus. Cell Microbiol 22:e13217. https://doi.org/10.1111/cmi.13217
3. Brandt P, Singha R, Ene IV (2024) Hidden allies: how extracellular vesicles drive biofilm formation, stress adaptation, and host-immune interactions in human fungal pathogens. MBio 15:e0304523. https://doi.org/10.1128/mbio.03045-23
4. Rizzo J, Trottier A, Moyrand F et al (2023) Coregulation of extracellular vesicle production and fluconazole susceptibility in Cryptococcus neoformans. MBio 14:e0087023. https://doi.org/10.1128/mbio.00870-23
5. Welsh JA, Goberdhan DCI, O'Driscoll L et al (2024) Minimal information for studies of extracellular vesicles (MISEV2023): from basic to advanced approaches. J of Extracellular Vesicle 13. https://doi.org/10.1002/jev2.12404
6. Freitas MS, Bitencourt TA, Rezende CP et al (2023) Aspergillus fumigatus extracellular vesicles display increased galleria mellonella survival but partial pro-inflammatory response by macrophages. J Fungi (Basel) 9. https://doi.org/10.3390/jof9050541
7. Piffer AC, Reis FCG, Moyrand F et al (2024) Protocol for separation of fungal extracellular vesicles using ultracentrifugation from solid medium cultures. STAR Protoc 5:103069. https://doi.org/10.1016/j.xpro.2024.103069
8. Meng F, Liu X, Duan H et al (2024) Aspergillus fumigatus-derived extracellular vesicles mitigate fungal keratitis by modulating the immune cell function. ACS Infect Dis 10:500–512. https://doi.org/10.1021/acsinfecdis.3c00454

Index

Natalia Mach (ed.), *Aspergillus fumigatus: Methods and Protocols*, Methods in Molecular Biology, vol. 3028,
https://doi.org/10.1007/978-1-0716-5222-0,

MIX
Papier aus verantwortungsvollen Quellen
Paper from responsible sources
FSC® C105338

If you have any concerns about our products,
you can contact us on
ProductSafety@springernature.com

In case Publisher is established outside the EU,
the EU authorized representative is:
Springer Nature Customer Service Center GmbH
Europaplatz 3, 69115 Heidelberg, Germany

Printed by Libri Plureos GmbH
in Hamburg, Germany